Operatic Geographies

Operatic Geographies

The Place of Opera
and the Opera House

EDITED BY SUZANNE ASPDEN

The University of Chicago Press   Chicago and London

The University of Chicago Press, Chicago 60637
The University of Chicago Press, Ltd., London
© 2019 by The University of Chicago
All rights reserved. No part of this book may be used or reproduced
in any manner whatsoever without written permission, except in
the case of brief quotations in critical articles and reviews. For more
information, contact the University of Chicago Press, 1427 E. 60th St.,
Chicago, IL 60637.
Published 2019
Printed in the United States of America

28 27 26 25 24 23 22 21 20 19   1 2 3 4 5

ISBN-13: 978-0-226-59596-2 (cloth)
ISBN-13: 978-0-226-59601-3 (paper)
ISBN-13: 978-0-226-59615-0 (e-book)
DOI: https://doi.org/10.7208/chicago/9780226596150.001.0001

Publication of this book has been supported by the Manfred Bukofzer
Endowment of the American Musicological Society, funded in part by
the National Endowment for the Humanities and the Andrew W. Mellon
Foundation.

Library of Congress Cataloging-in-Publication Data

Names: Aspden, Suzanne, editor.
Title: Operatic geographies : the place of opera and the opera house / edited by
    Suzanne Aspden.
Description: Chicago : The University of Chicago Press, 2019. | Includes
    bibliographical references and index.
Identifiers: LCCN 2018041795 | ISBN 9780226595962 (cloth : alk. paper) |
    ISBN 9780226596013 (pbk. : alk. paper) | ISBN 9780226596150 (e-book)
Subjects: LCSH: Opera. | Opera—Production and direction—History. |
    Theaters—History. | Cultural geography.
Classification: LCC ML1700 .O6835 2018 | DDC 782.109—dc23
LC record available at https://lccn.loc.gov/2018041795

♾ This paper meets the requirements of ANSI/NISO Z39.48–1992
(Permanence of Paper).

Contents

1 Introduction: Opera and the (Urban) Geography of Culture 1
SUZANNE ASPDEN

2 The Legal Spaces of Opera in The Hague 12
REBEKAH AHRENDT

3 Opera at School: Mapping the Cultural Geography of
Schoolgirl Performance 26
AMANDA EUBANKS WINKLER

4 London's Opera House in the Urban Landscape 39
MICHAEL BURDEN

5 Opera and the Carnival Entertainment Package in
Eighteenth-Century Turin 57
MARGARET R. BUTLER

6 Cockney Masquerades: Tom and Jerry and
Don Giovanni in 1820s London 74
JONATHAN HICKS

7 The City Onstage: Re-Presenting Venice in Italian Opera 88
SUSAN RUTHERFORD

8 Between the Frontier and the French Quarter: Operatic
Travel Writing and Nineteenth-Century New Orleans 105
CHARLOTTE BENTLEY

CONTENTS

9 *L'italiana* in Calcutta 119
BENJAMIN WALTON

10 Thomas Quinlan (1881–1951) and His "All-Red"
Opera Tours, 1912 and 1913 133
KERRY MURPHY

11 Empires in Rivalry: Opera Concerts and Foreign
Territoriality in Shanghai, 1930–1945 148
YVONNE LIAO

12 "Come to the Mirror!" Phantoms of the Opera—
Staging the City 162
PETER FRANKLIN

13 Open-Air Opera and Southern French Difference
at the Turn of the Twentieth Century 178
KATHARINE ELLIS

14 Pastoral Retreats: Playing at Arcadia in Modern Britain 195
SUZANNE ASPDEN

15 The Opera House as Urban Exhibition Space 213
KLAUS VAN DEN BERG

16 Underground in Buenos Aires: A Chamber Opera at
the Teatro Colón 234
ROBERTO IGNACIO DÍAZ

Acknowledgments 249   Contributors 251
Notes 255   Index 311

ONE

Introduction: Opera and the (Urban) Geography of Culture

SUZANNE ASPDEN

Musicologists and cultural historians have long acknowledged opera and opera houses' importance as forums for the performance of power, whether of the monarchy or the state.[1] Indeed, recognizable through their grandiose architecture and use of symbolic locations, opera houses are still often *lieux de mémoire*, repositories of collective memory, specially situated within the urban environment.[2] Yet this situatedness, and the way that opera and the opera house have interacted with and in the formation of their physical settings, has seldom been treated as a phenomenon worthy of examination in its own right. This is perhaps surprising, given the long-standing recognition of the theater's potential: in his 1989 *Places of Performance*, Marvin Carlson declared that theater historians were now as likely to consider theater as "a sociocultural event" (in the vein of the then-burgeoning field of performance studies) as they were to examine it as the representation of a dramatic text, and as such he proposed studying "how places of performance generate social and cultural meanings of their own which in turn help to structure the meaning of the entire theatre experience."[3] Carlson (following Eric Buyssens) invoked a visit to the opera house, in particular, as "the richest object available for semiotic analysis," not only for its multiple theatrical means, but for "how it fits into the social routine, where the opera house is located in the urban plan and how one arrives there, what preparations must be made for the operatic event."[4] Nonetheless, the opera house was not Carlson's primary focus (though it provided abundant pickings in the examination of theater-as-monument).[5] And while Anselm Gerhard's pioneering 1992 study of the "urbanization" of French nineteenth-century grand opera also followed this line of thought in its introductory discussions of developments in the Parisian operatic environment,[6] and the collection of essays on European concert venues, *Éspaces et lieux de concert en Europe, 1700–1920: Architecture, musique, société* (2008), has undertaken an examination of concert venues similar to Carlson's in approach,[7] few others have considered the ways in which opera's physical situation relates to and engages with the development of its social, cultural, and political functions, despite the attention that has been paid, in numerous studies, to the representation of urban, civic, or national life (or "place," more broadly) on the stage.

Yet since, as Carlson pointed out, theaters have been one of the most persistent features of the urban landscape, considering their place, agency, and representational mode helps uncover the "shifting meanings in the urban text,"[8] as well as in the nature of opera as a genre. We are thus enabled to see beyond the opera house as a mere receptacle for operatic events and appreciate it as a participant in negotiations of (urban) territory. Not only have opera and its urban environment developed side by side throughout the genre's four-hundred-year history, but the notion and performance of place (as itself an often-contested expression and experience of power) is as vitally important to opera as this most prestigious of art forms has been to the development of a sense of civic and national identity. This volume of essays thus sets about exploring something of that ever-changing relationship, from the peripatetic and contingent nature of late seventeenth-century opera and its venues, to the establishment of opera houses as defining civic spaces in their own right in the eighteenth century, to the opera house (and operagoing) as a cultural commodity and a source of regional, national, and international territorial definition in the nineteenth and twentieth centuries, to the challenges and disillusionments attending on that success and diffusion of the operatic (and opera-house) ideal in the twentieth and twenty-first centuries.

In exploring the opera house's potential meanings, cultural geography offers a valuable framework, its consideration of the ways in which the social and the spatial intertwine offering at least as much to opera as to other cultural forms. It is a field that has long engaged in the study of human interaction with (and formation of) the landscape—typically, in the earlier twentieth century, at a regional scale, focused on rural "field" sites. Since then, cultural geography has become increasingly concerned with problematizing the conceptual and historical separation of the human and the environmental, the natural and the cultural, productively calling into question "culture," "landscape," and "nature" as discrete concepts.[9] As Barney Warf explains in his introduction to *The Spatial Turn* (2009), "Geography matters, not for the simplistic and overly used reason that everything happens in space, but because *where* things happen is critical to knowing *how* and *why* they happen."[10] Recognition of these and other interdependencies has prompted considerable expansion within the field of cultural geography, such that it now seems almost all-encompassing; indeed, the editors of the 2003 *Handbook of Cultural Geography* describe it simply as "a series of intellectual—and, at core, politicized—engagements with the world."[11] Correspondingly, those (including musicologists) who have traditionally sought to understand the cultural "how and why" are coming to realize the importance of the "where," finding that (as with cultural studies more generally) seemingly normative or abstract patterns and principles make better sense when viewed in relation to their physical environments. In particular, understanding music through seeking to explore its place in its physical setting has brought insights into Western art music as diverse as Notre Dame organum and the twentieth-century symphony, as well as into contemporary popular music and world musics (the traditional stomping ground of geographers interested in music), providing one very concrete means of contextualization for a discipline still disentangling itself from the reification of score as work.[12] Indeed, for fields such as medieval chant and nonliturgical song (where the work concept has had less hold), reconceptualization of music around ideas of space and place has generated rich insights, and in the process regenerated interest in the repertoire itself.[13] Similarly, for opera, a genre still more dependent on a sense of place and the temporal inhabiting of space for staking its claim to status and potential meaning, a cultural-geographical approach offers an illuminating and fruitful set of perspectives. Although other— particularly French—studies have touched on aspects of the relationship between traditional human geography and opera, there is still much to explore, particularly in light of the expansion of cultural geography.[14]

While cultural geography in general provides a useful mode through which to approach opera, recent developments within the field hold particular relevance. Most importantly, for a genre often explicitly concerned with the expression of power relations, the cultural and spatial turns that have characterized developments in the social sciences and the humanities over the past twenty years or so have also encouraged the emergence of what in the 1990s became known as a "new" cultural geography, one committed to taking account of the political and the economic, the expression of power, and the function of physical environments as "texts" in its analyses.[15] Opera historians are no strangers, of course, to consideration of politics, economics, and power relations in their studies of opera companies and opera houses (whether these studies chime with the "new" musicology or not);[16] indeed, given the manifold contingencies of this expensive, multidisciplinary art form, political and economic interests are generally acknowledged as crucial to the genre's long-term viability. However, the interconnectedness of these issues with topographical or spatial concerns has rarely been considered—and yet the growing interest in music's (or, more broadly, sound's) relationship to space, while largely focused on modern experience, has obvious relevance for opera, as a genre traditionally associated with the political elite. As Jacques Attali pointed out in his influential book *Noise* (1985), music's capacity to connect the seat of power with its subjects—and, one might add, to perform that connection as itself a cultural expression—makes it "an attribute of power" and a means to "creation or consolidation of a community."[17] Thus sound's inherent capacity to travel (which made it integral to communicative and imaginative structures long before the technology of the printing press gave a presumed preeminence to the visual) is as evident in studies of medieval and early modern soundscapes as it is in those of contemporary popular culture, and demonstrates the centrality of music to the projection and negotiation of meaning within geographical as well as temporal space.[18] Exponents of opera certainly recognized its potential for such negotiations, associating the performance of the genre—even before construction of specific houses for it—with the centers of power.[19] So, as Warf's observation regarding the relation between geography and social politics suggests, the spatial turn is relevant to opera both with regard to the physical situation of an opera house and in examining opera's expression onstage.[20] Indeed, while the political slant to the new cultural geography might go hand in hand with the perception of opera and the opera house as forms of social and topographical texts, to be read for their expression of power, opera's phenomenological distinctiveness and visceral presence in performance mean the mode is also primed for analysis in terms of so-called nonrepresentational (or "more-than-representational") geography, which is grounded in the affective and the quotidian and in less hierarchical assessments of meaning.[21]

Recent work has certainly made clear cultural geography's potential for opera, although studies have tended thus far to focus on particular composers and/or works, exploring concepts of "landscape"—and, at times, connections to urban geography—expressed in opera: for example, Emanuele Senici's *Landscape and Gender in Italian Opera: The Alpine Virgin from Bellini to Puccini* (2005);[22] Daniel M. Grimley's "Carl Nielsen's Carnival: Time, Space and the Politics of Identity in *Maskarade*" (2010);[23] Christopher Morris on the alpine landscape and German modernist composers in *Modernism and the Cult of Mountains: Music, Opera, Cinema* (2012);[24] and Arman Schwartz on Puccini's use of sonic realism in *Puccini's Soundscapes: Realism and Modernity in Italian Opera* (2016).[25] David Charlton considers the aesthetics of landscape projected in late eighteenth-century opéra comique, in "Hearing through the Eye in Eighteenth-Century French Opera" (2014).[26] For these and other studies, "landscape" is primarily considered in terms of what we would call the "natural" environment, and valued for its metaphorical and symbolic functions within musicodramatic structures.

This volume—which had its origins in a 2014 conference in Oxford—takes a broader purview than that of most other studies of opera and geography. The aim of the volume as a whole is to offer a set of perspectives on the changing and often contested sociopolitical meaning and configuration of opera's physical context throughout the genre's history—in particular, examining the opera house's evolving place in its (largely urban) environment, from the seventeenth century to the present day. So Rebekah Ahrendt and Amanda Eubanks Winkler show that the establishment of opera in the Netherlands and England in the late seventeenth century was linked both to civic infrastructural growth and regulation, and to emerging discourses around opera's place in the urban economy, framed by concerns educational (Eubanks Winkler) and legal (Ahrendt). Their studies demonstrate equally the local contingency of early operatic development and the degree to which our artistic narratives of the genre are dependent on sociocultural and geographic circumstance: England's thriving suburban education market provided both impetus and financial security not otherwise available for English operatic ventures, perhaps helping to explain its distinctive style, while the Netherlands' geographical location encouraged a sophisticated system of peripatetic operatic circulation, contributing to a defining feature of the genre in the international movement of performers and repertoire. By the eighteenth century there was growing awareness of the opera house's potential to manifest and broadcast civic ambition, as essays by Michael Burden and Margaret R. Butler on late eighteenth-century London and Turin show. Butler demonstrates the care with which eighteenth-century Turinese impresarios engineered entertainment to construct their city's greater glory and restrict what they saw as less prestigious forms, showing how the opera house (like the later concert hall) served to mark out territory—to aestheticize the city in structural terms and thereby affirm elite control of the civic environment. In the case of London, on the other hand, Burden proposes that the everyday commercial demands of this, Europe's largest and most modern of cities, often thwarted civic aspiration in practical terms, no matter how clearly and ambitiously it was expressed.

The situation of the opera house in Europe's civic landscapes in the nineteenth century invoked both unabashed commercialism and a heightened (and commercially exploited) self-awareness, as is shown by Susan Rutherford's essay on the problematic touristic portrayal of Venice and Jonathan Hicks's on the insertion of Don Giovanni into the ebullient original "Tom and Jerry" craze in *perpetuum mobile* London. Whether in London or Venice, it seems, the weight of historical and cultural expectation framed narratives around the opera house much as they framed operatic narratives themselves—indeed, in this period, when the opera houses are well established, it is the stories told about opera and its locations that most emphatically enhance its edifice.

The sophisticated web of national, regional, and local self-definition and promotion that underpinned such narratives extended still further in the nineteenth century, of course: opera was part of an international cosmopolitan culture that, while centered on Europe, saw both colonial "outpost" and European center enhanced by the connections forged between them and the stories told about them, as Charlotte Bentley, Benjamin Walton, and Roberto Diaz show in essays on New Orleans, Calcutta, and Buenos Aires, respectively. The cachet of opera's Eurocentric and elitist associations made it attractive even in countries where the difficulties of its multimedia representation should have precluded it, helping to explain the attractions of operatic concert programs in factional Shanghai, and of touring opera round the "all-red" British colonies, as Yvonne Liao and Kerry Murphy show respectively. In fact, these difficulties were part of opera's appeal—a way of indicating that New Orleans, Calcutta, Buenos Aires, Sydney, Melbourne, and Shanghai had "arrived" economically as much as culturally in the late nineteenth and early twentieth centuries. The association of opera with power seemingly also made it an ideal means to project colonial or European might beyond the Old World— opera in India, the Americas, Asia, and the antipodes, whether in custom-built houses or nonspecialist theaters, could, through its attendant social rituals as well as its music, create a penumbra of exclusivity that appeared to reinforce social hierarchies just as effectively as military muscle might do. But, as all of these essays demonstrate in different ways, the compromises and contingencies of opera's adaptation to new environments inevitably called into question European hegemony, destabilizing narratives of cultural superiority. Indeed, because of their association with the performance of power, touring artists and operas often found themselves at the center of shifting and competing projections of imperialism and regionalism. Taken together, these essays suggest that, in its representational and performed complexity, colonial opera has the potential to add vitally to cultural geography's ever-shifting understanding of negotiations of power and meaning in colonial spaces.

The destabilization of opera's status was felt in Europe in the late nineteenth and early twentieth centuries too. The genre's populist potential in this period—which was perhaps the inevitable corollary of its place, geographically and ideologically, as a fulcrum of national (and nationalist) life—prompted growing concern about and interest in its function within the (variously defined) popular urban environment, as Peter Franklin shows for fin-de-siècle Vienna. Perhaps responding to such concerns, some early twentieth-century operatic endeavors sought to break opera's association with the city altogether, offering a new idealization and (in some senses) sanitization of the genre, whether in the French towns of the Midi that Katharine Ellis explores or in the English countryside and country-house ventures of Glastonbury and Glyndebourne that I examine in my chapter. The imbrication of opera and consumerism was not to be abandoned, however, and as Klaus van den Berg and Roberto Ignacio Díaz show (as much as Ellis and I do), the new opera houses of the twentieth century have all, in various ways, embraced the staging of operagoing as itself an act to be consumed. That the edifice of the opera house remains an important marker of opera's politicocultural status is demonstrated in the care taken over the design (always freighted with meaning) of modern opera houses, as well as those of the past—demonstrated by the Bastille Opéra in Paris and the Winspear Opera House in Dallas (van den Berg), as much as by the Teatro Colón in Buenos Aires (Díaz). But internationalism and cosmopolitanism could also give way to localism, in opera as much as in other popular genres:[27] just as opera's association with the expression of power lent itself early on to the building of overtly national houses (particularly in central Europe in the mid-nineteenth century),[28] interest in national or local identity also played itself out in operas themselves, whether in nineteenth-century Venice or twenty-first-century Buenos Aires. Indeed, Díaz suggests, it is up to smaller, modern venues and new works to offer a challenge to more traditional, officially sanctioned views of the social function and meaning of opera and the opera house.

As this chapter overview suggests, within the volume's broad historical examination of opera's place in its physical environment, other themes appear that are also informed by developments in cultural geography and that enrich our presentation of the interface between that field and opera studies. With the recent changes in cultural geography came a shift from a Marxist view of the cultural (and physical) landscape as one inevitably determined by dominant socioeconomic forces to an acknowledgment and accommodation of multiple forms of ownership and expressions of identity. This contributes to our approach in this volume by complicating the understanding of a genre traditionally seen as the plaything and projection of the elite. Indeed, there is increasing recognition both of the sophisticated and diverse ways in which hegemony (whether that of capitalist globalization or more traditional patriarchal modes) is resisted at the popular level, and, by way of counterbalance, of the thoroughgoing cultivation of everyday life as "spectacle" or performance, in ways that reinscribe and in some senses democratize (capitalist) hegemony through ubiquitous commodification.[29] From both perspectives, narratives of cultural geography in and about the opera house are enriched, as the essays by Hicks, Walton, Murphy, and Franklin, on the one hand, and those by Eubanks Winkler, Burden, Bentley, van den Berg, and myself, on the other, show.

Issues of identity politics that have become central to the field of cultural geography "through the less-than-tangible, often-fleeting spaces of texts, signs, symbols, psyches, desires, fears and imaginings" also have obvious resonance for opera studies.[30] The fluid cultural configurations bound up with a sense of place, and the mapping and classifying of identities contingent on that sense, should affect our understanding of opera, as a genre often treated as representative of different constituencies. The politics of identity features widely in this volume, whether considered from hegemonic perspectives (as with Ahrendt's examination of legislation in The Hague, Butler's study of eighteenth-century Turin and its all-powerful impresarios, or Rutherford's examination of mid-nineteenth-century representations of Venice) or from the point of view of a broader urban populace (as in Franklin's investigation of fin-de-siècle Vienna and Ellis's of provincial France), or that of idealized individuals (as in Hicks's essay on "Tom and Jerry" in London and Díaz's study on the representation of Victoria Ocampo at the Teatro Colón).

Recognition that identity is contingent on many factors, including on sense of place, and that it is therefore susceptible to mutation as circumstances change, has in turn over the past decade contributed to cultural geographers' increasing interest in the dynamism of their objects of study, expressed through encounters with mobility, migration, transnationalism and diaspora within a modern climate of globalization.[31] Scholars such as Mimi Sheller and John Urry have proposed a new "mobility paradigm" within cultural geography.[32] While our consideration of the geography of the opera house within the city might seem to conform to rather traditional ideas of (urban) space as a fixed "container" for social action, a more relational and mobile view of space helps illuminate the practices and interpretations of opera in the urban environment.[33] This is most obviously true for the earlier period of the genre's history, before fixed opera companies in designated opera houses became the norm; Ahrendt's discussion of the ways seventeenth-century contract law facilitated the movement of performers (and other artists) around Europe gives new resonance to the etymology of "entrepreneur" as someone responsible for carrying business not only between parties but between places. But equally, the solidity of establishment venues in later periods seems to have encouraged a renewed awareness of the value of mobility—in both physical and discursive terms—as Hicks's essay most clearly demonstrates but Ellis's, Diaz's, and my own also attest.

Within the mobility paradigm, cultural geography has joined with postcolonial studies to investigate not just "peripheral" reactions against hegemony, but also the fragmented and multiply contested projections of normativity coming from the supposed center.[34] Opera, as an art form that (like all the arts) both represents and tells truth to power, is increasingly recognized as having played (and as still playing) a significant role in networks of cultural transfer that helped to create and, equally, to contest national and regional identities. Opera's cultural mobility has allowed it to be used to negotiate, heighten, and at times transcend boundaries of identity. These processes occurred—and occur—within European countries, particularly (though not exclusively) in the self-consciously nationalistic nineteenth century—as Butler's, Rutherford's, and Ellis's essays all show. But they were just as potent when opera was exported to other territories, as a marker in various ways of European "civilization" and power, as well as a means of resisting and complicating it—in turn "provincializing" Europe.[35] The physical trace of colonial power in the urban environment, following the pioneering work of Jane Jacobs's *Edge of Empire: Postcolonialism and the City* (1996), reminds us that the opera house as a presence in the landscape could be a potent marker of colonial might. Bentley, Walton, Liao, Murphy, and Díaz all variously explore the ways in which opera outside Europe—and particularly opera on tour—contributed to a mobile transnational culture while also interacting (often in unlooked-for ways) with local cultural geographies. Attending the opera in such places, these essays remind us, could simultaneously serve to express personal taste, local status, and a delicate balance between alliance to and cooption of a presumed European cultural center, on the one hand, and subservience to that culture, on the other. Operatic listening—like contemplative listening more broadly— signaled not (or not simply) one's separation from the world but, as Richard Leppert put it, "one's control and domination of the world," as an activity that stops—and so is privileged over—all other activity.[36]

Placing opera and the opera house within the field (or viewing it through the lens) of cultural geography is in many ways also to understand it as one form of media among many within the urban environment. An interpretation of the history of opera and its venues in cultural-geographical terms then also invokes one branch of the field of media archaeology, in that it seeks to disrupt the traditional teleological and isolationist narratives that generally pertain to high art by instead situating these (and other) media in richer, more discursive social contexts, and acknowledging the interconnectedness of media in historical terms.[37] This now-venerable approach was pioneered by art historians such as Aby Warburg, Erwin Panofsky, and E. H. Gombrich, who were interested in placing artworks within history and linking recurring visual motifs across genres high and low, much like the cultural critic Walter Benjamin in his *Arcades Project*. The latter work, on nineteenth-century Parisian culture of all kinds, is used as a starting point by van den Berg in his essay on three modern opera houses and the ideology of their design and situation.[38] Butler's study of mid-eighteenth-century Turin shows that totalizing approaches to urban spectacle are not simply the projection of the modern historian, but were carefully stage-managed realities in environments where, because of the relatively confined nature of the urban setting and the hegemonic social structure, it was possible for one group to enact a uniform vision across a city's various forms of media. Hicks and Bentley demonstrate the importance of popular literature and theater to the cultivation and dissemination of operatic cultures within Europe (Hicks) and between Paris and New Orleans (Bentley), as Walton shows equally for visual representations of colonial outposts in mediating ideas of Italian opera. Franklin's examination of early twentieth-century Viennese opera also adopts an intermedial logic to justify the reappraisal of populist works previously deemed unworthy of attention. These and other studies' approach to opera as one medium within a network, the significance of which was (and is) partially constructed geographically, stimulates the connections between opera, media archaeology, and cultural geography, not least by offering a counterbalance to other approaches to media archaeology that have become popular within musicology (particularly German media theory, following Friedrich Kittler's work).

Operatic Geographies thus seeks to deploy aspects of the contemporary practice of cultural geography and related fields, in order to show that for opera the "new historicist" principle that "every expressive act is embedded in a network of material practices" applies not only to the materiality of performance, but also to the physical environment of production.[39] In so doing, this volume will address a significant gap in our understanding of opera's cultural meanings and encourage further study. Through it we also hope to stimulate thought within the fields of cultural geography and media archaeology about the interface between environment and an art form—opera—for which the intertwining of the social and spatial is integral, both on and off the stage.

TWO

The Legal Spaces of
Opera in The Hague

REBEKAH AHRENDT

Around 1717 Anna van Westerstee, also known as Anna Beek, published a newly engraved map of The Hague. Her *Nieuwe Platte Grond van 's Gravenhage*, dedicated to the town's magistrates, expanded on a feature of previous maps: "public buildings" were carefully numbered, with a tabular key provided. New with Beek's map was the number and kind of buildings so depicted—testimony to the town's great growth in status and population in the decades around 1700. And for the first time, the map included an opera house, on a site that had been unmarked on previous maps, here numbered 37 and circled in figure 2.1.

How the opera house ended up on Beek's map is a tale of urban renewal and spatial reorganization that is in line with many such accounts of opera's participation in civic life. But it is also a story of how opera and law interacted—of how savvy entrepreneurs worked within (and on the margins of) a legal system. To date, opera studies writ large has primarily considered law in terms of intellectual property rights or institutional organizations.[1] Legal documents, including contracts and lawsuits, have been read as evidence for who was employed where, how, and at what time, particularly in studies of early modern opera.[2] I propose in this chapter to read legal documents from a different angle: for what they can tell us about the gradual, historical integration of operatic troupes into the European urban landscape. In this I

THE LEGAL SPACES OF OPERA IN THE HAGUE

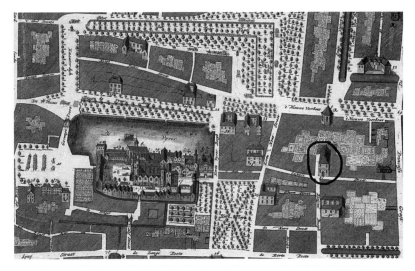

2.1   Anna van Westerstee, *Nieuwe Platte Grond van 's Gravenhage* (c. 1717), detail. Biblioteca Nacional de Portugal.

respond to recent calls for recognizing law and the legal as cultural constructions, as dependent upon and constructive of place as any other aspect of cultural geography.[3] Following Naomi Mezey, I intend not merely to show that law partakes of culture or that culture refracts law, but to demonstrate that they are mutually constitutive.[4] In the case of opera, legal agreements helped constitute the urban opera house, its inhabitants, and even its repertoire. Only through these agreements could the opera house become a site of performance and a feature of urban geography. In other words, it was through legal documents—themselves negotiated through acts of performance—that opera became a legitimized space. And opera in turn helped shape law: it caused cities to rethink urban planning projects, to regulate performance spaces, to legislate the identities of (foreign) performers in relationship to natural citizens, to reform tax laws to accommodate mobile populations and sporadic performance.

My concern is thus not just with opera's spaces, but with how space is itself inherently mobile and mutable.[5] Space, like law or culture, is a product of interrelations, of heterogeneity, of relationality—it is produced and productive.[6] Particularly salient here is the period around the turn of the eighteenth century when opera, at least in cartographic terms, was becoming *less* mobile. Indeed, one might view this period as the most significant for establishing a tradition of enduring urban opera houses. Yet, the fixed, purpose-built opera houses with stable companies that are

13

generally associated with opera's urban functions were still a rarity. Many itinerant opera troupes still performed on temporary stages, in the open air or in buildings originally intended for other purposes. Nor, when the new opera houses emerged, were the personnel fixed; rather, the opera house's integration into the broader European operatic labor market enabled personnel mobility. And like the constitution of the opera house itself, that mobile labor pool depended on shared concepts of law.

My primary thesis is that a shared European legal foundation—the *ius commune* based on Roman and canon law that persisted until the codification movements of the late eighteenth and nineteenth centuries[7]—enabled performer mobility and the establishment of opera. People carried their rights and privileges under Roman law with them, whether they worked on the Italian peninsula, in France, in the Dutch Republic, or elsewhere.[8] This transposability of legal norms across borders provided a common understanding of legal entitlements and obligations, which in turn shaped the foundation of opera companies. A common basis in law meant that performers and impresarios could expect to find the same terms in legal contracts across Europe. At the same time, as we will see below, the local ordinances and case law that overlaid the *ius commune* could provide a source of confusion or frustration for mobile populations, particularly those, such as performing troupes, who depended on the goodwill of local authorities.

Contract law in particular was vital to the establishment and maintenance of opera companies as regular business operations. As an aspect of private or civil law within the *ius commune*, participation in the formation, execution, and enforcement of contracts was not limited to citizens, meaning that even temporary residents or economic migrants could enjoy contracts' benefits and limitations. Accordingly, opera companies, which, like traveling theater troupes, were largely made up of noncitizens and even nonresidents, could conclude internal contracts of association and employment as well as external contracts with local venue owners, financiers, and artisans. Contracts made in any jurisdiction—whether domestic or foreign—were legally enforceable across borders, further contributing to performer mobility.

That is not to say that contracts were *exactly* the same everywhere: while European states shared the fundamentals of Roman law, local variants were legion. Hence, the ways in which contracts were enforced or even created were shaped by and subject to local jurisdiction. At times, local public law could limit the enforcement of contracts made in other jurisdictions; for example, an ordinance of the city of Amsterdam banned foreign performers from the city's stages in 1683, removing the capacity of foreign performing troupes to act on contractual relationships that had been formed abroad. In effect, the city's ordinance voided preexisting contracts concluded by foreign performing troupes with venues, service providers, and performing personnel.[9]

It was in part because of Amsterdam's ban on foreign performers— enforced well into the eighteenth century—that opera in the Dutch Republic more readily found a home in The Hague.[10] The Hague was not like other European locales, nor even like other Dutch ones. In what follows, I will examine the unique governmental and juridical structures of The Hague and their interaction with the establishment of opera in the town around 1700. From the first recorded opera performances of 1682 until the demise of a long-standing company in 1714, French-language opera was a significant part of urban society—unlike anywhere else in the Dutch Republic.[11] My examples here largely derive from a company that left an extraordinarily long paper trail thanks to the many legal documents its members created and attempted to have enforced. This was due in part to the unusual juridical landscape of The Hague, which fundamentally shaped the organization, location, obligations, and personnel of the company, as well as their litigious tendencies. And yet, as I emphasize, the opera company's participation in the town's legal geographies transformed what was essentially outsider space—theaters run by and for marginalized foreigners in a seedy entertainment district—into a signifier of prestige, a provider of social welfare, and a successfully redeveloped theatrical center.

Stages were regulated, and custom—based on the Roman concept of the *privilegium* granted to a corporation—dictated that performers obtain permission from the powers that be. The granting of such privileges in turn reinforced the sovereignty of local authorities, who were thus empowered to determine *where* performances could occur, and, by extension (or outright censorship), *what* could be performed in a given place. Thus, before an opera company could even begin to prepare for performances in The Hague, it had to seek permission to perform from the local authorities. This obligation was a tradition in the Dutch Republic for all public performing groups, dating back to the turn of the seventeenth century and following similar procedures developed in the French provinces.[12] Given that most itinerant troupes working in the Republic originated in France, and that many "Dutch" (though French-speaking) troupes performed in the French kingdom, it is unsurprising that the two lands' permissions processes developed in tandem. By the time opera arrived in the Dutch Republic in 1682, the process had been thoroughly formalized as part of local custom.

The first step in gaining permission was to submit a request—having found out to whom it should be addressed. A town like The Hague (it never received city rights) presented special difficulties for obtaining performance permission, for this "most beautiful village in Europe" was in fact *two* villages.[13] The central area of the town, or Hofbuurt (Court neighborhood), housed Dutch government institutions and associated elites, and was jurisdictionally and politically separate from everything else surrounding it—the true "village," with its mix of classes and occupations.[14] The Hofbuurt included lands originally belonging to the counts of Holland, whence the town's proper name, 's-Gravenhage ("The Count's Hedge"), derives. This area, roughly that featured in figure 2.1, and including the Binnenhof, the Buitenhof, the Vijverberg, the Hofstraat, the Plein, and part of the Kneuterdijk, came under the jurisdiction of the Hof van Holland (Court of Holland), the country's highest court, which also had authority over all Dutch nobility. The use of land and buildings within the Hofbuurt was regulated by the Hof van Holland and its chairman, the stadholder; in the absence of a stadholder, the States General took control. Outside the Hofbuurt, The Hague was administered by the local magistracy, led by the town bailiff (*baljuw*, a lifetime appointment) who oversaw three nominated mayors (*burgermeesteren*) and the members of the town council (*schepenen*). The magistrates had the legal authority to create and enforce town ordinances as well as to adjudicate local civil and criminal disputes.[15]

Because of the bifurcated jurisdictional nature of The Hague, an opera company had to know their performance site *before* obtaining authorization. If performances occurred within the magistrates' jurisdiction, as was most common, companies were subject to a number of regulations, including mandatory donations to local poor relief in order to appease the Dutch Reformed Consistory, which was generally opposed to public performances. The Hof van Holland, perhaps because of the stadholder, seems to have been more lenient than the town council in granting permission to performers, as the Consistory could not act against it; as a result, many theater companies set up their portable stages on the Buitenhof.[16] This was not a good option for opera companies, however, which needed a theater properly outfitted with a stage and—preferably—stage machinery. One such theater did exist on Court land, a former stable belonging to the stadholder known as the Manège or Piquerie (visible to the extreme left of fig. 2.1, below the square). However, this theater had been long occupied (rent-free!) by the French-language spoken-theater companies preferred by the stadholder's court, and as such was not an option for opera.[17] Rather, opera companies from 1682 on favored a former tennis court within the magistrates' jurisdiction, discussed further below.

Petitioning the wrong authorities caused at least two performing companies to fail in The Hague. In 1605 the town's first recorded French-language theater troupe was expelled for violating their permission. They had solicited the magistrates' consent but proceeded to perform on the Buitenhof without the Hof van Holland's authorization.[18] Rivalry for local power between town and Court persisted across the seventeenth century, causing the various authorities to closely guard their rights and jurisdictions. Such was the case in 1697, when the experienced opera singer and director Jakob Kremberg, formerly of the Hamburg Gänsemarkt Opera, received permission from the magistrates to establish an opera company in The Hague, but with the stipulation that a performance location had to be specified before performances could start.[19] This Kremberg seems to have been unable to do, for the mayors made no further mention of his project and Kremberg seems to have moved on to London. No operatic activities are again recorded in The Hague until three years later, when Kremberg's former Hamburg rival Gerhard Schott received permission to form a company.[20]

It is perhaps no accident that Schott, like many an Italian impresario before him, was a lawyer. Schott had previously utilized his legal talents and connections to ensure the success of Hamburg's Gänsemarkt Theater; for whatever reason, he decided to establish a franchise in The Hague in 1700. Schott's original request to form an opera company in The Hague does not survive, but the response from the bailiff and mayors clearly indicates that Schott made use of his training and titles in order to assert his authority. The magistrates mention that they had received a request in French, "in the name of and by order of Monsieur Schott, said to be 'Senator and first judge of the city of Hamburg.' "[21] Schott's self-identification in the request submitted by his associate François Colombel placed the Hamburg senator on an equal footing with the magistrates of The Hague.[22] Presenting himself as an equal would have served only to enhance the authority of the magistrates, for here was someone with great legal and social standing appearing as a supplicant. Schott surely recognized this and was likely confident of his success.

Schott's request was indeed approved, provided that the company donate two stuyvers (about 1 euro) per spectator per performance for the benefit of the poor. Poor relief—the Calvinist excuse for worldly entertainments—was a long-standing tradition in the Dutch Republic.[23] The practice had a history elsewhere too, particularly during times of war, need, or public outcry against theatrical frivolities.[24] For example, Louis XIV decreed in 1699, during a time of financial crisis, that a sixth of the revenues of the Opéra and the Comédie would go to the Hôpital Général.[25] In Schott's case, the mandatory donation was to be collected after every performance and delivered to the deaconate of the Reformed Church. Under the original terms of Schott's permission, the accounts of the deaconate record "gifts" of 140 guilders in 1702 and 150 guilders in 1703, indicating that the opera drew some 1400 paying spectators between late October 1701 and March 1702 (there being 20 stuyvers per guilder or florin) and 1500 spectators during a similar period the following year.[26] Though the sum received by the deaconate was fairly insubstantial, these "gifts" had the symbolic effect of integrating the opera company into the town's social norms.

Further evidence of integration is found in the many contracts that this French Opera of The Hague concluded over its years of existence. Contracts of association formalizing the structure of the company, rental contracts for the physical space of the opera house, financial agreements with creditors, and employment contracts concluded both in The Hague and abroad constitute further spaces in which opera's legal status was negotiated and formalized. Hence, these documents—in their material presence, in their delineation of spatial relationships and legal obligations, in the associations they created between persons—can be taken as part of the geography of opera in The Hague.

The first important legal step to be taken after the town council's permission was received was to formalize the company's structure: in order to be treated as a corporation—which in The Hague would provide exclusive rights to operatic performance—the opera company needed to *be* a corporation, participating in the Roman tradition of *societas*. Essentially, *societas* formalizes the pooling of resources for a common purpose; association members have independent charges and duties and share a community of property and assets, all of which are outlined within a contract of association.[27] Such documents were particularly important for companies that operated on an entrepreneurial model, as was true of most opera troupes by 1700.[28] The entrepreneurial or impresarial model accorded partnership rights and duties within the corporation only to the signatories of the contract of association, who tended (for opera, unlike for spoken theater) to be nonperformers.[29]

Creating a contract of association meant that the new company's directors needed to enlist a notary's services. The importance of notaries cannot be overestimated when considering the geographies of opera in the past. Notaries were the first point of contact to both domestic and foreign legal systems. Theirs was a respected and often well-remunerated profession, which carried with it the obligation to correctly instruct their clients in matters of law.[30] If they did not, notaries were in danger of losing both their accreditation and their standing in the community. Thus, it was in a notary's best interests, both professionally and socially, to keep up-to-date with legal developments, including the enactment of local ordinances or the progress of customary law in the courts. Notaries often had close relationships with local courts and lawyers, who provided channels of information legitimizing practice. Most notaries indeed signed their acts with reference to their accrediting court, particularly if they were associated with the higher courts of the land. The physical location of the notary's office, too, could provide a strong signal to potential clients about the relationship between the notary and local power structures. All of these factors contributed to the establishment of a notary's reputation, an essential quality on which depended his business and his client relationships.[31] Given these factors, it is unsurprising that opera troupes tended to choose—and choose very carefully—one principal notary for concluding the company's contracts.

Proximity to power seems to have been a deciding factor in the choice of notary made by the French Opera of The Hague, though reputation and confession may have played a role. On 31 May 1701 the opera company's would-be directors repaired to the office of the favored notary for Huguenots and well-heeled foreigners in The Hague, Samuel Favon. Favon was a respected member of the French Church, where Stadholder-King William III worshipped. His was the only notarial office on the Binnenhof, located between the apartments of the stadholder and the French Church, thus at the very center of power in The Hague, in the midst of the opera company's intended audience.[32]

The act of association reveals that the direction and proceeds of the company were to be divided into three parts: one-third for Schott; one-third for the Huguenot Jean-Jacques Quesnot de la Chenée, who signed the contract for both himself and Schott; and one-third for Louis Deseschaliers, to be shared with his wife and business partner, Catherine Dudard.[33] Each associate had certain responsibilities: Schott agreed to furnish the company with scenery, decor, and costumes in his possession that had been used previously in Kiel. Quesnot would travel to Hamburg to collect properties from Schott. Quesnot would oversee the day-to-day operations of the company and appoint a receiver to handle daily expenditures. Deseschaliers was charged with directing the music and engaging singers in Brussels, Rouen, or Paris. Dudard would serve as properties manager and conduct business on her husband's behalf whenever he was away. As stipulated, the contract would endure for three years, and the opera would open on 1 October 1701 with a production of Lully and Quinault's *Armide*.[34] The contract was duly signed in the presence of the notary and two in-house witnesses: Favon's son, Samuel Favon, Jr., and Philippe Beaujean.

Philippe Beaujean deserves special mention, for he provided a particular connection between the opera company and the legal landscape of The Hague. His work in Favon's office—witnessing requests of all sorts, perhaps assisting in document preparation or ensuring that Favon's scribal supplies were in good order—was a day job. Otherwise, he was a bass player in the opera company's orchestra. In an act passed by Favon on 2 February 1704, Beaujean's signature appears not as a witness, but as a symbol of his membership in a group of "actors, actresses, dancers, and others of the aforementioned Opera."[35] Beaujean was in fact a versatile theater musician who had worked with provincial theater companies in France in the 1690s.[36] Indeed, Beaujean's experience in negotiating multiple engagements with notaries across France may have qualified him to work in Favon's office.

Having made their association official, it remained to the new Société des Concerts (as they first called themselves) to find a theater. Since the town council, rather than the Hof van Holland, responded to Schott's original request, it seems it was assumed that the company would perform in the only suitable theater under the magistrates' jurisdiction: the tennis court (*jeu de paume*) of Pierre van Gool, located on the Casuariestraat, steps away from the Buitenhof. The Casuariestraat was historically a fringe adult-entertainment area of the town. The street was named for its most famous occupant, the storied brothel known as De Casuaris (a reference to the imagined sexual proclivities of the cassowary).[37] The Casuariestraat was marginal in jurisdictional terms as well: the north side of the street was within the jurisdiction of the Hof van Holland, while the south side was controlled by the local government. Two *jeux de paume* had been built on the street in the early seventeenth century, one on the north side and one on the south. That belonging to Van Gool, on the south side and thus under the control of the magistrates, had been the theatrical venue of choice for traveling opera troupes since at least 1682.[38] By 1700 it had been expanded, remodeled, and equipped with loges.[39]

On 6 June 1701 Quesnot and Dudard (on behalf of her husband, who was off in search of talent) signed a contract with Van Gool's widow,

Johanna van Rode, for the theater, a storeroom, a small garden, and two small houses, then occupied by a Dutch joiner and a French hatter (both potentially useful to an opera company).[40] This theater would serve the company well until 1709, when it moved across the street to the other indoor tennis court–*cum*–theater—the *jeu de paume* formerly of Pierre Noblet, commonly referred to as the "Royal Tennis Court," which is the building indicated on Beek's map.[41] Thus, for the first time, opera and the opera house moved into a different jurisdiction and a symbolically higher level of society, under the direct control of the States General.

Though a close relationship to the States General was no doubt the company's aim from the start, it first had to survive its infancy. Opera companies were expensive to run, so it is unsurprising that the company soon found itself in financial trouble. Even before the official opening (nearly a month late) on 28 October 1701, the company's directors had taken extraordinary measures to raise capital from creditors. What legal rules applied to such transactions varied from place to place and according to the gender of the borrower, as we shall see.

On 30 September 1701, Catherine Dudard, holder of a one-sixth interest in the opera company, appeared before the Amsterdam notary Hendrik de Wilde. She traveled from The Hague with her longtime friend, the singer Marie-Thérèse Charpentier.[42] Charpentier was a leading singer in the new company as well as a source of financial (and, probably, moral) support for Dudard and Deseschaliers. Dudard's mission was to obtain a loan of 2000 florins (a considerable sum), from "master surgeon" Élie Hébrard of Amsterdam in order to pay the travel expenses, salaries, and maintenance of the *opéristes* who had been hired in Paris by Deseschaliers.[43] The need was urgent, as the singers were expected in The Hague at any moment. But as Hébrard had already loaned 500 *Reichsthaler* (about 1250 florins at the time) to Quesnot in order to retrieve properties, scenery, and costumes from Schott in Hamburg, he reserved the right to place his own collection agent within the theater if the debt were not repaid by the deadline of 30 December 1701. And Charpentier, *jeune fille majeur*, provided "even greater surety" (*encore plus grande seureté*) for the loan by agreeing to treat the 2000 florins "as her own debt" in case Dudard defaulted.

Charpentier's support for Dudard's loan presented a legal difficulty, however. Charpentier, having reached her majority, unmarried, and likely orphaned, might have expected to be able to obtain loans with no hindrances, as she could in multiple French jurisdictions, including Paris and Toulouse.[44] There was no father, brother, or husband to speak for her, and she seems to have enjoyed financial freedom, based on subsequent personal loans made to Dudard. But De Wilde's contract makes clear that Charpentier, as a woman, was not authorized to stand surety for the loan. As he worded it, taking on this obligation meant

renouncing for this purpose all exceptions and benefits, especially those of *Senatus Consultum Velleianum* and the ordinances of this City, which state that no women or girls may stand caution or oblige themselves to others, of which she acknowledged having been instructed by myself, the aforementioned Notary, obliging to this her person and each and every one of her goods, present and future.[45]

De Wilde here refers to two separate traditions of law operative in the Dutch Republic: the heritage of Roman law and the city of Amsterdam's local laws and customs. He specifically mentions the *SC Velleianum* (46 CE), a Roman Senate decree prohibiting women from guaranteeing others' debts. This imposed legal protection was intended to protect a socially weaker group—in this case, to prevent the financial ruin of women who might not be acting of their own free will.[46] While Charpentier may have been familiar with *SC Velleianum* from French law, it seems unlikely that she would be aware of Amsterdam's local prohibitions, hence the necessity for De Wilde to "instruct" her, part of his obligation as a notary. Had Charpentier *not* waived her protections, Hébrard would have found it difficult to pursue any breach of contract in court, especially in Amsterdam. As this would place Hébrard in an unfair and potentially extralegal position, De Wilde would not have formalized this contract without her renunciation. Charpentier thus received a lesson in women's legal disabilities in the Dutch Republic—even though in practice, unmarried women in the Dutch Republic, like their Venetian counterparts, generally enjoyed independent legal capacity.[47]

As for Dudard, married women were not legally competent in most European locales; therefore, she should have been unable to perform independent legal transactions.[48] But Dudard and her husband had already recognized this disability and devised a remedy. Two years prior, when they were contemplating the journey to Warsaw, Deseschaliers had granted Dudard full authority to conduct business on his behalf in the form of a document known as a general procuration.[49] This document seems to have had legal effect outside of France, for when presented to de Wilde, the Amsterdam notary recognized it as the foundation for Dudard's contract with Hébrard. De Wilde remarked that Dudard appeared before him in person as the "wife and procurator of Louis Deseschaliers, director of the Royal Academy of Music of Lille in Flanders, based on her general procuration for all their affairs, concluded by Sieur Pierre Roland, Royal notary of the residence of the city of Lille, dated 10 February 1699."[50] We might see Dudard's procuration as a symbol of the trust Deseschaliers placed in her as an equal partner both in life and in the business of opera.

Dudard obtained her loan from Hébrard just in time to fulfill the contractual obligations her husband had initiated with the company's performers. Like the contract of association between troupe principals, employment contracts with performers stem from the practices of traveling theater troupes across Europe and became essential in the entrepreneurial business model. Yet few contracts for opera performers exist in the notarial records of The Hague. This could indicate that, like Venetian *operistas*, the *opéristes* concluded informal or verbal agreements with the company's directors.[51] Alternatively (and more probably), it could be because nearly every performer in The Hague was recruited abroad. The originals of contracts signed in Paris, Brussels, or Rouen would be stored in notaries' offices there; if needed, signatories could obtain copies to carry with them.

What is truly remarkable is that these legal documents enjoyed cross-border vitality—because, as I have emphasized, they were based on common law as it was understood across Europe. Thus, just as contracts signed in Metz established a company that would operate in Sweden,[52] so the agreements that Deseschaliers concluded in Paris were valid in The Hague. Remarkably, contracts drawn up in Italy, France, the Dutch Republic, and elsewhere during this period contain nearly identical legal language and obligations. An agent promised to arrange performers' travel, maintenance, and wages; performers in turn agreed to a number of formalized obligations, including attendance at rehearsals, being provided for in case of illness, and the length and nature of employment.[53] Perhaps this might exemplify the systems that enabled the free movement of labor throughout Europe long before the European Union was imagined.

One singer who signed a contract in Paris for employment in The Hague was Marie Rochois. Rochois, the niece and heir of the famous Paris Opéra star Marie Le Rochois, who created the title characters in Lully's *Armide* and Charpentier's *Medée*, was first hired by Deseschaliers in Paris on 2 August 1701. Her initial contract was drawn up in the most respected notarial chamber in Paris: the Conseillers du Roy, notaires au Châtelet du Paris, who, beginning in 1696, possessed the right to affix the seal of Louis XIV.[54] Rochois's agreement obliged her to sing whatever roles

Deseschaliers asked her to, in The Hague or "in whatever other places that [the opera] might be transported" (*en tels autres endroits ou il poura estre transporté*) and to be present at all rehearsals on pain of a fine. The contract stipulated that she neither quit the opera nor leave The Hague at all for at least one year without giving Deseschaliers at least three months' notice.[55] Curiously, the last stipulation essentially repeated an earlier one: Rochois was required to "follow the opera wherever it might be transported" (*suivre ledit opera par tout ou on le transportera*) at Deseschaliers's expense. Performer mobility was clearly of paramount concern, whether the company stayed at home or toured abroad.

In November of 1704 Rochois renegotiated her contract in The Hague. The contract bears every mark of a negotiation still in process, of the conversations that continued even in Favon's office. Nearly an entire clause is stricken; a broad left margin gave Favon space to insert additional phrases, neatly signaled by crosses or carets. And the closing formula, common to all such contracts across Europe, gives it further liveliness:

> For the accomplishment and execution of all that is aforementioned, the said appearing parties together and each one individually oblige and have obliged by these presents their persons and goods, nothing excepted, submitting them to the rigor of all Courts, Judges, and Justices. Made and passed in The Hague, in the presence of Philippe Beaujean and Samuel Favon, Jr., Clerk, witnesses to this legal demand.[56]

The phrase "ces présentes," from "ces présentes lettres," acquired its value within the medieval chancellery: the notion of an event that manifested itself at the moment it was spoken, representing something that happened before one's very eyes. The written record thus carried the memory of the act that passed between Deseschaliers and Rochois. By the act of appearing together before Favon in a space designated for the conclusion of legal agreements, these two were able to enact a legal relationship. They pledged not just their goods but themselves to the positive accomplishment of the acts formalized by the contract. Should the contract be breached, their bodies and goods would be subject to all justice—which in The Hague meant a lot of justice.

The Hague was (and is) a town of courts, as well as a (sometimes) courtly town. Besides the local magistrates' tribunal, The Hague was home to three different provincial high courts, the military high court, and the highest court in the Dutch Republic. Plaintiffs therefore enjoyed great freedom in finding remedies, practicing what is nowadays termed "forum shopping." There was a hierarchy and also a procedure to be fol-

THE LEGAL SPACES OF OPERA IN THE HAGUE

lowed, certainly, but the acknowledgment of multiple legal systems was itself a part of the language of contracts. Given the unusually high degree of legal pluralism, it is perhaps unsurprising that this opera company was inordinately litigious. Lawsuits among performers, directors, theater owners, printers, and creditors litter the legal landscape throughout the eighteenth century's first decade. The magistrates may have come to regret their decision to grant the troupe permission to perform, for that permission—like a contract—created obligations for the magistrates as well as for the company. The magistrates were responsible for policing the company and for being the court of first instance should difficulties arise. And difficulties were legion. Between 1702 and 1704 members of the company, local artisans, and hired help came before the court of the magistrates at least thirty times, with complaints ranging from trickery to failed payments to unfulfilled employment obligations to accusations of blasphemy.[57]

However, the integration of the company into The Hague's urban fabric had significant effects. First, it increased the town's prestige. Bringing local entertainments into line with those of European capitals like Paris or Rome increased The Hague's stature, in both image and fact. Second, the troupe's contributions to urban poor relief, no matter how small, had the symbolic result of assimilating popular entertainment into social welfare schemes, thereby transforming members of the opera troupe into upstanding members of society. Lastly, this integration brought the theater itself closer to the well-heeled bourgeois's concerns in The Hague. What had been a shady adult-entertainment district became gentrified, a locale for polite diversion. And by the time Beek's map was prepared, the opera had become a fixture—a destination, a dominant part of the urban landscape, inventoried along with other urban institutions such as churches, embassies, and the poorhouse.

THREE

Opera at School: Mapping the Cultural Geography of Schoolgirl Performance

AMANDA EUBANKS WINKLER

The seventeenth century in England was a time of enormous upheaval and uncertainty for musicians, as for society in general. But with uncertainty came opportunity, particularly as the national population grew and shifted to cities and towns, and people reaped the economic benefits of increased social mobility. England and Wales are estimated to have grown from a population of around 4 million in 1600 to 5.5 million in 1700, and London ballooned from 250,000 to 600,000 in the same period.[1] For musicians, the decline in stability of court patronage after the restoration of Charles II was accompanied by an increase in other kinds of cultural consumers. Not surprisingly, that newest and most high-profile of genres, opera, offered particular opportunities for such engagement, not least because it was not clearly sponsored by any single constituency, as it was in absolutist France. As such, musicians' operatic endeavors often thrived not only where subsequent history has looked for them—in high-profile, public, urban theaters—but also in venues of particular interest to musicians intent on a stable living: in schools. How this mutable but often licentious genre intersected with the world of the girls' school is my subject.

In a 2009 *Early Music* article, Bryan White describes a discovery: a "letter from Aleppo" written by Rowland Sherman, a factor of the Levant Company stationed in what is now Syria, to his merchant friend James Piggott in London.[2] Sherman asks Piggott to send a harpsichord arrangement of the symphony from "Harry's" masque "made" for a ball at Josias Priest's Chelsea boarding school—seemingly a reference to Henry Purcell's *Dido and Aeneas*. Predictably, the discovery of Sherman's letter reopened scholarly debate about *Dido*, but, from the vantage point of my study, the most interesting thing about White's discovery was not the possibility that *Dido* was first performed at Priest's school (rather than at court) or that the date of this performance had to be 1687 or 1688, not 1689, because of the timing of Sherman's departure for Aleppo. Rather, I was intrigued by the notion that concert-loving merchants such as Sherman attended these school "balls" and that such performances appeared to have been part of a larger tradition.

I am not the first to be interested in operatics at boarding schools for girls. As others have noted, the performance of *Dido* at Priest's school was not unusual; other musical works, including Thomas Duffett's *Beauties Triumph* (1676) and John Blow's *Venus and Adonis* (1684), were also performed at Chelsea. That merchants such as Sherman seemed to view these performances as just another entertainment option is intriguing, suggesting that such occasions were not purely pedagogical, but reached consumers beyond prospective students and doting relatives—indeed, Sherman's casual mention might indicate that they formed an established part of a region's (in this case, London's) musical offerings. The girls were entertainers as well as students, and, as we shall see, the social liminality of their position caused considerable anxiety for those who mounted the shows and for some audience members.

The liminality was not only social, however. In considering schoolgirl operatic performance in England in the seventeenth and early eighteenth centuries, I focus on two well-documented locations—the famous school at Chelsea and a lesser-known one at Besselsleigh (a village near Oxford)—both of which were on the geographical margins of burgeoning urban centers, and both of which seem to have relied equally on proximity to and distance from these centers to ensure the success and moral probity of their institutions. Locating early modern English opera within a school-based context enriches our understanding of canonical works such as *Dido and Aeneas* and *Venus and Adonis*; it also challenges our assumptions about amateur and professional musicians (or, to use Christopher Marsh's helpful formulation, recreational and occupational musicians).[3] And, most importantly in the context of this volume, it connects both of these topics with a re-examination of our preconceptions about where opera might be performed.

Just as recreational/occupational and schoolroom/stage distinctions were porous in early modern England, so too were generic designations, not least because performance circumstances were variable (even when associated with the "professional" theater). Music played a substantial role in English operatic entertainments in this period, even if there was also spoken dialogue. Such works go by various generic designations: masque, opera, dramatic opera, even "ball."[4] Thus, for the purposes of this essay, "opera" is defined capaciously.

Location, Location, Location

The operatic performances under consideration here took place at institutions that were in close proximity to—but not directly in—the middle of urban centers. Some of the most famous London-based schools were not in the City (within the bounds of the old Roman wall) or in fashionable Westminster or Covent Garden, although there were schools there too. Rather, they were in London's affluent suburbs or on its rural outskirts. Aside from Chelsea, other well-known girls' schools were located in Hackney and Putney.[5] This positioning may seem surprising, but the first half of the seventeenth century saw an increase in travel by coach, including the development of coach-hire ranks in London, which enabled social as well as physical movement and would have allowed girls, teachers, and parents to travel to these schools.[6] Some wealthier Englishmen, such as John Verney, whose niece attended Priest's school in Chelsea, owned their own coaches.[7] New roads were built to accommodate the heavier traffic flow, further facilitating travel.[8] Urban dwellers during this period engaged in "skirting" behaviors, moving regularly among city-center and emergent suburban and even rural spaces. Margaret Pelling explains that these skirters were "town-dwellers following patterns of living involving avoidance of, as much as commitment to, urban environments."[9] Why suburban and rural schoolmasters might have chosen such locations—whether the reason has to do with exclusivity, moral concerns, pastoral healthfulness, or even the cost and availability of suitable houses—is unclear, but given these habits and new modes of transportation, it would have been relatively simple for urban dwellers to attend school-based performances.

The location of these boarding schools was also redolent with cultural meanings; then, as now, neighborhoods had specific associations mutually created by their inhabitants and outsiders. As is now well-known, the physical location of public theaters across the river in Southwark in

Shakespeare's time reveals much about the anxieties of early modern English society.[10] The Rose and later the Globe were purposely located on the margins, outside the City walls.[11] On the other hand, the Chelsea and Besselsleigh schools' locations signified an aspiration to align with privilege and intellectual capital (in Besselsleigh's relative proximity to prosperous Oxford) and fashionability (Chelsea). Chelsea, in particular, was an extremely desirable location, known for its grand houses along the river and its popularity with courtiers in the late sixteenth and early seventeenth centuries, when many of these manor houses were built.[12]

These choices of locations also suggest negotiation of a balance between the need on the one hand to be accessible to staff based in urban centers (as I will explore below), and on the other to separate the girls from the urban environment's immoral temptations.[13] The implicit desire for removal from temptation was as relevant to the fortunes of Besselsleigh as it was to Chelsea. Oxford, the nearest urban center, was booming in the late seventeenth century, the University expanding significantly during this period with the building of the Sheldonian Theatre (1664–69) and the Old Ashmolean Museum (1678–83). Oxford residents were generally wealthy, thanks to a robust service sector, and enjoyed various entertainments and diversions, from which some may have wanted their daughters removed.[14] There were other young women's boarding schools in Oxford, but Besselsleigh seems to have been a rather famous establishment with strong ties to the University, as we shall see.[15]

Physical Spaces and Performers' Networks

Locations near London and Oxford proved particularly appealing for school-based performance for practical reasons: proprietors often had participated in the urban centers' performing arts themselves or had close associations with urban musicians, dancing masters, and playwrights. Usually, women ran the institutions' day-to-day operations, designed to educate girls in the "ornamental arts" and make them more marriageable: the (female) proprietor administered the school, sent out tuition bills, interacted with students and parents, and hired musicians and dancing masters to work as freelancers—at additional charge to students. Sometimes these administrators had close associations with or were married to a musician or dancing master, who may have viewed running a school as a sensible way of procuring a regular income. Hannah Playford (wife of the music publisher John Playford), for example, ran a school opposite the church in Islington. She advertised in her husband's publication *Choice Ayres* (1679), stating that pupils might "be instructed in all manner of curious work, as also reading, writing, musick, dancing, and the French tongue."[16] Franck (Frances) Priest, wife of the renowned dancing master Josias Priest, similarly ran the day-to-day operations at the Chelsea boarding school from 1680 onward.[17] Around the same time a dancing master named Mr. Hazard ran a school in Chelsea jointly with his wife. When Mrs. Hazard died in 1682, her husband made clear that he would hire a woman to look after the girls and the household, as it was apparently inappropriate for a man to run a girls' school without a woman.[18]

The most prominent boarding school in late seventeenth-century England with regard to operatic performance was undoubtedly the one in Chelsea, which (unsurprisingly, given its fashionable location) sponsored lavish events, described as "balls." Bryan White has suggested that these balls were probably held on a yearly basis in spring, at least in the years between 1684 and 1686.[19] It is unknown when the school opened, but by 1676 the professional musicians Jeffrey Banister and James Hart were running a boarding school "in that House which was formerly Sir Arthur Gorges," as described on the title page of the printed version of *Beauties Triumph*, a "masque" given at the school.[20] The house, a typical Elizabethan gabled structure (clearly large enough for a school),[21] was built by Sir Arthur at the west end of the village around 1617–19;[22] it was sold after his death in 1661 and rented (after Banister and Hart's tenure) to Priest in 1680.[23]

As was typical for staff at London-area boarding schools, Banister and Hart also held positions at court. Banister was a member of the Twenty-four Violins,[24] and Hart sang bass in the Chapel Royal.[25] After Priest and his wife took over the school in November 1680,[26] it appears, judging by a bill dated 20 January 1685 listing his name, that Hart continued on as a singing teacher.[27] Priest had previously run schools in Holborn and Leicester Fields; he may have also provided choreography for the London stage.[28] One presumes that, in addition to Priest's role as the school's dancing master, the couple hired friends and acquaintances from court and stage (always keen for additional income): the preface to Thomas D'Urfey's play *Love for Money; or, The Boarding School* (1691), for instance, tells us that he lived at the school for a time, although whether he taught students or simply penned an epilogue for Lady Dorothy Burke to declaim is unclear.[29] Similarly, the relationship between Thomas Duffett, the author of *Beauties Triumph*, and the Chelsea school is shrouded in mystery.

We know less about the boarding school operating in the late seventeenth and early eighteenth centuries (until 1717 at the latest) about five miles to the west of Oxford,[30] in the village of Besselsleigh, although music and dance apparently played a significant role there as well. Anthony Wood writes of "a solemn ball" held at a school between 23 and 25 October 1694 "in the house of Sir John Lenthall, performed by 50 or 60 maides, virgins of quality, that are sojourners under the government of. . . ."[31] The ancient Besselsleigh (or "Blessells Legh") manor, named for the "Blesells" who "hathe beene Lords of [the village] syns the tyme of Edwarde the First or afore," according to John Leland,[32] was acquired in 1630 by William Lenthall and passed to his son, John, upon his death.[33] It was pulled down in the nineteenth century, but before then was described as having "three principal fronts, with a quadrangular court in the centre, on one side of which was a Chapel. The walls were covered with ivy of lucsuriant growth, and opposite to each front were magnificent avenues of Elm and Lime trees. The old hall and kitchen are said to have been more adapted, from their size, to a College than a private mansion."[34] Perhaps this large hall served as a performance space for the "balls," apparently held in October of each year, but there is no conclusive evidence for this.

Evidence is lacking in other regards too: Wood apparently did not know the name of the schoolmaster and trails off into a frustrating ellipsis. Only scraps of information are known about the school's personnel, but it appears at least one or perhaps two members of the Oxford music faculty were involved. *Musica Oxoniensis* (1698) includes three songs from "the Mask of ORPHEUS and EURIDICE: Perform'd at the Boarding-School at *Besselsleigh*, in *October* 1697," composed by the then Heather Professor of Music at Oxford, Richard Goodson.[35] Similarly, the New College organist, John Weldon (appointed in 1694), published "Stop, o ye Waves" as "*Orpheus's* Song to the Vaves, in the Mask of *Orphus and Euridicy*" in *Mercurius Musicus* (May–June 1701).[36] At least one other professional may have been brought in for Besselsleigh's masque of *Orpheus and Euridice*: the *Mercurius Musicus* songsheet indicates that Mrs. [Mary] Lindsey performed as Orpheus. Lindsey's first known role was as a follower of Cynthia in Elkanah Settle's *The World in the Moon* (Dorset Garden Theatre, June 1697); she remained active on the London stage until 1713.[37] Perhaps, for this pivotal role, the organizers hired a professional to bolster the talents of the schoolgirl cast,[38] although the printed song may have recorded a London concert performance.[39] "Stop, o ye Waves" also exists in manuscript GB-Och MS Mus. 389. Many of the manuscript's pieces are from London stage works for the period 1694–98, but the manuscript "Stop, o ye Waves" may present an earlier version than the 1701 songsheet, which substantially recomposes the declamatory passages.[40] Indeed, the more highly ornamented printed version might preserve something of Lindsey's performance, whether at Besselsleigh or in Weldon's concert. Lindsey's possible participation in the Besselsleigh performance suggests just how permeable the boundary was between recreational and occupational musicians. Even if *Mercurius Musicus* records a concert performance it is still significant, for it shows the circulation of didactic musical repertoire beyond the school's walls into the concert room. If Oxford- and London-based musicians had some kind of association with Besselsleigh, it confirms a pattern seen in institutions such as Chelsea: occupational musicians in the Restoration period were (no doubt necessarily) involved in multiple projects and held multiple positions concurrently. Equally, schools wanted to employ well-known musicians in order to attract students.

The involvement of occupational musicians (both regular teachers and occasional hires) was almost certainly vital to the schools' operatic productions. Evidence from other contexts as well as from surviving sources associated with Chelsea and Besselsleigh school performances suggests that the girls often performed alongside their teachers or possibly other professionals. Earlier in the century, schoolmasters as well as music and dancing masters from girls' schools performed with their students, as did additional professionals on occasion. For example, in *Cupid's Banishment* (1617), girls from Robert White's school in Deptford performed at Greenwich alongside their schoolmaster, some court musicians, and sons of local dignitaries. Thomas Jordan's *Cupid His Coronation* (1654) was performed at Christ's Hospital by "Masters and yong Ladyes that were theyre Scholers."[41] Evidence suggests this mode of performance continued unabated after the Restoration. The printed text of *Beauties Triumph* (1676) calls for violins, recorders, and flageolets, but because woodwind instruments were considered inappropriate for women, some of Banister's and Hart's friends from court may have played at the event.

Opera at School: Moralistic Aims and the Problems of Performance

Boundaries could also be blurred onstage in these performances in ways that potentially problematized not only teacher-pupil relationships, but also gender roles. Sometimes girls took male roles: for instance, the Priests' daughter played the ill-fated huntsman in the 1684 *Venus and Adonis*, as John Verney recorded in his libretto.[42] On the public stage, women playing breeches roles (and wooing other women) had an erotic effect upon audience members;[43] one wonders how this registered in the environment of Priest's school, where the girls had preexisting relationships with each other fostered through cohabitation.[44]

Concern about appropriate behavior for young performers and the juxtaposition of moral concerns with pedagogical aims may be apparent in materials relating to *Venus and Adonis*. Given his association with Besselsleigh, manuscript fragments copied by Richard Goodson (GB-Och MS Mus. 1114) and arranged for treble voices may have been intended for performance by the girls there.[45] The text in Och MS Mus. 1114 is bowdlerized, softening the sexual content found in GB-Lbl Add. MS 22100 and Priest's school libretto: instead of "Absence kindles new desire: / I would not have my lover tire," we have "Thus you will the kinder prove / Since absence tunes the mind to love." If it was intended for Besselsleigh, perhaps the schoolmaster was less free-thinking than the Priests at Chelsea. (In contrast, GB-Och MS Mus. 360—a manuscript that may have been used by Goodson to educate choirboys—includes exercises copied by a student and has trebles of various songs, including John Eccles's bawdy "A Soldier and a Sailor" with no redaction.)

Given the undoubted anxieties about schoolgirl performance, especially in a context of concern about stage performance generally, it is somewhat surprising that the operatic entertainments given at Chelsea and Besselsleigh engaged with amorous love at all. Still, the purpose of sending a girl to a boarding school for "improvement" was to secure a good marriage, and in this context the concern over such subject matter becomes more understandable. Works performed by schoolgirls exploited the suggestiveness of the plots and enhanced opportunities for physical display (through dance and gesture) on the one hand, and maintained a veneer of sexual propriety on the other. Thus, *Venus and Adonis* and *Dido and Aeneas*, the two most famous examples of schoolgirl opera, explore stories of illicit passion, which result in death or suffering for one of the protagonists—a moral lesson of sorts.[46] *Beauties Triumph* (Chelsea, 1676), and *Orpheus and Euridice* (Besselsleigh, 1697) similarly offered warnings to the girls about the price of pursuing love, and, more specifically, sexual desire.

The Chelsea production of *Dido and Aeneas*, memorialized in a sole surviving libretto (GB-Lcm D144), a souvenir from the performance, offers a sense of the tightrope these schoolgirl operas walked.[47] The dances in the prologue and elsewhere would have showed off the girls' terpsichorean talents; indeed, dancing appears to have played a significant and memorable role in the entertainments at Priest's school, as John

Verney's letters to his brother, described below, attest. On the other hand, some of the dances seem potentially inappropriate for young women (e.g., the dance for "*2 Drunken Saylors*," in act 2 of the Chelsea *Dido*);[48] teachers or other professionals might have performed these anti-masque-style dances, in keeping with earlier practice.[49] Similarly, the libidinous satyrs in the Besselsleigh *Orpheus and Euridice* called for in the song "Let me, Ye Satyrs" might have been acted by male schoolteachers or professionals, in earlier anti-masque style.[50] Even the roles of the Sorceress and her coven in *Dido* may have been taken by teachers or professionals. If the schoolgirls did perform the roles of the witches and the sexually active Dido ("One Night enjoy'd, the next forsook" as Aeneas laments),[51] then, like the performance of Venus's erotic escapades in *Venus and Adonis* at the same school a few years earlier, it could have been understood as a moral counterexample.[52]

Equally, more prosaic reasoning may have helped justify performance: the cachet of court association—perhaps enhanced by parental connections—may have encouraged the production of otherwise inappropriate operas. So, while the Chelsea libretto's stage directions for *Dido and Aeneas* describe effects that may have been beyond the resources of Priest's school (e.g., "*Phoebus* Rises in the Chariot, / Over the Sea, The *Nereids* out of the Sea," and "*Venus* descends in her Chariot"),[53] these lavish descriptions might have encouraged the reader to align the work with the spectacle of the court masque, even if Purcell's *Dido* was never performed there. The standards of the court might even have influenced some operas' moral arguments: in Thomas Duffett's *Beauties Triumph*, the long-established English custom of using Paris's tale to indict those ruled by pleasure and lust was turned on its head.[54] Instead, Duffett's drama argued that love *should* triumph over Fate, a rhetorical move in keeping with contemporary discourses aligning the Stuarts with love, and with the goddess Venus specifically—another intersection between court culture and school-based musical entertainments.[55]

Undoubtedly, moral concerns shaped the school performances of *Beauties Triumph*, *Venus and Adonis*, and *Dido and Aeneas*; however, evidence, including the prologue and epilogue to *Beauties Triumph* and the epilogue to *Dido and Aeneas*, strongly suggests that the erotic charge of schoolgirls singing and dancing (perhaps with their teachers) was not entirely contained by the pedagogical frame. Indeed, the eroticism might have been necessary in order to showcase the girls as desirable marriage prospects. Such "shows" also formed a conduit between the "cloistered" suburban or rural schools and the social duties and obligations found outside their walls.

Between Public and Private

School-based operatic performances occupied an interstitial space, neither belonging to the public stage, accessible to all who could afford a ticket, nor, to judge from contemporary accounts, being entirely private: the merchant Sherman, with whom this essay began, might have attended or at least heard about a "ball" at Priest's school. For some, these performances were far too public and thus highly suspect, particularly given the long-standing negative associations suffered by musical and theatrical performance in England.[56] Mrs. A. Buck, writing to Mary Clarke (wife of a Whig Parliamentarian), lamented the state of London education, mentioning boarding schools at Hackney and Kensington as well as D'Urfey's play, *Love for Money; or, The Boarding School*: "Att present all schoolls are rediculd: they have latly made a Play cal'd The Boarding School." She then held up the performances at Priest's school for particular disapprobation: "Preists att Little Chelsey was one which was much commended; but he hath lately had an Opera, which I'me sure hath done him a great injury; & the Parents of the Children not satisfied with so Publick a show."[57] Although D'Urfey protested in the preface to *Love for Money* that he might as well have set the play in York as in Chelsea,[58] some of his contemporaries clearly interpreted his play—filled with amorous activity between students and their singing and dancing masters—as a *roman à clef*.[59] Another account, from 1717, indicates that men were drawn to boarding schools by the young ladies; attending an operatic performance could have provided an opportunity for suitors to get a full sense of the pupils' charms: "Bessilsleigh . . . is a mighty pleasant Place. Here was a few years since a noted boarding School, where were abundance of young Ladies, wch occasioned a more than ordinary Resort thither of the young Oxford Scholars."[60]

Indeed, the prologues and epilogues to works for schools revealed considerable anxieties. Duffett's prologue to *Beauties Triumph* apologized for the entertainment's public nature. The "young Lady" who spoke the text directed it exclusively to the other "Ladies" in the audience, establishing a communication among women, ameliorating the problem of an unwed girl speaking directly to men. Although actresses on the public stage addressed men frequently, they were sexually available in a way that these students (presumably) were not—a distinction the prologue carefully drew. This "young Lady," speaking Duffett's words, further distinguished the public stage from the entertainment the audience saw at the school:

They [the Masters] humbly bow,—and beg a kind excuse,
For straiten'd time, and a disorder'd House;
Hoping, the want of practice, fitting dress,
And glorious Scenes, may make our failings less:[61]

Continuing in the same vein, she asserted that "this [entertainment] was intended for our selves alone."[62] After the masque was over, the epilogue proclaimed the chastity of the students, drawing a clear distinction from the actresses of the public stage: the girls are "cloister'd Nuns with virtuous zeal inspir'd."[63] Indeed, performance was simply one of many activities the girls practiced at their school; others mentioned in the epilogue included embroidery, singing to the lute, and molding forms into "painted Wax."

D'Urfey's epilogue for the student performance of *Dido and Aeneas*, spoken by Lady Dorothy Burke, deployed much the same rhetoric regarding chastity and performance. The girls' singing did not inflame illicit passion in auditors; rather, their vocal performance was akin to something by a celestial choir: "All that we know the Angels do above, / I've read, is that they Sing and that they Love."[64] The girls' angelic love was chaste, their cloistered position at the school saving them from "these grand deceivers, men." Burke declared (in words D'Urfey framed for her) that she and her schoolmates were "bred to virtue," kept behind a "nunnery-door . . . charm'd to shut out fools," for they were "Protestants and English nuns." As in Duffett's epilogue, there was an assertion regarding the girls' spatial (and moral) separation from the pleasures of London, for neither "play-house wit" nor the excesses of the "vile lewd town" could charm them.

Despite clear anxieties regarding schoolgirl operatics, such events still took place. Why? Aside from any advantages accruing to the proprietors, such performances provided entertainment for those inclined or invited to attend. These activities also advertised the training received at the school and showed off accomplishments to parents, relatives, and other interested parties, as well as to prospective students. The Verney correspondence strongly suggests that the latter role was extremely important.[65] John Verney attended his niece's performances at Chelsea and reported back to her father, Edmund (Mun) Verney. On 24 April 1684 he wrote that "we [John and his wife] were both at the Ball & opera last Thursday where we saw my neece dance & act some part in it, and truly she doth it very well & I wish you had seen her dance twould (If it be possible) have sett her a degree higher in your affections, she is a good child & will doubtless deserve very well at your hands."[66] In subsequent years John and his wife took other relatives to see the "Grand Ball" as well. On 22 April 1686 he told Mun: "This day (as was last Thursday) is ye Grand Ball at Chelsey School, wherein your daughter is a greate Dancer, My wife carryes as many of the Stewkel[e]ys [their cousins] as her Coach can hold, last weeke 4 of that family went to it."[67] These comments indicate several things about the "Grand Ball" performances at Priest's: they were given more than once; groups of relatives—perhaps (if not in this case) including prospective suitors—attended together; and dancing was thought to be a highly desirable skill, a sign of "goodness" and perhaps filial obedience, and thus a spur to fatherly approbation.[68]

Schools also used public performance as a strategy for recruiting new students, although not always with success. While we lack evidence about this practice for girls' boarding schools, performance played a key role in student recruitment at young gentlemen's academies. Much earlier in the century Francis Kynaston's failed academy for gentlemen, Musaeum Minervae, located in Kynaston's own lodgings in fashionable Bedford Street, Covent Garden, put on public presentations of music and other "shewes."[69] The school trained the nobility in military pursuits as well as the arts; to advertise the latter, in *The Constitvtions of the Mvsaeum Minerva*, Kynaston mentioned that "the day of publick Musick is Tuesday, beginning at two of the clock in the afternoon" and mandated that "to *Publick shewes*, *Presentations*, and *Musick meetings*, none shall be admitted but such as shall bring with them the character [i.e., proof of gentility] which shall be given forth by the Regent for that time."[70] At least in one instance a masque, *Corona Minverva*—presented before Prince Charles, the Duke of York, and their sister, Lady Mary, in 1635—was performed in an effort to procure patronage and to demonstrate the talents of students and teachers.

In 1649 Sir Balthazar Gerbier, a Huguenot émigré, opened a similar academy for gentlemen at his house at Bethnal Green (a decidedly pastoral location then) offering music and dance instruction.[71] Like Kynaston, he sought support through music-making, but this strategy backfired. According to *Severall Proceedings in Parliament*, a newspaper published by Robert Ibbitson, the free public lectures his academy sponsored sometimes included "an Accademicall entertainment of Musicke" or "an intermixture of pleasing music."[72] But, although he provided ticketed reserved seating for "all honourable persons" and notwithstanding his removed location,[73] too many people of the "lower sort" attended these events, and his academy began to lose elite students as a result.[74]

John Banister, another musician-*cum*-master, sponsored the performance of *MUSICK: OR A PARLEY of Instruments* (1676), featuring music masters at his gentlemen's Academy in Little-Lincoln Inn Fields, possibly in an attempt to generate publicity and draw prospective students. According to the wordbook for *Parley* (GB-Lbl 11621.f.31), the "Scholars Teaching" dictated that performances could not begin until "six a clock in the Evening." The "scholars" were not the primary performers of *Parley*, however; a contemporary advertisement indicated that it was "perform'd by Eminent Masters," indicating that professionals were employed for the occasion.[75] Furthermore, a ball was held after the performance at the request of "some Persons of Quality," and for that "twenty violins" were on hand "to attend them until ten a clock at Night."[76] It is doubtful that twenty violins were consistently on hand to staff the school; Banister was clearly drawing upon colleagues from the king's musical establishment to pull off this lavish performance.[77] Thus *Parley*, in its printed and performed iterations alike, served to bring attention to Banister's educational enterprise, facilitated through his contacts with the London professional music scene.

In all these cases, attendance at entertainments was made possible by the schools' proximity to urban centers. But this proximity fomented moral anxieties, particularly where girls were concerned. As the prologues and epilogues to works given at Chelsea and accounts of the "Scholars" from Oxford visiting Besselsleigh make clear, temptations of the "Town" (either directly or via staff or audience members) might render the girls sexually vulnerable. At Chelsea there was a clear desire to separate the girls' performances from those of morally dubious stage professionals, while Anthony Wood's description of Besselsleigh emphasizes the students' virginity. Yet, given the somewhat "Publick" nature of these performances and the fact that the teaching staff was drawn from the professional musicians circulating between schools and urban centers, claims about the girls being safely "cloistered" were clearly untrue. Perhaps the rhetoric of the nunnery—in a country for which the cloistered Catholic votary was exotic or historically distant—was itself an anxiously overdone attempt to address that concern. Such discourse, like the works themselves, and the networks within which they were developed and circulated, suggests the contested and heterodox environments in which English opera developed.

FOUR

London's Opera House in the Urban Landscape

MICHAEL BURDEN

In about 1900 it was possible to stand at Trafalgar Square and look west up Cockspur Street to see the newly completed Carlton Hotel on the corner of Pall Mall to the left and the Haymarket to the right (fig. 4.1). To the right could also be seen Her Majesty's Theatre, designed by Charles Phipps, the final and present incarnation of London's eighteenth-century house for Italian opera. Since its opening in 1705 the building had been successively called the Queen's Theatre (under Queen Anne), the King's Theatre (under the Georges and William IV), and Her Majesty's (under Queen Victoria); until the mid-nineteenth century it was the home of London's elite opera and dance, restricted by a largely annual license to the staging of foreign works, for the most part by foreign performers. The story of this opera house, and its uneasy relationship with its environs, is the subject of this essay, for whatever the perceived grandeur of the theater building by 1900, its status as a monument to London's cultural wealth had not always been obvious.

The new hotel in figure 4.1 (run initially by the fabled duo, Swiss manager César Ritz and French chef Auguste Escoffier) is also significant to the story of the opera house. It took its name, "Carlton," from the nearby Carlton House Gardens in Pall Mall, built on the site of the Prince Regent's

MICHAEL BURDEN

4.1   Her Majesty's Theatre (institutionally, formerly the Opera House), far right, with the Carlton Hotel on the corner of the Haymarket and Pall Mall. York and Son, photograph, 1899–1900. Reproduced with permission of Historic England Archive, NMR DD97/00353.

Carlton House.[1] When the Prince Regent moved into the house in 1783, he and the architect Henry Holland had set about remodeling it into a startling and controversial building.[2] This reconstructed residence was also the reason that the architect John Nash embarked on building the thoroughfares that together formed the ceremonial route from Carlton House to Regent's Park (seen in figure 4.6): Waterloo Place, Lower and Upper Regent Streets, Portland Place, and Park Square.[3] These works, started in 1811 and completed in 1825, formed one of the most ambitious and successful pieces of town planning London has ever seen and placed the Italian opera house in a prominent position in London's urban environment for the first time. The significance of this moment in the capital's operatic history, and how it might be (or might have been) linked to opera's place in the urban environment, is the chief focus of my narrative. To appreciate this, however, we need a larger perspective, stepping back to the origins of London's opera house.

St. James's and the Haymarket

Until the 1890s reconstruction described above, the King's Theatre stood almost on the corner of the Haymarket and Pall Mall. Pall Mall ran from the bottom of the Haymarket through to the front of St. James's Palace and for much of its life was one of the most fashionable streets in London. The Haymarket, which formed a major route from Pall Mall to Piccadilly, was (as it remains) a commercial thoroughfare: it is found on a plan of around 1585 and appears in its present-day form on the Fairthorne and Newcourt map, surveyed in 1643–47 and published in 1658.[4] However, as a site for the new theater in the 1700s (one originally proposed in 1703 by architect and playwright John Vanbrugh, for the staging of both opera and spoken drama), it attracted much criticism, largely from those with vested interests. Colley Cibber's later commentary sums up much of the argument:

At that time it had not the Advantage of almost a large City, which has since been built, in its Neighbourhood: Those costly Spaces of *Hanover*, *Grosvenor*, and *Cavendish* Squares, with the many, and great adjacent streets about them, were then all but so many green Fields of Pasture, from whence they could draw little or no Sustenance, unless it was that of a Milk-Diet. The City, the Inns of Court, and the middle Part of the Town, which were the more constant Support of a Theatre, and chiefly to be rely'd on, were now, too far, out of the Reach of an easy Walk; and Coach-hire is often too hard a Tax, upon the Pit, and Gallery.[5]

Cibber's narrative has colored subsequent views of the circumstances. As Judith Milhous has pointed out, the street is not at all remote: "Anyone who has walked through the theater district in London, or looked at a map, knows that the Haymarket is just a mile from Drury Lane," and while it may not have been a desirable trip on all occasions, a short chair ride would get one there easily.[6] It should be further added that if, as seems likely, the aim of the theater's promoters was to distinguish the opera house from other theaters, then whether the pit and gallery could afford the fare was not necessarily the first priority, and may even have been an advantage in that it excluded the less well-heeled clientele. Indeed, although the project was headed by Vanbrugh, it was supported by twenty-nine aristocratic subscribers, many of whom were members of the Kit-Kat Club, with its literary connections and Whiggish political aims.[7]

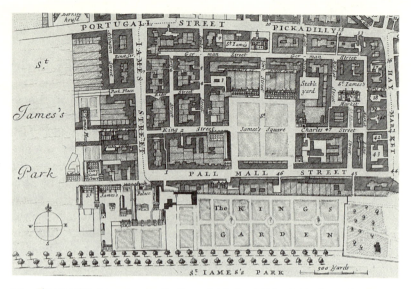

4.2 The area of St. James, showing the Haymarket to the right-hand side. The yard at No. 43, known as Phoenix Yard and later as King's Yard, was to be the site of Vanburgh's theater, then the Queen's Theatre and later the Opera House. William Morgan, *Map of the whole of London*, 1682. Detail from a facsimile published by Harry Margary in association with the Guildhall Library, 1977.

But perhaps even more to the point was the character of the district of St. James's, in which the theater stood. In 1660 the land had been part of the royal park that encompassed both St. James's Park and Green Park, but it had been shorn off by Charles II, who granted a lease to Henry Jermyn, the first Earl of St. Albans. St. Albans's plan was to develop the area as an aristocratic enclave—as indeed it remained until the Second World War—and he designed a street grid with the fashionable and aristocratic St. James's Square near the center and the open St. James's Market up toward the junction of Haymarket and Piccadilly. As the map in figure 4.2 shows, by 1681–82 the area was laid out and well developed: by the 1720s seven dukes and seven earls were in residence in St. James's Square alone. And as part of establishing the area's new identity and ensuring adequate provision for the residents, there was an early move to create a new parish. When the lease was granted, it was part of the parish of St. Martin-in-the-Fields; pushes for separation came in 1664, 1668, and 1670. Anticipating eventual success, the new church's foundation stone was laid in 1674, and construction began in earnest; St. James's, Piccadilly, was consecrated in 1684 and in 1685 (with grant of the freehold) was made an independent parish, bounded by Pall Mall, Oxford Street, and the center of Haymarket. Building the theater was, then, neither an automatic financial risk nor was it hopelessly situated: by 1703 the area had a clear identity. The Haymarket was a well-established thoroughfare, now running alongside a developed and fashionable neighborhood.

Nevertheless, it would be a mistake to sanitize the district. The aforementioned general produce market of St. James's (see fig. 4.2) was still active, patronized by the lower classes of the area and, one assumes, the servants of the grander establishments. And as discussed below, the market was still going strong when John Nash improved the area the following century. He was greatly concerned by the loss of the amenity and proposed a suitable replacement. Further, there was another, long-standing market from which the Haymarket—"a spacious Street of great Resort, full of Inns and Houses of Entertainment; especially on the West Side"—took its name: in 1720 it was noted that "the Market for Hay and Straw is kept every Tuesday, Thursday and Saturday."[8] In 1766 *A New and Accurate Survey of London* observed that among the "remarkable things" in the parish of St. James's was "an opera-house, and a play-house in the Hay-market, in which there is also a market for hay and straw three days in a week."[9] The figures for the use of the market are staggering: in the month of February 1774 it "was attended by over thirteen hundred hay and straw carts," and "between Lady Day 1827 and Lady Day 1828 over 26,000 loads of hay and straw were registered by the toll collectors."[10] The market remained in the street until 1830, when it was moved to St. Pancras. The three market days a week, which remained Tuesday, Thursday, and Saturday, were also the days on which the Opera performed, almost guaranteeing that the elite audience attending the theater would have encountered a street strewn with straw, horse dung, and other market detritus. The street was perpetually busy, which was a problem for the citizens and theaters alike; successive managements frequently requested that the horses have their heads turned toward Pall Mall in an effort to ensure that the Opera House traffic flowed one way only: "To prevent Inconvenience to the Nobility and Gentry getting into their Carriages, they are most respectfully intreated to give positive Orders to their Servants, to set down and take up with their Horses Heads towards Pall-Mall. The Doors in Market Lane for Chairs only."[11] The street itself was lined with shops: the range of businesses is evident in surviving trade cards: the British Museum's eighteenth-century holdings include cards for the Café de la Europe, the chemist Richard Sidall, the ironmongers Elwell and Taylor, the vintners B. Vale and Brother, and the Music Warehouse of Corri, Dussek and Co. By the early nineteenth century, Haymarket's mixture of houses, theaters, and retail shops can be clearly

MICHAEL BURDEN

4.3 The row of shops in Pall Mall East, at the junction of Pall Mall and the Haymarket. The Opera House encased with John Nash's colonnades is on the right-hand side. Thomas Hosmer Shepherd, watercolor, 1829. © Trustees of the British Museum 1880, 1113.2240.

seen in schematic street maps, and, as seen in figure 4.3, the street's retail image had also moved upmarket, for there were distinctly smarter shops at the bottom of the Haymarket than could have been found previously. To the right can be seen the arcades of the Opera House, juxtaposing the entertainment of the stage with the entertainment of shopping; this is, in fact, one of the few images that includes this juxtaposition.

A "Hidden" Opera House

The history of the building itself is, like much of the history of London's Italian Opera, a confused shambles, involving theft, litigation, and skullduggery of all kinds.[12] Originally intended for both plays and operas (as noted above), the building was first mentioned in a letter dated 5 June 1703 from John Vanbrugh to the publisher Jacob Tonson, and soon the project was being described thus:

The ground is the second Stable Yard going up the Haymarket. I give 2000. for it, have laid a Scheme of matters, that I shall be reimburs'd every penny of it by Spare ground; but it is a Secret lest they should lay hold on't to lower the Rent. I have drawn a design for the whole disposition of the inside, very different from any other Other House in being. but I have the good fortune to have it absolutely approv'd by all that have seen it.[13]

Vanbrugh's boast that the design was "very different from any other Other House in being" has perplexed scholars, although it has usually been considered to refer to the building's interior. The most convincing and elegant solution to the conundrum of the auditorium's form, offered by Graham Barlow based on some sketches by Vanbrugh, is a design of classical proportions using three overlapping ovals.[14] From a technical point of view, the design may also have included a permanent orchestra pit, seen in the engraving *Coupe prise sur la longueur du Théâtre de l'Opéra de Londres* prepared for Gabriel Dumont.[15]

The interior, of which no illustration survives other than Dumont's, may well have been as elaborate as Cibber's mention of "vast columns," and "gilded cornices" suggests, but the acoustics of the new theater were, Cibber noted, initially troublesome: "For what could . . . their immoderate high roofs avail, when scarce not one word in ten, could be distinctly heard in it?"[16] Alterations finished in 1709 appear to have been an attempt to put this right and may also have been an effort to improve the house for musical performances: after this date, the theater was a house devoted to opera, for Cibber reports that a "flat ceiling is now over the orchestra," a modification that can be seen in Dumont.[17] The theater's problems, as described by Cibber, made their way into later accounts about spoken theater,[18] but there were no complaints about the space as far as opera was concerned. After 1709 the house served up opera and dance, with few structural changes until 1778, when remodeling began in earnest: there were extensive works in 1782, a complete rebuild with a new design after the fire in 1789, and further alterations in 1796, 1799, and 1865.

Opening a theater in 1705 was doubtless an "event," but there are no contemporaneous illustrations of the outside of the new structure, an inexplicable lacuna given the status of the building and the position of those who subscribed to its construction. Or was it? As mentioned above, Vanbrugh's site was in "the second Stable Yard going up the Haymarket," known then as Phoenix Yard and later as King's Yard; this was a fact seized upon by Daniel Defoe in his satirical prologue "On the New Playhouse in the Hay-Market," published that year:

Your own magnificence you here Survey,
Majestick Columns stand, where Dunghills lay,

MICHAEL BURDEN

And Cars Triumphal rise from Carts of Hay.

. . . .

A *Lay-stall* this, *Apollo* spoke the Word,
And straight arose a *Playhouse* from a Turd.[19]

Defoe's ridicule not only emphasized that the site had been a stableyard, but proposed that the new building had been erected from horseshit. Turning back to the ground plan of St. James's (see again fig. 4.2), the theater's site at the bottom of the Haymarket can be seen in the lower right-hand corner. The area, the second courtyard on the left side of the street (marked 43), is entirely surrounded by buildings, leaving the narrow entrance as the principal way into the new structure. The fronts to both the Haymarket and Pall Mall were shops, which remained in one form or another into the nineteenth century. The absence of a ceremonial entrance to the theater might go some way to explaining the lack of an early visual record of the building's exterior. There could be no picture, because there was nothing to produce a picture of.

It is not until the 1714–22 topographical view of London and Westminster by Johannes Kip that a meaningful illustration of the exterior appears, although quite *what* it means is unclear. Kip's engraving shows the theater as an independent building located at the bottom of the Haymarket. It is quite inaccurate, but in that very inaccuracy we may be seeing an attempt to present the theater as a building of public consequence in the urban environment, for the distant representation offered by Kip appears to be in the form of a semi-independent classical temple. The first surviving (and oft-reproduced) illustration of the theater's entrance is given in figure 4.4, which originated in a 1783 watercolor by William Capon. This was the façade to the foyer, a two-story brick building with an arcade below and a single large room above. The room, having served as a dancing school in the 1781–82 season, was at this stage a fencing academy, Ridaut's, and remained so in the hands of Henry Angelo until the whole theater was destroyed by fire in 1789, an event which saw Capon's watercolor engraved and circulated.

The fire of 1789 was devastating. It broke out "five minutes before ten o'clock" in the evening, when the performers were "practicing a repetition of the dances" on a nonperformance night,[20] and began at the top of the building in the flies or the scene room: the "first notice [the dancers] had of the mischief was sparks falling on their heads."[21] Angelo arrived to find the building in flames and hastily rescued his portrait by Mather Brown of the fencing master Joseph Bologne, Chevalier de Saint-Georges, but lost much else through theft: his fencing room, however,

46

4.4 The Opera House façade to Haymarket, when the room over the portico was Rigaut's Fencing School. After William Capon, pen, gray ink, and watercolor, 1783 (1818). © Trustees of the British Museum 1880,1113.2219.

was to be the only fragment of the theater remaining when the flames died down.[22] In the aftermath of the blaze, which also destroyed shops and other parts of the site, at least two proposals emerged for the theater's rebuilding. One was a spectacular design by Robert Adam, which included the rest of the block bounded by Pall Mall, Haymarket, Charles Street, and Market Lane, with a small bridge across Charles Street to a little pavilion. It was conceptually a wonderful design, particularly giving a sense of consequence to the corner of the Haymarket and Pall Mall, and offered a grand entrance from Pall Mall itself. Adam also offered a version of the design with a frontage to Haymarket the approximate length of the old building. Little is known about the circumstances in which Adam drew up the plans: there was no formal competition, but public discussion was widespread, and Adam may have responded to it. With somewhat more authority than Adam, Michael Novosielski, an architect and the theater's scene designer, produced a plan with arcading and a rusticated ground floor, similar to Adams's second version, but in a more

MICHAEL BURDEN

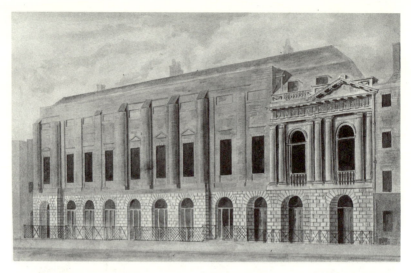

4.5 The Opera House façade to Haymarket, designed by Michael Novosielski, started 1790 but left uncompleted until John Nash encased the building in 1816–18. Anonymous watercolor, after 1794. © Trustees of the British Museum 1880, 1113.2221.

severe classical style and with a low central dome. Both Novosielski's and Adam's designs relied on the purchase and demolition of the seven or eight small houses that faced the Haymarket, which would give the Opera House a ceremonial street front for the first time.

Not surprisingly, the insider Novosielski's design was adopted, and work began in 1790. What he originally proposed was a building with a grandiose eleven-bay front facing the Haymarket; what London got when the money ran out and Novosielski died (in 1795) was the frontage seen in figure 4.5, with only the first two bays completed. This fragment of a frontage was all there was to look at, a façade that attracted the scorn of commentators for many years. It was described in 1802 by the critic James Pellar Malcolm as being

> fronted by a stone basement in rustic work, with the *commencement* of a very superb building of the Doric order, consisting of three pillars, two windows, an entablature, pediment, and balustrade. This, if it had been continued, would have contributed considerably to the splendour of London; but the unlucky fragment is fated to stand as a foil to the vile and absurd *eminence* of brick pieced to it, which I have not patience to describe.[23]

And there things stuck, with an ungainly and unfinished structure that conveyed little of the glories therein and did not represent in any obvious way the role of an opera house in London's urban environment on one of the most frequented thoroughfares in the district.

At the same time, a competing opera scheme focused on Leicester Square for which monumental designs were drawn: two by John Soane, one by Thomas Sandby, and a fourth, painted by Thomas Hodges, that was clearly based on those of Victor Louis for the Grand Théâtre de Bordeaux.[24] Anthony A. Le Texier also used the Théâtre de Bordeaux as a model in his March 1790 essay *Idées sur l'Opéra, presentées à messieurs les souscripteurs, les actionnaires, et les amateurs de ce spectacle.*[25] His text offers a range of advice to the opera managers, but it is clear that he also produced plans. There may additionally have been a three-dimensional model; he does refer to the model's "building," and the press commented: "LE TEXIER'S model for a new Opera House, has been much admired for ingenuity of contrivance, and elegant machinery. It is not certain how much of the plan shall be adopted."[26] Both the model and *Idées sur l'Opéra* came under attack from at least one critic:

M. Le Texier, not content with the praise given to his Readings, has lately attempted some new and curious adventures to fame!—He first forms a model of the theatre of Bordeaux, the most defective structure in Europe—and offers this as his own design for a New Opera House! He next translates a few desultory passages from Count Algarotti, and calls them his *"Ideas on the Opera!"*—Such is the opinion which Gaelic *modesty* can form of English judgment.[27]

Whatever the pros and cons of Le Texier's scheme, what was fundamental to the (unrealized) Leicester Square proposals was the notion of an impressive, independent building, signifying the grandeur of its purpose.

The Opera House in the Haymarket stood as the eyesore identified by Malcolm until the arrival on the scene of John Nash. Nash, the architect of the Prince Regent's Carlton House, was also able to convince the prince of the merits of the extensive works required to develop Regent Street (see fig. 4.6). These were enshrined in an act of Parliament, the New Street Act of 1813, which was important to the shape and development of the Opera House, for although the act's priority was the linking of Carlton House to Regent's Park, it also provided for the creation of "a more convenient communication from Mary-le-bone Park [. . .] to Charring Cross." To achieve this, the commissioners were empowered, where necessary, to enact the "diverting, altering, widening, and improving such parts of the present Streets, as will form entrances into such new Streets or into the Streets, Squares and Places connected therewith."

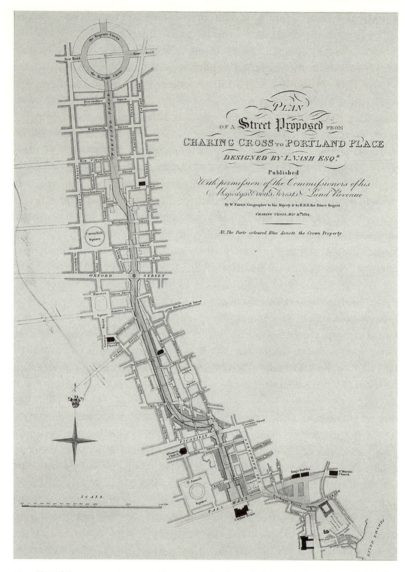

4.6   Plan of the proposed ceremonial way from the Regent's Park to Carlton House, Pall Mall. The shape of the Opera House auditorium can be seen at the corner of Pall Mall and the Haymarket. Surveyed by William Faden, engraved by Michael Thompson, 1814. © British Library Board, Maps Crace Port. 12.1.

In the works required to bring the Opera House into the scheme, there were three provisions that affected the building. First, the widening of the east end of Pall Mall involved the demolition of the buildings on the south side of the Opera House. Secondly, in the creation of Waterloo Place, the houses behind the Opera House on the other side of Market Lane would be cleared. Thirdly, in order to extend Charles Street through to the Haymarket, the buildings to the north end of the Opera House would have to be demolished. If these works were carried out, then the Opera House—consisting then largely of Novosielski's 1790 auditorium, his concert room of 1793–94, and the ugly-duckling Haymarket façade—would be left in isolation but capable of further development and improvement in the context of the grander street schemes envisaged by Nash.[28]

Nash's solution to the problems presented by these works, seen in figure 4.7, was to encase the building and site in a parapetted façade with sash windows, and then to construct colonnades surrounding the building from Pall Mall, up the Haymarket, and along the new line of Charles Street. Despite all the demolition and street widening, some

4.7 The Opera House façade to Haymarket, from Pall Mall East, with encasing by John Nash, 1816–18. Drawn by Thomas H. Shepherd, engraved by M. Fox, 1828. Collection of Michael Burden.

MICHAEL BURDEN

shops remained on the perimeter of these sides of the site, and these were included inside the colonnading. The remains of Market Lane later became the still-extant Opera Arcade. A square tower at each end of the block articulated the Haymarket façade of Nash's new building, and the entrance to the theater was framed by two small projections at first floor level. The entrance to the theater had Doric columns rather than arches, and a frieze representing music and dance, by James Bubb after drawings by John Flaxman, was placed below the cornice. The Doric columns continued along the Pall Mall frontage.

The works began in 1816, and by 1818 Nash's building, with Novosielski's auditorium, stood in splendor on the corner of the Haymarket and Pall Mall. It was destroyed by fire in 1867 but swiftly rebuilt in the shell of the old building, only to be demolished in its entirety by the Crown Estate after 1891 to fulfill their planning ambitions, which resulted in Phipps's Her Majesty's.

The King's Theatre as Monument

That Robert Adam and Michael Novosielski (and probably Anthony Le Texier) offered elaborate designs that imposed themselves on London's urban landscape was wholly to be expected. By the time of the 1789 fire, eighteenth-century discussions of the role of theater in the urban environment presented it—as represented by its building—as a "cultural monument" rather than a "private possession," a function of the theater that was also felt in France.[29] Voltaire remarked, "Our dramatic works have given France a superiority no longer disputed; but the stranger or the citizen who regards the monuments which embellish the capital will seek in vain for a theater worthy of Corneille, Racine, Molière, Crébillon, Voltaire."[30] The repeated criticisms of theaters in Paris, which became a chorus with the two burnings of the Opéra, spawned a number of French treatises.[31] Of these, there was one manifesto, *Observations sur la construction d'une nouvelle salle de l'Opéra*, published in Paris and Amsterdam after the 1781 fire at the Opéra, that was penned by a practitioner, Jean-Georges Noverre, a ballet master and composer of dances who had worked in theaters all over Europe. Noverre's main theme in *Observations* was the necessity for an island site, both in case of fire and to improve the practical functioning of the theater. For example, he stated: "A theatre building must be absolutely separate from other structures. It is desirable that the roads that approach the building be sufficiently wide and spacious so that pavements can be designed to prevent the greater

52

part of the theatre-going public from being run over by carriages."[32] He followed this with a clear, practical outline of what might be gained in the management of the building's internal spaces from this external situation:

A theater sited in this manner would provide the architect with the space to increase the size and number of lobbies and exits. He would be able to create beautiful corridors, magnificent staircases, and exterior galleries, and neglect nothing that can contribute to the comfort and safety of the public. This sort of internal layout is impossible to achieve on a small site that is closed in on all sides by buildings. The theater must have nearby a similarly isolated building to which it is linked by two bridges, or covered galleries, that run to the right and to the left of the theater. The main part of this second building would serve as dressing rooms for the actors, actresses, dancers, and all the performers employed in the staging of the opera. One could build a great foyer, one large enough to be adequate for the practice of ballets, at moments when the building might be pressed for space.[33]

Noverre also argued that the theater should serve as an embellishment to any capital and picked up on the notion of the French minister responsible for authorizing the Académie d'Opéra in the seventeenth century, Jean-Baptiste Colbert,[34] that such an edifice would play a role in the economy of a city:

As for the cost, it would be considerable, no doubt. But it is not for me to calculate the resources and to determine the use of the public purse. The citizen, one who is a friend of the arts, has accomplished his task when he has proposed the visions and ideas that he believes to be useful. It is up to the wisdom of the men of state who govern us to combine the circumstances and the time, the effort and the advantages, the goal and the means. Colbert regarded a sum of money well-spent when it was destined to embellish the capital, encourage the talents and the arts, and attract foreign businessmen there by the fascination of pleasure.[35]

Little of what Noverre had to say was entirely original: for example, the "bridges" he proposes were planned (and probably under construction) for what became the Théâtre de l'Odéon, the Théâtre-Français du Faubourg Saint-Germain, built between 1779 and 1782;[36] his comments on the use of lighting appear to be drawn from Patte's 1781 article in the *Mercure de France*;[37] and his closing call for a theater on the Place du Carrousel was decidedly a case of Johnny-come-lately. But although Noverre and others commented little on the interiors of theaters, limiting their remarks to an emphasis on the patrons' convenience and a vague idea of excellent decoration, there is no doubt there was an

MICHAEL BURDEN

underlying assumption that the builders would create an interior worthy of a grand exterior.

At the time *Observations* was published, in 1781, Noverre was starting work managing dance at London's King's Theatre. After his first season the building was extensively remodeled internally. I have argued elsewhere that this was primarily to do with accommodating the needs of the new type of dance Noverre's season had made the norm: his 1781–82 season was a spectacular success and cemented what came to be called the *ballet d'action* as the preferred style of theatrical dance on the London stage; Noverre himself intended to return for the next season.[38]

The timing of *Observations*, Noverre's presence in London, and the interior alterations to the King's Theatre all suggest that London's Opera House was under the influence of a man not only with supreme skills as a ballet master, but with determined views on the form of the opera houses in which he worked. While Noverre promoted the island-site approach on practical grounds, the attendant opportunities for physical magnificence contributed to the potential for the theater to act as a cultural monument. Noverre's ideas had impact in France, his standing being acknowledged by at least one Parisian architect, Bernard Poyet:

I must say again that if [the project] brings together several advantages, I shall be indebted to a brochure entitled *Observations on the construction of a new opera house* by M. Noverre. His name alone provides his elegy, and I do not fear to say that it is difficult that an artist of such a distinguished merit might be able to mistake himself on the facts of a monument of this sort.[39]

But however strenuously Noverre promoted the idea of a theater on an island site, in London it was not to be. As a city of merchants, in London such a structure could only be achieved with royal or parliamentary intervention, neither of which was forthcoming. And therein lies the essential difficulty: London's imperatives were essentially commercial, and most eighteenth- and early nineteenth-century commentaries on the London opera were as deeply concerned with how the Opera House functioned financially (mostly it did not, so this is perhaps not unreasonable) as they were about the building: discussions concerning the site of an opera house were bogged down in problems with leaseholds or squabbles over money, rather than the possible aesthetic statement of a building in the capital of an empire.

In contrast, Parisian debates were concerned not just with the building, but, as Christopher Mead has argued, with the Opéra's institutional place within the changing social structure of the city.[40] Administratively, the Opéra had moved from the king's direct patronage to oversight by the Ministère de la Maison du Roi; the Ministère transferred part of that control to the City of Paris from 1749 to 1757, when it then again took control until handing it back to the City in 1790, during the Revolution. During that time, the Opéra occupied four different sites, moving progressively from the courtly environs of the Palais-Royal on the rue Saint-Honoré, where it was first housed in 1673 and rebuilt (after a fire) in 1763, to a temporary house on the site of the theater's scenery warehouse on the boulevard Saint-Martin after another fire in 1781, to the Palais Garnier in 1875. As Mead notes, these moves signaled subtle shifts in orientation. Thus the 1781 move to the scenery warehouse on the boulevard Saint-Martin "recognized the boulevards' increasing integration into the city, while also suggesting by its location a previously unimaginable commonality of purpose with the nearby vaudeville houses."[41] And although the Palais Garnier was part of a planned development of the area that included a number of high-end shops, there can be no questioning the building's independent presence in the urban landscape.[42]

In the end, in London, it was Nash who, in an effort to complete the ceremonial (but fundamentally commercial) Regent Street scheme, created the sort of "public" building that Kips, Noverre, Adam, Novosielski, and others imagined. And finally, in 1818, it stood resplendent on the corner of Pall Mall, surrounded by arcades, a monument worthy of London's urban landscape. Or was it? Can we in fact make any substantive claims for Nash's achievement? Despite the reconfiguring of the site, Nash's building is architecturally weak; indeed, his input is essentially a tidy-up job, marzipan applied to an unsatisfactory cake: Nash took an existing building, surrounded it with shops and loggias, and added a shopping arcade. Far from projecting the opera house into the urban landscape, he had ensured that it was cloaked with a design that blended into the general run of nineteenth-century commercial developments being undertaken in the West End. And despite the efforts of some Nash scholars to aggrandize his work on the Opera House, Nash's own text in a report he prepared in 1810 gives the game away. In the section headed "Beauty of Metropolis," Nash wrote: "The beauty of the Town, it is presumed, would be advanced by a street of such magnificent dimensions; by the Colonnades and Balustrades which will adorn its sides; by the insulating the public building of the Opera; by the effect of the Monuments in the centre of the crossing streets."[43] The opera house is "insulated," not celebrated; monuments will instead stand at the center of crossing streets. Although he had the Haymarket street front to develop, Nash had not created the sort of temple to the arts or monument discussed by continental authors, or the independent building of standing proposed by Noverre. Perhaps, indeed, the commercial priorities of London had the last laugh, since the noise from the busy shopping streets required some kind of "insulation" for the theater. And the theater, as Her Majesty's, did not escape commercial concerns. Returning to figure 4.1, we can see that in the newly completed Carlton Hotel and Her Majesty's Theatre, the commercial imperative was such that in Phipps's design, the theater building has been moved from the corner of fashionable Pall Mall to the less prominent site on the corner of Charles Street. This street scene emphasizes the extent to which Her Majesty's was reabsorbed into the commercial landscape; London's former Italian opera house, for all the pomp of its architecture, is now visually just the same as the Carlton Hotel.

FIVE

Opera and the Carnival Entertainment Package in Eighteenth-Century Turin

MARGARET R. BUTLER

Musicologists have long acknowledged opera's power to illuminate contemporary society in its role as a complex "theatrical conversation" involving performers, venues, audiences, and related art forms.[1] In Italy, carnival season provided the temporal frame for this conversation by affording a city's residents myriad opportunities to engage in acts of sociability and spectatorship before, during, and after opera performances. Carnival opera's connection with diversions in and outside the opera house, and audiences' engagement within a broad system of urban entertainment, are thus recurring themes in Italian opera studies.[2] The relationship of that engagement to broader goals, and the links between the carnival season's cultural products and the spaces where they occur, can also shed light on the nature and development of the genre.[3] Understanding city spaces as meaningful wholes that strongly influenced cultural processes is essential for achieving a richer view of carnival opera. This is particularly true for the eighteenth century, when theatrical venues all over Italy underwent profound changes and increasingly became the focal point of the urban experience.

Turin, the eighteenth-century capital of the Savoyard state, within the principality of Piedmont, exemplifies the interconnectedness of performance-based arts in Italian cities of this period. The Savoy dynasty, whose roots go back to the eleventh century, had formed its state by gradually acquiring "a mosaic of territories straddling the western Alps" through its long negotiations between France and Spain in their struggles for Italian dominance.[4] Piedmont developed as the state's economic and political heart. Having earned a royal title in 1713 (through acquisition of the kingdom of Sicily), Victor Amadeus II instituted centralizing reforms and urban developments in the ensuing decades, aiming to assert his state's status. His reorganization of governmental structures and creation of new, politically unifying institutions ultimately led to Turin's rise as a thriving cultural center.

Turin's large, sophisticated, and lavishly decorated opera house, part of this chain of developments, dates from 1740 and is unusually well documented.[5] Turin therefore provides fertile ground for studying the institutional operatic experience within its urban environment.[6] The connection to the house's surroundings was particularly acute where the Regio was concerned, because of its management structure. The Regio's administrative directors, the Nobile Società dei Cavalieri, functioned as a collective impresarial group,[7] managing all aspects of the city's theaters, and more: they facilitated diverse activities occurring before, during, and after the carnival operas, such as gambling, shopping, masked balls, marionette plays, acrobatic shows, wax-figure exhibits, circuses with exotic animals, fireworks, and other spectacles.[8] An examination of Turin's cultural life must, therefore, begin with them.

The Cavalieri's administrative statutes testify to the sovereign's policies around cultural activities and to these events' civic importance, detailing their obligation to provide two carnival-season opere serie with high-quality music and poetry, but also other types of entertainments; indeed, performance of these other entertainments without Cavalieri oversight was prohibited.[9] The group's secretary kept detailed records of its plans and decisions about artistic production, much of which survives.[10] Contracts and other materials show that Turin's entertainments were highly regulated in terms of both place and time of performance. The carefully prescribed physical spaces, operating schedule, and emphasis on visual spectacle formed a network of meanings for the sovereign, nobility, and middle classes. Because of the Cavalieri, Turin's audiences experienced opera as part of what we might term a carnival entertainment package that was place- and space-specific and possessed of a distinctive character. The opportunities for sociability these activities provided demonstrated cosmopolitanism, assisting the sovereign's quest to gain recognition for Turin as a leading European capital.

Engineering Place and Time

Among the several factors unifying Turin's eighteenth-century carnival package, the most important were regulations prescribing the geographical and temporal organization of the city's entertainments and a city-wide culture of spectacle. Although extant contracts for the extra-operatic carnival entertainments reveal no details of their content, one clarifies the Cavalieri's engineering of the city's entertainments. Marionette shows were offered by a tailor, Antonio Vinardi (and associates), in unspecified locations around the city, starting from at least 1762. By 1776 Vinardi had become sufficiently successful to have moved from performing in a general area to having his own 'Teatro Vinardi'—subsequently one of the city's important minor theaters for popular entertainment. Vinardi and his partners' 1762 contract (the earliest extant for any of Turin's non-operatic performances) clarifies that they could only present their spectacles during the day, and had to conclude by 5:00 p.m. Geographical boundaries beyond which the associates could not extend were specified; they were to notify the Cavalieri when they had chosen a location, which had to be approved.

The contract's clause identifying the boundaries reads: "These contracting parties may not present the aforementioned entertainment [in the area extending] below the tower in the direction of Piazza Castello, but [may perform] above it in the direction of Porta Susina, that is, in [the area above] a straight line [extending] from the tower of the city to Porta Palazzo."[11] These boundaries kept the marionette shows well away from the Teatro Regio—in the central Piazza Castello—and Carignano, the city's comic theater and venue for balls, just outside the city center, as figure 5.1 illustrates.[12] The black lines indicate an area to the right of Via Dora grossa (the vertical line) and above Via S. Francesco d'Assisi (the horizontal line). The city's tower was located where the two streets intersect. The Teatro Regio is several blocks below that point, at I, and the Carignano to its left on the map, at T (in the Palazzo Carignano). This location for Vinardi's puppet shows evidently remained fixed: the 1776 carnival entry for Turin names a "Teatro Vinardi in Strada Dora grossa"; elsewhere it is called "Teatro delle marionette in Dora grossa alla Torre della Città," where it was adjacent to the tower. The injunction of 1762, and subsequent information on Vinardi's activities, seem to suggest that the Cavalieri strove to maintain a strict separation between entertainments—and presumably classes of people patronizing them.

The concluding hour of 5:00 p.m. is also significant because Teatro Regio operas began at 5:30 p.m.[13] The public attending the marionette

5.1   Nuova *Pianta della Reale Città di Torino* (Milan: Carlo Giuseppe Ghislandi, 1751), I-Tac, Simeom, Serie D, inv. 56. Boundaries added. Reproduced by permission of the Archivio Storico della Città di Torino. Reproduction or duplication by any means prohibited.

shows probably included some of the patrons occupying the Teatro Regio's *platea*, the ground floor, for those who did not own boxes. The location and timing of entertainments available to the middle classes, therefore, probably reflected the needs of the Teatro Regio's high-ranking audience members. It seems, then, that patrons were ushered to the Regio, and later to the Carignano, by means of the city-wide temporal organization of a carnival evening's multiple entertainments. Indeed, on the third renewal of Vinardi and his partners' contract (December 1765), a clause was added prohibiting them from organizing masked balls, privileging the Teatro Carignano, in the center of the city, as the only place to go to dance. Turin's carnival entertainment package thus reinforced class distinctions and created an expanded and multi-tiered space for sociability.

Marketing the Package

The Teatro Regio's prominence in the European operatic circuit, in terms of the leading performers, stage designers, choreographers, and composers it attracted, is now well-known.[14] Its repertory reflected leading European trends and stylistic features, thanks to the international trading of personnel and its organizers' contacts. News of its operas appeared in publications with international circulation.[15] Turin's carnival-season activity as a whole also gradually expanded over the second half of the century and became more widely known, as it—along with the city itself—was increasingly marketed to visitors as well as residents, attesting to Turin's burgeoning reputation as a cosmopolitan cultural center. It also paralleled a broadening of the areas designated for performances—a shift in the city's cultural landscape.

The printed *Almanacchi de' teatri di Torino*, published from 1780 to 1789, provided full information on the city's upper-class entertainments, with performance schedules and the names of box holders.[16] They emphasized particularly the Carignano's interior space and its transformation into a great dance floor during carnival. Still more notable was the *Almanacchi*'s predecessor, the Milanese *L'indice dei teatrali spettacoli*, which demonstrates the unified nature of a city's carnival season entertainment, especially for Turin.[17] Directed to agents hiring performers for theatrical seasons around Europe, the *Indice* appeared annually starting in 1764. In later decades it announced cities' total entertainment offerings, including operas, plays, and balls, perhaps taking on the secondary role of publicity tool. The publication demonstrates the theatrical industry's growth throughout the second half of the century.

The *Indice* shows that the theaters operating in Turin during carnival grew from two to five: to the Teatro Regio and the Teatro Carignano were added the marionette house, Teatro Vinardi, the Teatro Gallo (or Gallo Ughetti), and the Teatro Guglielmone (Teatro D'Angennes), in which itinerant troupes gave spoken-word plays. These three minor theaters have long been known for their non-operatic offerings,[18] but their histories are sketchy, with repertories and many details about when and how they were established unknown. Their eighteenth-century locations, however, are known and reflect the Cavalieri's evident desire for separation between high and low art forms: the Guglielmone and the Gallo were both located south of the Piazza Castello, not far from the city walls and the Po River, the Guglielmone in Piazza Carlina (point Q on the map) and the Gallo one block to the east (at S. Bonifacio, circled on the map).

The Teatro Guglielmone/D'Angennes is more amply documented than the Vinardi and Gallo theaters, though its eighteenth-century activities are still largely unknown.[19] Lorenzo Guglielmone was a tailor (like Vinardi) who began to give performances in the palazzi of the Marquis D'Angennes, after whom the theater subsequently created was named.[20] Guglielmone received written permission to present spoken-word plays in the Teatro D'Angennes for the carnival season of 1765 (with reference made to his having done this the previous year) but was denied permission when the forthcoming 1769 carnival season was unusually short (although he was permitted to offer performances elsewhere).[21] The Società's directors' treatment of Guglielmone suggests that the length of carnival season influenced the availability of Turin's non-operatic entertainments. The directors also closely monitored the respective schedules, profits, and productions of Vinardi and his partner (Trossarello, a painter), and of Guglielmone, comparing their activities in 1769, as the theaters were required to pay a fifth of their profits to the Cavalieri.[22] There may have been additional financial incentives for the Cavalieri to permit these lowbrow theatrical activities: Vinardi and Guglielmone both rented a coffee-shop space from the Cavalieri (probably near or inside the Marquis D'Angennes's palazzi, but also near Guglielmone's enterprise); their theatrical activities' success presumably affected the profits they (and by extension the Cavalieri) were able to turn at the coffee shop.

The activity of Turin's three minor theaters during carnival, alongside the prestigious Regio and Carignano, has generally escaped notice, but understanding it is important for a comprehensive view of Turin's theatrical life. The *Indice*'s entries for Turin between 1764 and 1774 list only the Regio's operas (although payment records verify that the Carignano balls were also given in those years). In 1775 the Teatro Gallo joined the Turin entry. In 1776, for the first time, all five theaters operated during the season and were publicized as doing so. These additions represented a significant broadening of the theatrical landscape and of urban activity.

Masked Balls and Their Role

Where the upper classes' experience of carnival entertainment was concerned, the center of gravity lay in the Piazza Castello. The Carignano's masked balls and the Regio's operas must have been enjoyed as a pair: their schedules were closely coordinated, and their venues were only about four hundred meters apart in the city center, near the Palazzo Reale, making the area the center of socially exclusive activities. The theaters bordered open piazzas linked by a short street, so the space for promenading and communicating between them was clearly favorable to social display, matching the magnificence of the artistic and architectural display. The new materials ordered to ornament the theater for the balls—including, in 1769, twenty-five chandeliers, ten gambling tables, and a mirror—show that they were privileged events.[23] Varying in schedule and number from season to season, the balls were held after some operatic performances, on non-performing nights during the operas' run,[24] and during the period when the Regio was dark, between the runs of the two operas presented each season.[25] They seemed to build in frequency toward the end of the season (with none at all sometimes given during the first opera's run) and were always held on the second opera's closing night and the two nights prior. They began either after the operas concluded (operas usually taking around five hours), or earlier on other nights, and lasted until early the next morning. The reveling so exhausted the instrumentalists who played the dance music that in 1781 their director petitioned the Cavalieri for an earlier end to the evenings.[26] As we have seen, the Cavalieri enforced the Carignano's monopoly on dancing. The balls might also have influenced alterations to the operas' musical components: evidence suggests the directors cut the operas' content significantly, removing sections of some arias and deleting others completely—on at least one occasion, it seems, so that the balls could begin earlier in the evening.[27] The operas and balls were valued by the institution for the revenue they generated through

5.2   Ignazio Sclopis del Borgo, *Passeggio della Cittadella con l'Elefante venuto in Torino l'anno 1774*. I-Tac, Simeom, Serie D, inv. 856. Reproduced by permission of the Archivio Storico della Città di Torino. Reproduction or duplication by any means prohibited.

entrance fees and gambling. Games of chance were available at both the Regio and Carignano, and the Cavalieri used funds generated from them, together with other annual sources of revenue, to help offset the annual losses incurred in running the opera—although the contribution made by gambling was nowhere near enough, and the annual "bailout" from the royal treasurer increased significantly over time.[28]

The publicizing of the balls in the *Indice*—including their specific dates, from 1783—attests to their importance for visitors.[29] Serving as an extension and expansion of the opera house's dynamic environment, the balls provided another type of visual spectacle. Traversing the short distance from the Teatro Regio to the Carignano for post-opera dancing capped off a Turinese carnival evening; the operas and balls formed a place-specific entertainment combination that was unified both geographically and aesthetically.

A Culture of Spectacle

Numerous sources document Turin's general interest in the visually thrilling, and not only during carnival. Throughout the year in this period, the Turinese patronized exhibits of exotic animals, including an elephant, camels, porcupines, and monkeys; they also enjoyed oddities such as a creature with two heads encased in a glass bottle and a dessicated crocodile.[30] When a live elephant was exhibited to the public, the exhibit's opening and closing times were specified in detail, and the show required an attendant appointed by the Cavalieri who himself received a percentage of the daily revenue from the elephant's appearance.[31] A contemporary engraving depicting an elephant's exhibit in Turin demonstrates the public's fascination with bizarre things particularly well (see fig. 5.2).[32]

Diversions offered specifically during carnival also capitalized on visual spectacle, with the Cavalieri drawing up contracts for non-operatic entertainments on a seemingly ad hoc basis. Turin was a lively place during carnival and in the weeks prior, as described earlier, and during the intervals of the five-hour opera performances—indeed, while the opera was being performed—theatergoers bought jewelry and other fineries from vendors who sold them along the theater's main staircase, as well as refreshments at the shop.[33] Gambling at the opera house's casino and masked balls after the operas ended, or on nights when operas were not given (Fridays, and sometimes Tuesdays),[34] rounded out the entertainment. Carnival in Turin must have been a feast for all the senses. But it enchanted the eyes above all else.

Given Turin's overall fascination with spectacle, evident from its sponsorship of these varied diversions, it is not surprising to find spectacle featuring prominently in the Teatro Regio's operas. A clause from the statutes approved in 1770 confirms the dedication to lavishness: "It will be the Società's great honor to produce the most beautiful and sumptuous operas possible."[35] Turin was admired for its technically sophisticated stage machinery; Giacomo Durazzo, director of Vienna's theaters from 1754 to 1764, solicited models of Turin's theatrical machines in 1762, and Turin was lauded in a Florentine publication for the variety of its machines in 1775.[36] Numerous works given throughout the century highlighted the manmade spectacle typical of Metastasian opera, which featured battle scenes, conflagrations, and processions (often involving several live horses). But in Turin the stage spectacle consistently extended beyond this type, reflecting the views Francesco Algarotti expressed in his famous manifesto of operatic reform, *Il saggio sopra l'opera in musica* (1755): that opera should consist of a varied spectacle with numerous components to delight the eye and should favor remote and exotic locations.[37] Turin's operas consistently featured visually striking elements such as airborne gods and goddesses, sea monsters, and exotic locales such as Mexico, China, India, and Africa, thanks in large part to the Teatro Regio's house poet, Vittorio Amedeo Cigna-Santi. His *Motezuma* of 1765, concerning Mexico's conquest by Spain, drew on an operatic narrative first written by Alvise Giusti as *Motezuma* and set by Vivaldi (Venice, 1733) and continued with Graun's 1755 Berlin setting of *Montezuma* by Tagliazucchi (Cigna-Santi's teacher). Italian theaters found Cigna-Santi's libretto the most attractive of those on the subject, and it was the source for a number of later productions.[38]

Animals and imaginary creatures made frequent appearances at the Teatro Regio, apparently serving as central components of Turin's emphasis on spectacle. Among the live animals gracing Turin's opera stage were an ox in *Ifigenia in Aulide* of 1762,[39] a parrot in *Motezuma*,[40] fifteen horses—unusual for the specified number—in *Creso* of 1768,[41] and caged wild beasts in *Tamas Kouli-Kan* of 1772.[42] Supernumeraries appeared in some operas costumed as lions, tigers, and bears—even, in one case, a dragon.[43] Dancers portrayed a wild boar in the ballets of carnival 1755, monkeys in those of carnival 1756 and 1766, and a hippogriff (a mythical creature with the front quarters of an eagle and the hindquarters of a horse) in carnival 1766.[44] Large numbers of supernumeraries were involved in some of the operas: 112 in *Tigrane* of 1761, and 92 in *Ifigenia in Aulide* of 1762, for example.[45] The second opera of carnival 1762 (which became *Ifigenia in Aulide*) was budgeted by the directors to be "un opera delle più spettacolose" in an attempt to recoup losses incurred from the three previous seasons, though, as a comparison of revenue over time indicates, it seems to have been ultimately unsuccessful.[46] The choice of Montezuma was likewise important in this regard, as the directors thought its spectacular component would guarantee its operatic success.[47]

Beyond the events onstage, the theater itself struck visitors for its grandeur, as oft-cited accounts of travelers demonstrate. Charles Burney, for example, stated, "The great opera house [in Turin] . . . is reckoned one of the finest in Europe. It is very large and elegant; the machinery and decorations are magnificent." And according to Joseph-Jerôme Le Français de La Lande, Turin's theater was "the best designed, the best planned, and the most complete that is to be found in Italy; it is the most richly and nobly decorated that there is in the modern style."[48] Its physical and cultural centrality were further enhanced when it played host to events such as those around the 1750 wedding of Vittorio Amedeo, Prince of Piedmont, to Maria Antonietta Ferdinanda of Spain. Aside from a *festa teatrale*—*La vittoria d'Imeneo*, set to music by Baldassare Galuppi, the scene designs for which (by the renowned Galliari brothers) illustrated the theater's capacity for spectacle[49]—the auditorium itself contributed to an aura of magnificence. Striking in the following account is the use of materials that produced and reflected light (candles, jewels, glass, brocade, and mirrors):

On this occasion a sumptuous illumination was given, "with approximately three thousand torches and wax candles distributed along the five tiers of boxes in candle holders adorned by golden leaves, with such beautiful symmetry that together they came to elevate the majesty and the richness of the theater," while the scene designs, illuminated by the "crystal lamps" [chandeliers] refracted the radiance of the festive ornaments.[50]

On nights when operas were given, and on many others, the city itself was illuminated, offering at once reassurance and visual enhancement: "But even apart from these occasions the city offered, by night, a delightful spectacle: 630 lanterns suspended from metal supports broke the darkness and lent newly made reassurance to the passers-by."[51]

The environment around and beyond the theater—the beauty of the city and its architecture—contributed to a broad culture of visual spectacle that was understood by those residing both within its walls and beyond them. Visual beauty was a recurring theme in eighteenth-century visitors' accounts of the city. Turin was widely admired internationally for Filippo Juvarra's and Benedetto Alfieri's sophisticated architectural designs, part of the large-scale urban modernization project undertaken in Turin earlier in the century. Juvarra, brought to Piedmont by Victor Amadeus in 1714, assembled a team of assistants, including Alfieri, and developed a "sumptuous, refined style [that] made Piedmont a centre of late baroque architecture."[52] He transformed Turin's urban landscape, creating numerous churches, piazzas, and palaces over the following decades. Charles de Brosses in 1740 judged Turin to be "the most beautiful city in Italy and perhaps in Europe, for the straight streets, the consistency of the buildings, and the beauty of the piazzas."[53] Near the century's end, visitors such as Élisabeth Vigée Le Brun were still admiring the city, and from her language, reminiscent of de Brosse's, we can infer that a trope of Turinese style had developed: "The city is very beautiful in itself; the streets are all perfectly straight and the houses constructed with great consistency."[54]

Turin's beauty was a source of pride for Savoy's dynasty and was evidently marketed as a concept to the broader public at mid-century, consistent with Victor Amadeus's aspirations to achieve recognition for his state as an international leader in art and architecture as well as in politics. Two publications in particular, aimed not at residents but at visitors, contain descriptions and images of the city that emphasize beauty in tangible ways: its first guidebook, Giovanni Gaspare Craveri's *Guida de' forestieri per la Real Città di Torino* (Turin, 1753), and the map shown in figure 5.1, published in Milan in 1751.[55] Craveri's *Guida* was evidently published in connection with the aforementioned royal wedding in 1750;[56] the map's publication may also have related to the wedding (although the circumstances of its creation are unknown).

Craveri offers the most detailed contemporary description of the Teatro Regio's magnificence:

> This theater is judged by all to be the most majestic and best furnished in Europe, and is justifiably the object of admiration of visitors for its vastness and expansiveness, for the architecture and conveniences of the building, and for the beauty of the interior's decorations, most of which are ornamented in gold. The painting of the ceiling is remarkable. Here every carnival season they perform operas with spectacle as magnificent as is appropriate to the grandeur of the royal court, who attend in the spacious balcony reserved for them, that is then illuminated.[57]

The book's numerous foldout engravings both enticed potential visitors to experience the city's beauty for themselves and served as an aide-mémoire to those who had already visited. The pages containing the images might themselves be viewed as a form of spectacle; figure 5.3's Torre della Città, for instance, folds out to more than twice the size of the book itself. The guide's descriptions highlight each building or monument's visually distinctive features. Figure 5.4 shows the Castello Reale (today the Palazzo Madama, located in the city's central Piazza Castello); its description emphasizes the Castello's grandeur, enhancing the reader's estimation of Turin because of its possession of an architectural marvel the equal of any in Italy:

> Castello Reale . . . today serves as the residence of the royal princes. It was built in 1416 by Amadeus VIII, First Duke of Savoy, with four towers, that is to say, one per corner. And at this point the boundaries of the city terminated. It was subsequently decorated, in 1720, thanks to the munificence of Madama Reale, mother of King Vittorio Amedeo, by a superb stone façade, which completely covers the western side, and is adorned with vases, statues, large windows, and passageways. At the entrance there are three large uniform gated doors, which do not obstruct the view of the superb atrium and the two great, spacious marble stairways, the whole of which is a construction of great magnificence. The celebrated Filippo Juvara [*sic*] designed it, and one can safely say that this is the best work of architecture in the city and that it is comparable to any of Italy's most beautiful structures.[58]

OPERA AND THE CARNIVAL ENTERTAINMENT PACKAGE

While Craveri's *Guida* is often cited,[59] the map's brief text is also illuminating (and has heretofore gone unnoticed). A short statement on the city's essential qualities appears in an index alongside the map in figure 5.1, using language (*bellezza, regolarità*) that, as we have seen, became a trope for the city's visitors:

> This city is well-known for its beauty and the uniformity of its buildings and level streets. Its citizens revel in many languages, particularly French; are polite, attractive, civil, lovers of virtue, courteous toward visitors, and faithful subjects of their sovereigns. It is located in the heart of Piedmont, which is overflowing with all things necessary for the nourishment of the human soul, where it is called the Garden of Italy.[60]

5.3 "Torre della città di Torino," in Giovanni Gaspare Craveri, *Guida de' Forestieri per la Real Città di Torino* (Turin: n.p., 1753), unnumbered page. I-Tac, Simeom, Serie G, inv. 1. Reproduced by permission of the Archivio Storico della Città di Torino. Reproduction or duplication by any means prohibited.

69

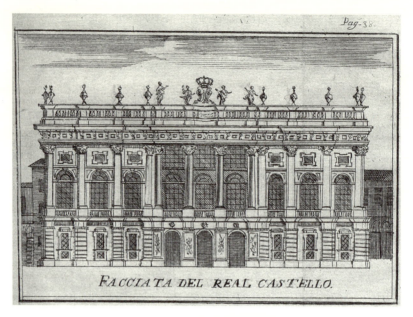

5.4 "Facciata del real castello," in Giovanni Gaspare Craveri, *Guida de' Forestieri per la Real Città di Torino* (Turin: n.p., 1753), 38. I-Tac, Simeom, Serie G, inv. 1. Reproduced by permission of the Archivio Storico della Città di Torino. Reproduction or duplication by any means prohibited.

This idyllic picture is revealing in that the citizens, as much as the city itself, seem to form part of Turin's tourist attraction, as "regular" in their behaviors as the streets themselves.

While Turin's Nobile Società dei Cavalieri offered the public a tightly integrated set of diversions, designed to convey the sovereign's munificence, and the image of Turin as a sophisticated capital city, the convergence of their activities with concerted efforts to project the public experience of the city as beautiful and striking and the targeting of international visitors point to a broader quest for the city to be seen as a modern urban cultural center in a burgeoning Europe-wide tourist economy. Considering a theater's operas as one entertainment form among many, within an urban environment that was itself something of a theater, helps us imagine more vividly the view contemporary audiences might have had in the mid to late eighteenth-century, and how they might have experienced opera during carnival.

Appendix 1. I-Tac, Libri conti della Nobile Società dei Cavalieri, vol. 32 (1756–57), p. 114.

Stato degl'Utili avuti sopra li Spetacoli dati nelli Teatri Carignano e Grondana, e per altre Esazioni

Spetacoli dati nelli Teatri Carignano e Grondana	1752 in 1753	1753 in 1754	1754 in 1755	1755 in 1756	1756 in 1757	Totale
Opere Buffe nella primavera		12389.12.2				12389.12.2
Dette nell'Autunno	6717.1		2580.1.4	4163.19.10	2359.9	15820.11.2
Commedie Italiane nella primavera	3010.17		2751.19		4120.16.8	9883.12.8
Dette nell'Estate	1736.4					1736.4
Dette nell'Autunno		1867.10.8				1867.10.8
Commedie francesi nella primavera ed'Estate				2686.10.6		2686.10.6
Equilibri nell'Estate		43.18.6				43.18.6
Forze d'Equilibri nell'Autunno					94.10	94.10
Balli di Corda e forze d'Equilibri nell'Estate		300.19				300.19
Saltatori, e Balli di Corda nell'Avento			912.19.2	472.11.10	150.16.6	1536.7.6
Balli di Maschera nel Carnovale		1985.19	770.0.2	1993.10.3	2904.1.6	7653.10.11
Totale degl'Utili	11464.2	16587.19.4	7014.19.8	9316.12.5	9629.13.8	54013.7.1
Si deduce il fitto di Teatro e spese straord.e	5847.18	6228	5187	5000	5038	27300.18
Resta d'utile sopra li suddetti spetacoli	5616.4	10359.19.4	1827.19.8	4316.12.5	4591.13.8	26712.9.1
Più per Esazioni diverse		6	43.19.3	87.15	145.10	283.4.3
Totale degl'Utili	5616.4	10365.19.4	1871.18.11	4404.7.5	4737.3.8	26995.13.4

Stato dell'Esatto, e speso, e perdite fatte sopra le opere del Teatro Regio ed altre

Opere Teatro Regio	1752 in 1753 Lucio Papirio, e Medo	1753 in 1754 Bajaset, e Demofoonte	1754 in 1755 Sesostri e Andromeda	1755 in 1756 Ricimero, e Solimano	1756 in 1757 Antigono, e Lucio Vero	Totale
Speso per le opere del Carnevale	94255.3	97050.5.9	91080.9.2	92231.8.8	91055.5.8	465672.12.3
Esatto per le opere suddette	84941.9.8	79652.12.6	64484.7.2	74192.16.1	75058.7.0	378329.12.5
Resta di perdita sopra le sud.e opere	9313.13.4	17397.13.3	26596.2.0	18038.12.7	15996.18.8	87342.19.10
Più per altre perdite e spese straordinarie			696.12.10	2029.9.8	472.2.6	3198.5
Totale delle perdite suddette	9313.13.4	17397.13.3	27292.14.10	20068.2.3	16469.1.2	90541.4.10
Deduzioni degli utili suddetti	5616.4	10365.19.4	1871.18.11	4404.7.5	4737.3.8	26995.13.4
Perdita Totale E deducendosi il Total dell'Esatto dal Tesoriere generale Signor Buttis	3697.9.4	7031.13.11	25420.15.11 12000	15663.14.10 24149.19.2	11731.17.6 15663.14.10	63545.11.6 51813.14
Si riduce la total Perdita						11731.17.6

Appendix 2. Contracts that document performances of entertainments during Carnival seasons specifically, or periods just prior to them. In Archivio storico della città di Torino, Carte sciolte 5499, Teatri e pubblici spettacoli (except where noted).

Marionette Plays:

For carnival 1763: 1762, 20 December. Pietro Maria Deferari del fu Eusebio, Michele Mazzolino del fu Franco, Pietro Dematuri del fu Gio Domenico, Vinardo.
1st renewal: 6 December 1763 [for carnival 1764]
2nd renewal: 7 December 1764. "Per il carnovale 1764 in 1765."

OPERA AND THE CARNIVAL ENTERTAINMENT PACKAGE

3rd renewal: 17 December 1765. Extra clause: "Si conferma l'avantiscrita delli 20 Xbre 1763 [*sic*] a favore del Sr. Vinardo, e compagni che non possino dar Balli, e rapresentino come anno fatto per il passato ad osservino li capitoli contenuti in d.a scrittura. (addition by Sig. Marchese D'Angennes)

For carnival 1768: 24 December 1767, Pietro Pastorini di Piacenza Dentista, Marionette.

For carnival 1778: 23 December 1777, Giacomo Agrizio and Antonio Ferrari, Marionette.

Wild Animals (Circuses) and Miscellaneous Entertainments:

14 November 1778 through all of December: Gio. Batta Bersegol, "un cavallo vivo, far vari giuochi e conti d'aritmetica" (a live horse, [and to present] various arithmetical games).

November 1781 for 20 days [leading up to carnival]: Nicoletti Vida, "Un Barraccone esisten.te in piazza castello, le ondici Bestie descritte . . . , fra quali un Leone, una Leonessa, ed una Tigre" ([to construct] a temporary stage in Piazza Castello [to present] the eleven animals listed . . . among which [are] a lion, a lioness, and a tiger).

Tightrope Walkers, Acrobats, Dancers, and Fireworks:

For carnival 1765: 17 November 1764, Gennaro Bologna, "balli di corda, fuochi d'artificio, e altri divertimenti in un Barraccone" (tightrope walkers, fireworks, and other entertainments on a temporary stage).

For carnival 1780: 2 December 1779, Domenico Merli, "balli di corda e salti sul filo di ferro in un barraccone, compagnia di saltatori, e ballerini" (tightrope walkers and acrobatics on an iron bar in a temporary stage, [and a] troupe of acrobats and dancers).

Concerts: 15 November 1784 until the opening of the Teatro Regio [on 26 December]: 1784, Franco Ferrarj, "dare concerti in musica da giorno d'oggi, sino all'apertura del Regio Teatro" (to give musical concerts from today until the opening of the Teatro Regio).

Lifelike Wax Figures:

9 November 1776 for thirty days: Francesco Marchetti Veneziano, "figure di cera al naturale" (lifelike wax figures).

Spoken-Word Plays (Comedies and Tragedies):

For carnival 1768: Libri Ordinati della Nobile Società dei Cavalieri, vol. 6 (1767), 204–6. Contract for Guglielmone, 22 December 1767. "Commedie e tragedie, nel teatro Sig Marchese D'Angennes" (Comedies and tragedies in the theater of Sig. Marquis D'Angennes).

SIX

Cockney Masquerades: Tom and Jerry and Don Giovanni in 1820s London

JONATHAN HICKS

Mozart's Don Giovanni seems to live his life—at least within the confines of the opera—perpetually on the run. But what to us is symbolic of his precarious emotional and sociocultural position had other associations in the early nineteenth century. In 1820s London Giovanni's popularity coincided with the brief but phenomenal success of Tom and Jerry, the protagonists of Pierce Egan's *Life in London*, first serialized in 1820.[1] These characters—and their period—had in common something of an obsession with mobility, an important topic in urban history and cultural geography, but one that has hitherto received little attention in opera studies.[2] Although a number of writers have considered nineteenth-century operas and opera audiences in relation to their places of performance and/or the urban landscapes they evoked, these studies have typically treated the city either as a synonym for social context or as a relatively static assemblage of sites, sounds, and atmospheres.[3] Less attention has been paid to the manner in which urban life was experienced and characterized not only as transitory—an observation for which Baudelaire, writing long after the Georgian period, is sometimes given credit—but also, more literally, as an experience in transit from one place to another, at speed or at leisure, by carriage or on foot. Attending to the history of operatic mobilities might help to animate established themes of locality and identity while opening new avenues for the study of opera and the city.

My focus here is on a cast of imaginary young men who possessed a good deal of social privilege and no less sexual curiosity, and who would have been easily recognized in Georgian London. The reckless aristocratic libertine so familiar to the public via the engravings of Hogarth and the playboy lifestyle of the Prince Regent (later George IV) was exaggerated to the point of caricature in the case of Don Giovanni, a high-ranking nobleman with sufficient confidence to seduce or assault whomever he desires—not to mention the chutzpah to invite a statue for dinner. There is even a cartoon from 1820 that puts the prince in the part of Mozart's character (with bare-breasted women upstage), surprised by the return of Donna Anna (Queen Caroline), while Leporello (Lord Castlereagh) unfurls a list of his master's conquests.[4] The social status of the urbane "Corinthian" Tom and his country cousin Jerry Hawthorn is less clear. Gregory Dart counts this once-famous pair among the poster boys of an emergent Cockney culture, which blurred established class boundaries and placed London at the center of the imaginative universe.[5] Whatever the differences between my three title characters, they had in common an obsession with youthful masculine vigor as well as a will to shape life around the pursuit of one pleasure after another.[6] Their journeys, you might say, were integral to their identities. And they crossed paths (and exchanged clothes) on more than one occasion.

The clearest sighting of all three together comes in the Green Room at Drury Lane, courtesy of an engraving by Robert and George Cruikshank that accompanied Egan's text (see fig. 6.1). The caption describes "Tom and Jerry introduced to the Characters in Don Giovanni," but this is not the opera familiar to audiences today. In 1820s London only the King's Theatre was licensed to stage Italian opera; Mozart's work had premiered there in April 1817. Yet the Cruikshanks' image shows us backstage at Drury Lane, a venue ostensibly reserved for spoken-word drama but increasingly home to "illegitimate" genres such as pantomime, melodrama, and burlesque.[7] The presence of Pluto and Mercury relaxing in the background, and a woman in the part of Giovanni, offers clues that we have wandered some distance from Da Ponte. In fact, the central figure here is the English actor, singer, and dancer Lucy Vestris, who appears costumed for the leading role of *Giovanni in London*. This burlesque (or *burletta*) after Mozart's opera was written by William Thomas Moncrieff and first staged at London's Olympic Theatre on Boxing Day of 1817, only months after the operatic original had made it to the capital.[8]

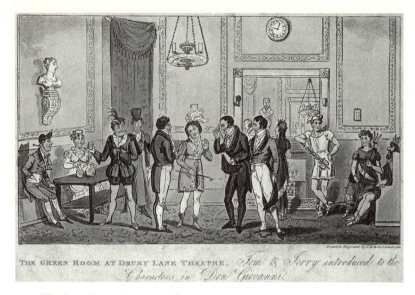

6.1 "The Green Room at Drury Lane Theatre. Tom & Jerry introduced to the characters in Don Giovanni." Drawn and engraved by Isaac Robert Cruikshank and George Cruikshank. Published by Sherwood & Co., London (1 June 1821). © Trustees of the British Museum. Tom and Jerry are the figures standing arm in arm, next to Vestris; Jerry is on the right, identified by his green coat, a marker of his origins in the country before he embarked on life in London.

When the Olympic's manager, William Elliston, took the lease on Drury Lane, he revived the *burletta* and chose Vestris—who had played the hero in Thomas Arne's setting of *Artaxerxes* earlier that season—for the breeches part in *Giovanni in London*; her first outing was on 30 May 1820. Such was her success in the role (with which she was to become identified) that she was still playing Giovanni at Drury Lane when the Cruikshanks' print was published, a year and a day later.

The image of Tom and Jerry in the Green Room provides a timely reminder of opera's cultural mobility in Regency London: within a relatively short space of time, *Don Giovanni* had first been mounted at the high-status King's Theatre, then burlesqued on the "minor" stage of the Olympic and at the patent house of Drury Lane. Before the burlesque was off the boards, Egan had written his fictional cousins into the action by showing them in a fashionable setting, one that would have been recognisable (if not familiar) to many of those who purchased *Life in London*. Of course, the notion of operatic culture as a locus of "fashionable acts" is well established.[9] But rather than drawing attention to the social

COCKNEY MASQUERADES

life of the opera house, the availability of the (travestied) Giovanni for literary and pictorial allusion suggests that opera in Regency London was not confined to the stage, let alone to the King's Theatre. Indeed, the story of Tom, Jerry, and Don Giovanni is one of a soon-to-be-canonic opera put to work in the context of a fantasy of urban freedom, a fantasy worth exploring further.

Movable Types

Egan's Tom and Jerry made their debut in September 1820, in the first issue of Egan's new monthly magazine, *Life in London*. Less than a year later the text (and images) were available in a single-volume format divided into two "books," the first comprising a discussion of Egan's ambitions as a metropolitan guide as well as introductions to the principal characters, and the second relating the cousins' adventures in the city.[10] At the time Egan was best known for sports journalism, particularly reports on prize-fighting. His accounts of "ancient and modern pugilism" provide some context for the rough-and-tumble picaresque of Tom and Jerry, in which the city functions as a playground–*cum*–fighting ring. This sense of the capital as a social space available for exploration and adventure—and not merely for omniscient overview—is central to Egan's urban vision.[11] With each passing chapter Tom and Jerry travel wherever fashion or happenstance takes them, requiring only the company of good friends, plenty of energy, and a willingness to dress for the occasion. When the single-volume version was published, the idea of social mobility was established with a "Corinthian Capital" frontispiece; it shows the British population arranged around a column according to their "ups and downs" and "ins and outs." The crowning "Flowers of Society" are far removed from the human vegetables propping up "The Base." The relative size of the figures in the image draws attention to the middle, the slipperiest part of the social scale. It is here that we find the book's principal characters—Tom, Jerry, and their Oxonian friend Bob Logic—enjoying a toast with one another while in touching distance of the "Noble," the "Respectable," the "Mechanical," and the "Tag Rag & Bob tail."[12]

Part of the pleasure of their middling position, the image seems to imply, is the access it affords to all walks of society—a point reinforced by the book's subtitle, advertising the trio's "Rambles and Sprees through the Metropolis." Readers turning to the first chapter, an "Invocation" of life in London, are promised men "upon the *strut* in Hyde-park" as well as "dissipated ramblers *touched* with the potent juices of Bacchus."[13]

77

In chapter 2, a "*Camera Obscura View*" of the city, they are introduced to those "whose whole evenings are continually occupied in *toddling*, as it is termed, from one lodge to another."[14] And so on. Egan's Londoners are rarely static, and a novel lexicon is brought out to accentuate their various comings and goings: "A *peep* at Bow-street Office—a *stroll* through Westminster Abbey—a *lounge* at the Royal Academy . . . a *strut* through the lobbies of the Theatres, and a *trot* on Sundays in Rotten-row."[15]

The penultimate chapter demonstrates the sheer variety of the scenes the main characters visit and the frequency with which they move from one place to the next. As the opening summary informs the reader, "The Trio" of Tom, Jerry, and Bob begin by "making the most of an Evening at Vauxhall." This location was (and had long been) recognized for its social diversity, as Bob Logic attests: "To me . . . Vauxhall is the festival of Love and Harmony, and produces a most happy mixture of society . . . Every person can be accommodated."[16] After Vauxhall, they attend the "Exhibition of Pictures at the Royal Academy" in Piccadilly, another fashionable destination, but far more upmarket. The three then head north, where we read of them "'masquerading it' among the Cadgers in the 'Back Slums' in the Holy Land."[17] Egan's studied, not to say exhaustive, use of informal vocabulary helped proclaim the series's authenticity, with explanations readily provided for Jerry, the "green" country cousin, and thence for the reader;[18] but it also helped draw attention to the telescoping of high and low spaces in the story, as when he compares "masquerading it" in the north London slums with a "Grand Carnival, or the Masquerade at the Opera House." And while there is a frisson of physical danger in the first of these places, there is never any indication that descent into the realms of the city's lowlife will cause long-lasting harm. (Moreover, when fights do ensue, they are treated with the light touch you might expect from one with Egan's positive disposition toward sports of violence.[19])

Needless to say, Egan's titular protagonists are both able-bodied sportsmen; Tom in particular is the epitome of the fleet-of-foot guide to the modern city. We find this expressed in his song, "A Description of the Metropolis," delivered near the close of the first book of the two-part text. Following a brief "rustication" at the country seat of Hawthorn Hall (where we meet Jerry), Tom resolves to take his cousin to London, a visit that will occupy book 2. Buoyed by the prospect, Jerry invites Tom to entertain the after-dinner party, and he obliges:

London Town's a dashing place
    For ev'rything that's going,
There's *fun* and *gig* in ev'ry face,

So natty and so *knowing.*
Where Novelty is all the rage
    From high to low degree,
Such pretty *lounges* to engage,
    Only come and see![20]

There follows a versified list (almost four pages long) of metropolitan novelties interspersed with a sing-along chorus: "Dancing, singing, full of glee, / O London, London town for me!" When read in the context of Egan's narrative, immediately following a hard day's hunting and a long night's drinking, the song can be taken as Tom's gift to good company. Singing was a common feature of Regency masculine sociability, and it is indicative of Tom's character that he should bring such merry verse to the table. The cue line for each chorus—"Here all dash on / In the fashion!"—typifies Tom's (and Egan's) vision of the city, where speed and style go hand in hand. And the exploration promised seems to take on the character of the event or place visited: "To the Opera prance," reads one line in verse 3, "See Vestris dance." We might also note the contrast to Jerry's partial rendition, on the previous page, of Charles Dibdin's "Labourer's Welcome Home." Where the country cousin sings of the plowman, the hedger, the woodman, and the shepherd, all kept in time by the "sweet" sound of the village clock, the city cousin invokes an alternative rhythm of distractions and delights: gala nights, grand parades, masquerades, and gaslights. Tom's song thus offers a reminder of the rural/urban opposition that defines the cousins' relationship and provides a sign of things to come: as both inventory and invitation, this breathless list hails the reader as a fellow traveler on a promised journey to the capital of fashionable life.

Egan ends book 1 in anticipation of arrival, with the cousins racing along the road to the metropolis while "Jerry, in raptures, frequently burst out, humming the last line of his Cousin's song of '*London, London town for me.*'"[21] It is telling that the mode of expression Egan chooses for such excitement, with the carriage moving "as fast as the horses could go," is not verbal but musical. Although the particular song in question seems to have been incidental to the original *Life in London*, which did not include it,[22] later editions came interleaved with a piano score credited to A. Voight.[23] Voight's song, like Tom himself, was well traveled: the opening of each verse is set to an adaptation of "O London is a fine town," an old tune long associated with picaresque London adventures.[24] Balancing the familiar and conventional opening, Voight's newly composed material offered a succession of effects and decorations, placing the emphasis on moment-by-moment communication of the text. The song incorporates both direct word painting (the appearance of the tightrope walker Madame Saqui occasions an octave ascent, while a "Peep into Hell" leads to the lowest note of the song) and opportunities for vocal mimicry (a hawker selling "rare hot pies!" and a "Matchgirl's bawl" on a sustained high G).

One fleeting impersonation is particularly worthy of note: on the words "Ambroghetti's squall" we find three full bars of sixteenth notes in an unprepared key area of B-flat with a crescendo to *fortissimo* marked in the piano score. Voight's melismatic setting is a rudimentary example of the (often more ambitious) English practice of parodying Italianate vocality. But there is more to it than that: Giuseppe Ambroghetti, a celebrated bass, was associated for much of his career with Mozart's *Don Giovanni*. He made his Paris debut in the title role in 1807 and played Giovanni again a decade later at the work's London premiere. Indeed, readers of *Life in London* had been reminded of this pairing of singer and role in book 1. During an introduction to the Italian opera, Egan remarks on "the languishing 'die away' strains of Ambroghetti's *Don Giovanni* [which] may almost cause an earthquake in the ear of the tasteful critic, and call forth 'bravo!'"[25] He even quotes Don Giovanni's act 2 paean to pleasure:

Vivan le femine [sic]
Viva il buon vino,
Sostegno e Gloria,
D'umanità
[Long live women
Long live good wine
The substance and glory
Of humanity]

There is doubtless something opportunistic about Voight's setting of the squall: among the various city sights referred to in Tom's song, those that also *sound* have an obvious appeal to a composer's imagination, and the Ambroghetti moment is not the only vocal effect committed to the score, as noted above. But the connections between Tom and Giovanni are more than circumstantial.

Early in Egan's text there is a passage describing the Corinthian's physical appearance, which was "the theme of universal admiration . . . his attractions were far above mediocrity, and demanded for him a degree of consideration and respect that must always cause him to appear CONSPICUOUS."[26] The supporting evidence for this claim comes in the form of a string of situations in which Tom could not help but be noticed: his "box at the Opera," for instance "caused many a languishing eye and palpitating heart to wander towards its elegant owner," while his "*stroll* through the Theatres caused a different sensation among another class of females, and no lures were neglected to decoy and entrap him."[27] If the sexual subtext already brings Mozart's rake to mind, it is not without cause. At the same point in the text Egan makes explicit reference to "the *dramatic* Don Giovanni" in the context of Tom's "descending into the *Hells* [gambling dens]."[28] Indeed, these allusions prepare the ground for Tom's choice of costume when attending a masquerade ball at the King's Theatre: whereas Jerry dresses as a huntsman and Bob appears as a domino, Tom is "attired as the gay libertine, *Don Giovanni*."[29] Later in the book, when Tom raises the prospect of a visit to the Green Room, Jerry replies that "it is Don Giovanni to night, and the numerous characters that piece contains will afford us plenty of fun! Besides, if it were nothing else but the introduction of one Don to another." Tom, we read, "smiled at this compliment, but made no answer to it."[30] He did not have to: the elision of Tom and Giovanni was sufficiently established by this point to require no further comment. And when *Life in London* received the same burlesque treatment as Mozart's opera, the overlap between the characters found even greater prominence. Let us turn, then, to the appearance of Tom and Jerry (and Don Giovanni) at London's Adelphi Theatre.

Musical Stages

William Moncrieff's Adelphi version, entitled *Tom and Jerry; or, Life in London, an Operatic Extravaganza* (which premiered on 26 November 1821) was the third adaptation, after those of one Mr. Barrymore and Thomas Dibdin (son of Charles).[31] In 1881 John Camden Hotten, introducing his edition of Egan's text, emphasized the impression this play made on its audiences:

[It] created a furor amongst play-goers, the like of which never occurred before, and has never occurred since. It ran for upwards of three hundred nights, and only gave over because the actors were tired out; the audiences were as mad for it as ever. It made the fortune of the house, and everybody connected with it—except the author.[32]

Despite theatergoers' propensity to revisit productions, for a play to stay on the boards for such a long time meant that a lot of different people attended. Moncrieff's own preface to the play is striking for the emphasis it places on the audience's social variety:

From the highest to the lowest, all classes were alike anxious to witness its representation. Dukes and dustmen were equally interested in its performance, and peers might be seen mobbing it with apprentices to obtain an admission. Seats were sold for weeks before they could be occupied; every theatre in the United Kingdom, and even in the United States, enriched its coffers by performing it.[33]

The appetite for the Tom-and-Jerry format extended beyond Moncrieff's production. And as with the *Don Giovanni* spin-offs, the sheer number of versions that emerged within a short period of time is remarkable. Hotten supplies a list demonstrating that this multimedia phenomenon was repackaged for all sorts of different markets: "Besides the authors already mentioned . . . Farrell, and Douglas Jerrold each produced dramas upon the popular theme; and in the summer of 1822 'Life in London' was being performed at no less than *ten* theatres in and around London."[34] The fact that these adventures were being played out across so many venues—from the patent house of Covent Garden to minor theaters across the Thames—says something about the cultural mobility of Egan's invention. The variety of productions encompassed a pantomimic version at Bartholomew Fair in 1823, and, at Astley's Amphitheatre, near Westminster Bridge, a "New Pedestrian, Equestrian Extravaganza, in Three Acts of Gaiety, Frisk, Fun, and Patter," by Egan himself. Even before the work had reached the stage, there were sixpence renderings of Egan's shilling stories, such as John Badcock's *Real Life in London; or, The Rambles and Adventures of Bob Tallyho, Esq., and his cousin the Hon. Tom Dashall, &c. through the Metropolis*, in which the adapted names emphasize the importance of movement to the characters' identities, with Tallyho (the Jerry character) ready for the country chase and Dashall (the Tom character) defined by his speed at moving about town.[35] Many found novel ways to depict London life via emphasizing different sorts of movement: Edward Fitzball's *The Tread mill; or, Tom and Jerry at Brixton*, for instance, was distinguished by its representation, at the Surrey Theatre, of the notorious south London prison wheel that forced the human body into a grotesque spectacle of perpetual motion.

Moncrieff's adaptation may not have gone to quite the extremes of Fitzball's "serio, comic, operatic, milldramatic [*sic*], farcical, moral burletta" but it is nevertheless an essay in Regency mobility. His take on the Tom and Jerry phenomenon uses Egan's basic outline: a tour around various sights in London with the three friends and the acquaintances they meet en route. The principal alteration to the story concerns the prominence given to three female characters: Kate, Sue, and Jane. The women frequently appear in disguise—first as fashionable men, then as female street singers, then as the three Trifle sisters—both to trick their beaux into trouble and to keep an eye on them along the way.[36] When we first meet the female trio, in act 1, scene 2, their gendered identities are made clear by their modes of mobility: they do not presume to partake of the masculine, hunting activity of the "chase" but follow their would-be partners with the speed of the mail coach. Soon after, Tom and Jerry join Sue in a song about "sprees and rambles night and day" entitled—and so evidently to the traditional tune of—"Over the Hills and Far Away." This number draws our attention to the practical reasons for augmenting the female parts, which reflect the contrasting generic requirements of a picaresque novel and a musical play: whereas the former can meander freely, the latter requires the balance of different voice types and, as a corollary to that, a pairing off of characters for marriage. This is precisely what happens in Moncreiff's version, in contrast to Egan's text, where the topic of domestic union is scarcely broached.

Music helps articulate the play's structural logic in other ways too: the unusually large number of scene changes for a piece of this period, which are necessary for the transfer of Tom and Jerry's hectic, cyclopedic tour from book to boards, are often marked by music. Whenever the characters discuss where to go next, the moment of decision is likely to be heralded in song, as a dramatically crude but theatrically effective expression of the non-linear and non-developmental design of the play. Music is, indeed, designated as a structural device from the outset: the first scene of Moncrieff's burlesque—variously labeled a "drama" and an "operatic extravaganza"—takes place at Hawthorn Hall in the country, with an optimistic chorus welcoming the coming day and celebrating those "who take, in the chase, and the glass, a delight." The scene ends with a request for further songs from the assembled company, thus setting the pattern for the rest of the play, in which few scenes pass without singing of one sort or another.

Music also serves as an expression of both pleasurable sociability and movement. When the men decide to make their way to Almack's ballroom to enjoy its entertainments, rapid movement is the theme: "Come along, then," says Bob to Jerry. "Now, Jerry, chivey!" As ever, Jerry is bemused by the use of slang terminology: "Chivey?" he enquires, to which Logic replies: "Mizzle!" Again, Jerry does not understand. "Tip your rags a gallop!" exclaims Logic, as if this would be clearer. Then "Walk your trotters!" And, finally, "Bolt!," which Jerry does understand. In the song that follows—a moment of performance that stops the characters in their tracks so the audience can enjoy the many styles of their (im) mobility—Logic sings, "Run, Jerry run, all London are quadrilling it."

The verb "to quadrille," also found in Egan's text, makes music, dance, and fashionable mobility indistinguishable. If such linguistic sleights of hand pass in an instant on the page, they are more involved on stage: the song that follows Logic's line is indeed a quadrille and references the names of two of the dance's movements: "La Poule" and "La Finale." This is also a song about the sheer energy and momentum of fashion: "Jerry, Tom, and Logic must not be behind; Come Jerry come, now for toeing it and heeling it." Once the singing has stopped, the stage direction stipulates there is no time to lose: "Run, Jerry, &c. Exeunt L."

The momentum of fashion in London is made audible not only in the act of singing, but in the choice of music. Prior to the scene described above, in act 1, scene 5, Kate, Sue, and Jane announce that they intend to save their men from the city's "vortex of riot, folly and, ruin" by anonymously delivering tickets to Almack's ballroom, where they will be waiting, in disguise. There follows a trio, sung to Mozart's "Military Waltz," in which the women resolve to "haste to the battle, haste, haste, to love's field." At Almack's itself, the scene is given over to the women, including a solo song by Jane (Bob's partner, and seemingly the principal vocalist) to the tune of "Di tanti palpiti" from Rossini's *Tancredi*, which had premiered in London, at the King's Theatre, just the previous year. In the following scene, a comic City character attempts to waltz to Rossini's unwaltzable tune. This is the cue for a whole series of waltzes—staged as part of the entertainment at Almack's—in which the love interests sing to one another: Jane to Logic, Kate to Tom, and Sue to Jerry. Without a score, it is difficult to get a sense of how these six singers worked together, but we do have this performance indication in Moncrieff's text: "All these Tunes harmonize together—the whole of the Parties fill in the front of the stage, and staying, sing the different Airs at the same time." This notion is a clear allusion to the famous ballroom scene in *Don Giovanni*, with its multiple interlocking tunes; Moncrieff was intimately familiar with it, of course, having burlesqued it for the Olympic in 1817.

The fashionability of Moncrieff's operatic quotations subsequently takes a pointed thematic turn, as the parallel to *Don Giovanni* is developed. In the next act, Kate, Sue, and Jane—dressed as men, and at Temple Bar in moonlight—encourage Tom, Jerry, and Logic to assault a night watchman. This was one of the features of Egan's text and Moncrieff's burlesque most commented on: violence toward upstanding officials was clearly an affront to decency and social order. Appropriately enough, all of the characters from this scene enter the next "confusedly" singing an air "from the spectacle of 'Don Juan.'" Although the words of the song do not mention Don Juan, we can assume that some among the theater audience would have caught the reference. And when the intrigue comes to a climax in the third and final act, Giovanni's presence is clear for all to see (and hear). Moncrieff's burlesque ends in Leicester Square with a masquerade ball, an "illuminated carnival" in which Tom makes his entrance in the costume of—who else?—Don Giovanni. According to the printed play text, the finale of the whole piece is sung to the tune of Mozart's "Giovinetti," with words expressing the spirit of the piece:

Tom, Jerry, and Logic, have made Life in London one holiday,
Bidding frolic and merriment reign;
In larks, sprees, and rambles, have sported thro' many a jolly day
And many will sport thro' again.
    Fal la la, &c.

That final "Fal la la" seems to confirm that the music was taken from act 1, scene 3 of Mozart's opera, in which Masetto and Zerlina express similar sentiments about *giovinette* and *giovinetti*, and the heady pleasure of "spinning around." The presence of Tom on stage in the clothes of Don Giovanni makes the connection unmissable.

The affinities between these two characters would not have come as a surprise to London audiences. As I suggested at the start of this chapter, the absorption of operatic culture into urban entertainment at large was a feature of Regency London. Moreover, the plays *Giovanni in London* and *Life in London* were written by the same dramatist and performed in the same city at the same time—sometimes on the same stages and with the same actors.[37] These burlesques, which made modern London more than a mere context for the intrigue, catered to a growing appetite among audiences for close-to-home locations and up-to-date geography. Some, like one contemporaneous editor, justified this taste in moral and practical terms, seeing Tom and Jerry's adventures as useful guides to the city: "After having made the grand tour of London, accompanied in his peregrinations by his 'guide, philosopher, and friend [Corinthian Tom]' . . . [the viewer] may be safely trusted to walk without leading-strings; and, having satiated his curiosity, retire with an abundant stock of useful experience."[38] But others were less complimentary about the connection between art and life: "It is impossible to say how long the days of the *Giovannis* and the *Toms* and *Jerrys* will last,—but, certainly, our theatrical taste is becoming as depraved and disorderly as our streets."[39] The conjunction of moral laxity and mobile, urban fashionability was indeed at the heart of Tom's and Don Giovanni's popularity and consanguinity in the 1820s, but it did not long remain so.

"Point non plus"

In Egan's literary *Tom and Jerry* the threat of arrest—in both senses of the word—is indicated by the slang term "point non plus," the jail sentence that would put a halt to any frolics around town. In Moncrieff's stage version the same threat is articulated musically, by a "Stop Waltz" in which Logic considers his fate while moving round the stage, presumably with comic interruptions. The possibility of going no further is also important if we are to historicize this late Georgian fixation on movement. While Tom and Jerry and Don Giovanni were racing round the metropolis at a devastating pace, they were already living on borrowed time. Although Don Giovanni had stalked the streets of European fiction for centuries and Egan's dynamic duo continued a tradition of urban explorers stretching back into the British eighteenth century, the particular mode of engaging with the city that these characters embodied could not survive into the Victorian era in either physical or aesthetic terms.

The change—historically and materially—might be dated to the decade following the cessation of major European conflict in 1815, which ushered in a new era of capitalist speculation. Where the previous generation of London financiers—most notably the Rothschilds—had made their fortunes providing credit for continental battles, the smart money was now in bricks and mortar. The leading figure in this early nineteenth-century building frenzy was the architect John Nash, whose series of royal commissions stamped the name of the Regent both on an expanding commercial network (Regent's Canal and Regent's Canal Docks) that linked the country and its capital to the ports of the world, and also on elegant shopping and leisure zones (Regent Street and Regent's Park) that included the most fashionable place of all, the opera house (and the new opera arcade) on the southern reaches of Regent Street. The promise of peacetime prosperity was not to be found "in the air," but on the newly fashionable sidewalks. However, we ought not to lose sight of the blunt fact that Regency London was a playground for only a tiny, wealthy elite. By the middle of the 1820s, and especially after the banking panic of 1825, the ideology of endless adventure was already looking threadbare. Nowhere is this clearer than in Egan's own publications, which included a "Finish to Life in London" and the death of two of his principal characters.

Tom and Jerry were now subject to the laws of mortality: their "rambles and sprees" became inseparable from recollections of a period that was fading further and further into the past. Only fifty years after their heyday, they were widely seen as little more than a footnote in the literary history of Georgian London, valued principally for what they might reveal to later generations about how the city used to be, at a time when—at least for men of a certain class—physical mobility also described a form of social mobility. As time dated Egan's cousins (and their slang terminology), they ceased to be avatars of Cockney urban experience and became instead a pair of curiosities, jellied in Regency aspic. By the same token, as Mozart's opera assumed the mantle of greatness, the stone guest had the last laugh: what had been a picaresque tale, a window onto the contemporary city, now became an opportunity for philosophical reflection and an ossified aesthetic object. For a brief moment, though, Tom and Jerry and Don Giovanni had been theatrical fellow travelers. The parting that followed is not necessarily something to mourn, but their conjunction allows us an opportunity to understand better how the operatic geographies of 1820s London stand as monuments to an impossible city, one in which the pursuit of pleasure knew few bounds and the last word in realism was a masquerade ball.

SEVEN

The City Onstage: Re-Presenting Venice in Italian Opera

SUSAN RUTHERFORD

The scenography of nineteenth-century Italian opera has been described as providing a kind of tour guide of the various regions, cities, and monuments of the peninsula (and elsewhere) for its audiences.[1] At a time when travel was restricted or at best difficult, the operatic stage, with its emphasis on spectacle and innovative technology, could offer a glimpse of a world beyond. Opera and the opera house also functioned as a means of drawing tourists into a city as well as of enabling the local population to engage with aspects of their heritage and history.[2] Centered on Venice, this essay explores notions of place and the city in relation to operas both "about" Venice as well as "of" Venice, notions that reflect tensions between how a city perceived itself and how it was perceived by others. Works such as Vincenzo Nolfi's *Il Bellerofonte* (1642) and the prologue of Giuseppe Verdi's *Attila* (1846) exemplify different ways in which the image of a city could be inserted into operatic narrative and how (in the case of *Attila*) that image could then be disseminated more widely.

The titles of operas produced by just one mid-nineteenth-century Italian composer—Errico Petrella—are enough to emphasize the preference for far-off locations: *I pirati spagnuoli* (Naples, 1838), *Le miniere di Freinbergh* (Naples, 1843),

Il carnevale di Venezia, ossia Le precauzioni (Naples, 1851), *Elena di Tolosa* (Naples, 1852), *L'assedio di Leida, o Elnava* (Milan, 1856), *Jone, o L'ultimo giorno di Pompei* (Milan, 1858), *Il duca di Scilla* (Milan, 1859), *Il folletto di Gresy* (Naples, 1860), and *La contessa d'Amalfi* (Turin, 1864). Drama, it would seem, generally happened to someone else, somewhere else. Strict censorship of theaters was probably a factor, ensuring that any potentially subversive elements (moral, religious, or political) in the narrative could be safely attributed to another period and locality. Such an approach was a legacy from much earlier opera. Monteverdi's *L'incoronazione di Poppea* (1643), written for Venice but set in ancient Rome, was one example: the location of its amoral narrative was perceived as demonstrating everything that Venice, in comparison, was supposedly not, at a time when the rivalry between the two cities was intense.[3]

There were, of course, exceptions to the siting of opera's narratives in distant environs. Verdi's *I Lombardi alla prima crociata*, composed for La Scala in 1843, revealed something of the history of the local Lombard crusaders, using the city's surrounding milieu (act 1 opens in Milan's Piazza Sant'Ambrogio) before exploring the more exotic settings of Antioch and Jerusalem. Other operas simply blurred the edges between "here" and "there." Petrella's *Il carnevale di Venezia*, staged by the Teatro Nuovo in Naples in 1851, is ostensibly set amidst the Venetian carnival, yet its characters are largely based on Neapolitan archetypes. For the most part, though, nineteenth-century serious opera explored a geographical and/or historical Other—other cities, regions, countries, or epochs—in its fictional narratives. In so doing, it accorded with older ideas voiced by Francesco Algarotti in 1755, who counseled that opera's inherent lack of verisimilitude (created by characters who sang rather than spoke) could be ameliorated by geographically and temporally removed settings, with their rich opportunities for presenting the visual "marvelous."[4]

Venice, with its distinctive architecture poised magically on the lagoon, provided one such useful setting, figuring in a number of Italian and foreign operas. As the Neapolitan author Stanislao Gatti commented in 1845, if there was one city in the history of the world that provided "abundant material for the theater," it was Venice, with its canals, islets, and marble palazzi, and the indomitable spirit of its people, with their affection for the urban landscape they had created for themselves: "Everything here is poetic: men, deeds, nature."[5] Not surprisingly, perhaps, it was here that the operatic practice of a visual incarnation of a city-on-the-stage first emerged, in Francesco Sacrati and Vincenzo Nolfi's *Il Bellerofonte* for the Teatro Novissimo in 1642,[6] a spectacle designed by

Giacomo Torelli that contributed to public opera's growing role in the Venetian tourist industry.[7]

The staging of a city was a practice borrowed from prose theater, beginning with the arrival of perspective sets, with their cityscapes, in Ferrara in the early sixteenth century.[8] In 1513 Baldassare Castiglione pursued the idea even further in his set design for Bernardo Bibbiena's *La calandria* in Urbino, where the whole room, used as the performing space, was transformed into a representation of a city, with the spectators encompassed within it.[9] Although set in Rome, Castiglione's design was both generic ("any city") as well as an "idealized" particular image.[10] In contrast, Torelli's design for the production of *Il Bellerofonte* over a century later was much more specific. During the prologue, Neptune causes a model of the future city of Venice to rise from the waves as Astrea and Innocenza watch, enchanted. The scene concludes with the trio's praise for Venice's forthcoming prowess:

Città sopra qualūque il mondo ammira
Saggia ricca e gentile,
Son de le tue grandesse un'ombra vile
Sparta Atene, e Stagira
Quindi vedranno i secoli future
Correr à i lidi tuoi gonfio di lumi
Per tributarti il Ciel conuerso in fiume.

[City admired by the world above any other as wise, rich, and noble, Sparta, Athens, and Stagira are but a modest shadow of your greatness. Henceforth the ages to come will see heaven, bathed with light, rush to your shores as a river to pay you tribute.]

The machinery required to achieve this spectacle received particular commendation;[11] yet it not only facilitated the stage picture but was a reminder of the technical ingenuity required to build the city of Venice itself. As Catherine Keen reveals, medieval images of the city in Italian literature had contrasted "city, culture and community" with the "opposing chaos of countryside, wilderness and division," ideas that lasted well into the Renaissance and beyond.[12] The emergence of Venice from the waves in *Il Bellerofonte* not only acknowledged the city's illusory suspension on water but also emphasized the continuing distinction of the security and civilization offered by the urban environment against the untamed expanse of nature. The archetypal elements of the imagined city identified by Paul Zumthor along with their corresponding affect— "enclosure/isolation," "solidity/security," "verticality/importance"— were not only represented visually by that model of Venice, but were experienced simultaneously by the spectators within the structure of the theater itself. The enclosed boxes ensured separation from the masses, the solidity of the large building and its furnishings assured security, and the verticality of the tiers conferred status and afforded a view from above not only of the stage but of other spectators.[13]

In seeing the city in represented form, citizens and tourists alike were thus also actively participating in or even embodying the idea of the city in their collective, ordered presence within the theater environment. Some spectators drew similar conclusions. Cristoforo Ivanovich wrote in 1681 that while the *platea* of the Venetian theaters held benches on which sat a masked audience (anonymity thus ironing out social distinctions), a number of boxes had been created mainly "for the convenience of the nobility, for the ladies like to stay unmasked there and to feel at liberty."[14] In this way, both *platea* and boxes differently afforded spectators freedom corresponding to the city's own image: "Venice, born free, wishes to preserve freedom for all."[15]

The Ottocento

Venice's freedom ended in 1797 with her defeat by Napoleon; the city was almost immediately passed over to the Habsburg empire until 1805, when renewed hostilities resulted in French rule until the Congress of Vienna in 1814.[16] The city lost governance not only of her present, but also of her past. Encouraged by both French and Austrian authorities, her earlier glories were retold in different guise. The workings of the city's former system of governance by the Council of Ten were now held to be tyrannical and unscrupulous, and its proud reputation for liberty was refashioned as moral licentiousness.[17] As James Johnson reveals, this "dark legend" of corruption and deception increasingly pervaded Ottocento operas about Venice.[18] Donizetti's *Lucrezia Borgia* (La Scala, 1833) opens before the Palazzo Grimani:

The festival by night. Masked revelers cross the stage from time to time. From the two sides of the terrace is seen the Palace, splendidly illuminated: at the back, the Giudecca canal, on which some gondolas pass at intervals in the shadows; in the distance, Venice by moonlight.[19]

From one such gondola Lucrezia Borgia emerges to recognize her sleeping son, Gennaro, setting in train the events that will lead to both his death and the deaths of his friends. That gruesome collective demise takes place in Ferrara, not Venice; but the latter's carnivalesque atmosphere and practices of disguise facilitate the first fateful encounter between mother and son. Other operas notably associating Venice with duplicity and amorality include Donizetti's *Marin Faliero* (1835), Mercadante's *Il bravo* (1839), Verdi's *I due Foscari* (1844), Sanelli's *Il fornaretto* (1851), and, much later, Ponchielli's *La Gioconda* (1876). Such works suggest that the city had ceased to be regarded as a refuge from nature, or indeed a triumph *over* nature, and had become instead a place where man's savagery and brutality had found new ways to thrive.

How effectively such a well-known environment as Venice (Byron wrote that the city was "one of those places which I know before I see it") was reproduced on stage raised other questions,[20] particularly given the Ottocento's growing emphasis on theatrical realism. The librettist Felice Romani opined that historical opera should be properly authenticated by sets depicting local sites and monuments;[21] Giuseppe Mazzini similarly insisted on the representation of *"historical* individuality" through the "character of places in which the opera is set."[22] Such acts of verification were also arguably bidirectional: the use of theatrical, impassioned narratives relating "real" events brought cold monuments and mute buildings to life. Yet only rarely did theater reviews provide extended coverage of scenic design. From a historian's perspective, the premiere of *Lucrezia Borgia* drew a frustratingly laconic comment from one critic: "As for the design of all parts of the production, we will not say a word. The audience expressed their opinion all too clearly."[23] Whether this was a positive or negative reaction it is difficult to say (more likely the latter). In contrast, a production designed by the Bolognese artist Luigi Martinelli of Donizetti's *Marin Faliero* provoked explicit plaudits for its verisimilitude in the Venetian settings:

The sets painted by signor Martinelli of Bologna, and in particular those of the Arsenale, and of San Giovanni e Paolo, are so well treated and faithful to the places that they represent that one might believe oneself to be an observer transported into the Adriatic capital—for which Martinelli obtained repeated "evvivas" and stage calls.[24]

Clearly, this was one production that (according to the critic) offered the spectators a seemingly authentic representation of the Venetian cityscape, providing a form of stay-at-home tourism. Confirmation of Martinelli's accuracy might be assumed from his subsequent hiring (along with that of Francesco Bortolotti) by the Teatro La Fenice the following season for

THE CITY ONSTAGE

the theater's own production of *Marin Faliero*.[25] Venetian critics, however, were harder to please in their assessment of the scenic fidelity of their city's panorama and culture.

The costumes by Antonio Ghelli were derided for various lapses from historical accuracy—not least that a beardless Faliero contradicted many of the local sources, as well as provoking laughter on his first entrance because he reminded the audiences more of Mamm' Agata (the infamous stage mother of Donizetti's *Le convenienze ed inconvenienze teatrali*, sung by a bass) rather than a doge.[26] Martinelli and Bortolotti similarly drew rebuke for the "hovel" (*sala pitocca*) in which a supposedly magnificent ball takes place at the end of act 1, deemed unfit for hosting the doge and the "first gentlemen" of the Republic. Nonetheless, the scenographers were commended for "a beautiful backdrop of SS. Giovanni e Paolo, where the colors and the background of the canal, which truly lengthens to the eye, are very natural and lifelike."[27] Even if the opera itself presented a less-than-flattering portrait of Venice's history, the beauties of her panorama could still be defended.

And what of Venice's own efforts to represent the city onstage? La Fenice, as the city's foremost theater, found itself compelled to reflect the city's changed political circumstances in operatic encomiums to foreign rulers on their visits or birthdays. This cluster of works (usually cantatas) followed a mythological theme, with the city invariably enshrined as "Adria," the spirit of the lagoon, in dialogue with other deities. The tone was submissive and placatory: *Adria consolata* (1803, for the birthday of Franz I of Austria), *Italia, ed Adria liberate, ed unite dal genio di Marte* (1806, for the visit of Eugène Napoleon, the new King of Italy), or *Il vero eroismo ossia Adria serenata* (1815, for the visit of Franz I). One such work was devised in 1838, following the crowning of Ferdinand I of Austria as king of Lombardy-Venetia in Milan. The emperor then journeyed to Venice, where further ceremonies included a grand regatta as well as a gala at La Fenice on 6 October. The main event, Viotti's ballet *Adelaide d'Aragona*, was prefaced by a "Prologo" composed by Giovanni Battista Ferrari to Giovanni Peruzzini's libretto, with scenography by Francesco Bagnara.[28] The cast included Carolina Ungher (Adria), Napoleone Moriani (Genio della giustizia), and Domenico Cosselli (Giove).

The work begins in Olympus, where the "spirits" (*genj*) of Peace, Justice, and the Arts ask Giove to bestow his favor on Adria, beloved by all since her birth. The next scene opens on an "Isola ridente sul mare," with a "Panoramica di Venezia" in the background. Fishermen are throwing their nets and singing in praise of the lagoon. Adria and the gods descend from the clouds. Adria agrees with the fishermen that here in this "giardin della natura" the wind sighs more sweetly and the sun shines more brightly. But, she advises, they will be happier still if they pray with her. All kneel and sing a prayer asking God to endow the king with his own image ("Donale nel suo Re / L'immagine di te"). A sudden light appears on the horizon as Giove accepts the prayer. The chorus declare that this day will be "dearer" than any other glory in Venice's history. Led by Adria, the spirits of Peace, Justice, and the Arts utter their blessings and predictions for a wonderful future: footnotes in the libretto explain that Adria's rhapsodic allusions to progress refer to the restoration of the Palazzo Foscari as a national museum, the building of a dam, and the forthcoming railway linking Venice with Milan. The mood is slightly undercut by the implicit threat in Justice's determination to root out any dissidents ("Se in grembo ti sorge degenere un figlio, / Strappartelo ardita sarà mio consiglio"). The light brightens yet further, and from the sea arise nereids, tritons, and other mythological denizens of the ocean, who dance as the chorus sing to the king: "Fra quest' acque non v'è core, / Che non palpiti per Te" (Among these waters there is no heart / That does not beat for You). The sun itself now begins to appear until finally, according to the stage directions, it "shows itself fully on the horizon. On its disc is written the name of Ferdinando." Adria and the chorus of fishermen enthusiastically exclaim: "FERDINANDO! . . . IL NOSTRO RE." The curtain falls.

Fabio Mutinelli's description of the evening in his album of the royal visit claimed that this "unusual and very brilliant" lighting effect provoked clamorous applause.[29] Lighting was a prized technology at La Fenice, given that it had been the first major theater in Italy to introduce gas lighting in 1833. Yet while in *Il Bellerofonte* stage machinery had been used to proclaim the advantages of the republic, now it merely enacted the city's servitude to Austro-Hungarian dominance.

Attila

Eight years later at the Teatro La Fenice, another sunrise introduced a very different image of Venice. The prologue of Verdi's *Attila* (1846) contains a famous scene of dawn rising above the Rio Alto. This scene was an addition to Friedrich Ludwig Zacharias Werner's original source, *Attila, König der Hunnen* (1809). There, the action opens in the ruins of Aquileia and then moves directly to the outskirts of Rome. Verdi's opera follows the same pattern, except that between the two locations lies a further scene set in the Venetian waters:

> A mudflat in the Adriatic lagoons. Here and there stand a few huts supported by piles and connected with each other by long boards held up by boats. In the foreground, similarly raised, stands a stone altar dedicated to San Giacomo. In the distance there is a bell attached to a wooden shed, which afterwards became the belfry of San Giacomo. The darkness begins to disperse amid stormy clouds, and then, little by little, a rosy light increases, until (toward the end of the scene) the sudden ray of the sun, flooding over all, re-adorn the firmament with the most serene and limpid blue. Slow strokes of the bell greet the morning.[30]

Foresto, the opera's hero, arrives with other survivors of Aquileia's devastation; lamenting the capture of his lover, Odabella, he plans his revenge.

At the opera's premiere, the scene was praised for the "profound impression" created by the combination of Verdi's music with innovative lighting effects, and has since been much analyzed as a turning point in Verdi's "scenic imagination."[31] Little is known, however, about the circumstances surrounding its inclusion. As Marcello Conati noted, the documentation relating to the compositional process of *Attila* is sparse compared to that for Verdi's other Venetian operas.[32] The first extant letter from Verdi regarding the opera's content was written on 12 April 1845 to the Venetian librettist Francesco Maria Piave (who had supplied libretti for *Ernani* and *I due Foscari*, both 1844).[33] Verdi's outline here makes it plain that at this stage no additional scene was envisaged:

> I am thinking of having a prologue and three acts. The curtain should rise revealing Aquileja on fire with a chorus of people and a chorus of Huns. The people pray, the Huns threaten etc. Then Ildegonda's entrance aria, then Attila's, etc. etc., and the prologue ends.
>
> The first act opens in Rome and instead of having the celebrations onstage place them offstage with a pensive Azzio onstage meditating on the situation etc. . . . etc. . . .[34]

Shortly afterward the commission for the libretto was offered instead to Temistocle Solera, with whom Verdi had worked on his very first opera, *Oberto* (1839), then on *Nabucco* (1842), *I Lombardi della prima crociata* (1843), and *Giovanna d'Arco* (1845). Having produced an initial draft, Solera abruptly departed for Barcelona in September 1845, and the completion of the libretto reverted to Piave.[35]

Reference to the additional scene first occurs at the end of that same month, in a letter dated 29 September 1845 by the impresario Alessandro Lanari to the president of La Fenice, Giovanni Berti. Lanari enclosed the libretto of *Attila* for approval and alerted Berti to a late change in the prologue—dawn rising on the Rio Alto—for which Verdi had given specific instructions. He quoted the composer directly: "What I really want to be sublime is the birth . . . of the city of Venice. Let the sunrise be particularly well done, since I intend to express it in the music."[36]

When, how, and why the idea emerged to introduce a new scene between those letters of 12 April and 29 September remains obscure. One factor might have been the image of Venice in Verdi and Piave's earlier opera, *I due Foscari* (1844). Written for the Teatro Argentina in Rome, this setting of Byron's poem shadowed many aspects of the "dark legend" in its revelation of manipulation and treachery among the Council of Ten. As Verdi himself later admitted, it was a bleak opera with a single *tinta* or mood.[37] At one point he had toyed with offering the work to the Teatro La Fenice but was warned that it might offend those families still bearing the names of the central characters.[38] Indeed, La Fenice only staged the opera in 1847, after Verdi had already supervised a production himself at the smaller Teatro Gallo (San Benedetto) in 1845 to warm reception.[39] Nonetheless, during the compositional process of *Attila*, it might have been thought expedient to present a more favorable image of Venice.

Another intriguing contextualization is provided by that earlier depiction of Ferdinand I in October 1838 as the "sun" basking in his rule of Venice. First, it might have constituted an initial experiment with the technology necessary to achieve a slow increasing intensity of light—three minutes is required in *Attila*, and from the libretto of the gala we can deduce that a similar if not longer period was necessary. But the political purposes to which that light had been put raise other questions. Verdi (then in Busseto) probably knew little if anything of the theatrical homage to Ferdinand, but Piave had returned to his native city of Venice in early September. He might even have approved the flattery of the La Fenice event. Three weeks later, in a letter to Jacopo Ferretti on 24 October, Piave described the emperor's recent visit to the publishing firm where he worked and his own reading of a fulsome sonnet he had penned in the monarch's honor: Ferdinand and his retinue had apparently been delighted and recommended its publication. Piave justified his action to Ferretti by referring to the political amnesty Ferdinand had issued in Milan, thus "conquer[ing] the love of this poor Italy."[40] Other authors later associated with the Risorgimento wrote similarly profuse odes. Temistocle Solera's poem for Ferdinand's visit to Milan was entitled "L'amnistia": it reportedly "spoke of the gratitude of the youth, ready to give their life for their beloved sovereign."[41] Andrea Maffei, who composed the verses for a flattering cantata set by Alberto Mazzucato, subsequently rationalized such blandishments by arguing that the Habsburg empire had then seemed "unstoppable."[42]

By 1846, however, the atmosphere had changed. Venice had not only been systematically stripped of her cultural treasures but also burdened by heavy taxation and treated as a food bank by the Habsburg Empire, despite the hunger of her own citizens.[43] While armed insurrection was but a vague aspiration for a few (the execution of the Venetian Bandiera brothers in 1844 had amply demonstrated its seeming futility), shivers of discontent had nonetheless begun to ruffle the lagoon. Whether intentionally or no, the new sunrise for *Attila* offered an opportunity to reclaim the dawn—at least symbolically—for Venice.

The first major difference between the two auroras is that mythology was replaced by history—albeit one that remained contested and uncertain. Although the origins of Venice lay hidden deep below the existing buildings, the stage directions specify a particular location: the belfry of San Giacomo. Rooting an ancient, amorphous vista of mudflats and the lagoon in the vicinity of a known landmark placed the spectators in a precise locale of Venice and thus drew on the discourses of scenic realism—a way of stimulating a double image of the imagined panorama against its actuality in the audience's recall. Some thought plainly went into the selection of an appropriate location. On receiving Lanari's news of the additional scene, the president of La Fenice wrote to the Direttore della Scenografia, Francesco Bagnara.[44] His letter first cited Verdi's comments about the new scene and instructed Bagnara to pass these on to the set designer Giuseppe Bertoja; a second and longer section in different handwriting comprises all the stage settings for the opera, presumably as they appeared in the libretto submitted by Lanari. The stage direction for the dawn scene is identical to the published version with one intriguing exception: the belfry eventually attributed to the church of San Giacomo is given here as the "campanile di S. Paterniano."

The critical edition of *Attila* makes no reference to this variant, nor does Conati's study of documents relating to Verdi's operas for La Fenice. Assuming that Berti's letter represented the original intention of Verdi and his librettists (rather than an error by the copyist), what might have guided the initial selection of San Paterniano? Why was that option subsequently abandoned for San Giacomo instead?

San Paternian's unusual pentagonal campanile, an architectural legacy of the Byzantine era, was supposedly the oldest bell tower in the city, dating back to the tenth century.[45] It also figured on the earliest map of the city, drawn up in the twelfth century and reproduced in 1781 by Tommaso Temanza.[46] As a noted landmark in Venice, it was included in

Giovanni Pividor's *Souvenirs de Venise*, a series of illustrations of the city first published by Joseph Deyè in 1836 (see fig. 7.1).

The history of the fourth-century Saint Paterniano, a refugee in Fano who escaped persecution by adopting a hermit's life along the banks of a river, bore faint resonance with Foresto's enforced exile. More importantly, the location of San Paternian supported a key myth attributing Venice's origins to Attila's destruction of Aquileia, whose citizens supposedly fled to the environs of the lagoon and there began the construction of a new city.[47] The mercantile area of San Paternian was considered one of the first to be populated, and was associated with prominent families in Venetian history: Andreada, Muazzo, Dandolo, Pisani, and Contarini—indeed, the Palazzo Contarini del Bovolo, with its beguiling external spiral staircase, was only a few meters away down the Calle Locanda. Derelict since 1810, the campanile and church of San Paterniano were demolished in 1871, when much of the piazza was reconfigured and a statue to Daniele Manin erected.

The campanile of San Giacomo, however, offered primarily religious connotations. Foresto arrives at the center of the other competing legend surrounding the birth of Venice: one that Edward Muir states had become "more powerful, influential and popular" than that of refugees fleeing Aquileja around 452.[48] In this account, a raging fire broke out among the houses of earlier settlers on the Rivalto; a shipwright named Entinopo vowed to build a church if he received divine aid in extinguishing the blaze. At once rain poured down from the heavens, and Entinopo subsequently erected Venice's first church, dedicated to San Giacomo at noon on 25 March 421—the feast day of the Annunciation of the Blessed Virgin. Celebrated in annual public festivals, this tale had threaded through various well-regarded histories of the city from medieval times to Domenico Crivelli's *Storia dei Veneziani* (1837).[49]

In 1844 a wider readership could instead consult Jules Lecomte's new guidebook, *Venise ou coup-d'œil littéraire, artistique, historique, poétique et pittoresque, sur les Monuments et les Curiosités de cette Cité*, which had been translated into Italian and published in Venice. As Margaret Plant notes, it was the "most stylish and wide-ranging guide to Venice in all its aspects" among the plethora of such works emerging in the 1840s.[50] Lecomte simplified the early history of the city to proclaim that "Venice was founded by Attila,"[51] but he also nominated the erection of San Giacomo as the first church in Venice.[52] Verdi's opera, in relocating the dawn scene to San Giacomo, similarly combined both myths: Venice is shown to have already been founded by religious hermits by the time of Attila's entry to Italy, but the "flagello di Dio" nonetheless spurred the city's development by provoking a new influx of political refugees.

7.1 "Campo di S. Paterniano" (Giovanni Pividor, 1836). Biblioteca Nazionale Marciana, Venice.

SUSAN RUTHERFORD

In recollecting the city's birth, the scene also evoked modern Venice. Perhaps the most famous pictures of a city long represented in art were Giovanni Canaletto's exact reproductions of vistas and monuments, much favored by tourists in the eighteenth century. In the 1830s and 1840s, however, another Venetian artist, Ippolito Caffi, brought a different approach by focusing not so much on the buildings themselves but on the play of light (both natural and artificial). His *Notturno con nebbia in Piazza San Marco* was much praised in *Il gondoliere* on 23 August 1845 for offering an uncommon perspective of a familiar landmark that, as the critic testified from his own experience, was in its rendering of a rare atmospheric phenomenon nonetheless "only the truth" (see fig. 7.2).[53] Curiously, that reviewer turns out to be none other than Piave, in his other guise as art critic. Did Piave's admiration for Caffi's unusual image contribute to or even inspire the imagining of the dawn scene in *Attila* a year later? Whether that was the case or no, the combination of Bertoja's design, the lighting effects, and Verdi's music corresponded with this new emphasis on perceiving Venice (quite literally) in a different, almost fantastic light—yet one which was at the same time a refiguring of reality.

The most obvious difference between the sunrises in Ferdinand's gala and *Attila* lies, of course, in what this new dawn brought. The opera opens with the burning ruins of Aquileja, destroyed by Teutonic hands. The subsequent scene change to the image of a nascent Venice played well into Risorgimento allusions to progress and regeneration—something that might have influenced the (by then) politically committed librettists Solera and Piave, if not Verdi. Still grieving for Aquileja, Foresto's forecast of a more glorious future recalls the optimism that we saw in *Il Bellerofonte*:

> Ma dall'alghe di questi marosi,
> Qual risorta fenice novella,
> Rivivrai più superba, più bella
>> Della terra e dell'ond stupor! [Prologo, scene 7]

> [But from the seaweed on these breaking waves,
> As a newly arisen phoenix,
> You will live again as the proudest, most beautiful wonder
> Of land and sea!]

The reference to a "newly arisen phoenix" was surely a nod to the Teatro La Fenice (the Phoenix Theater, recently rebuilt after a fire in 1836) as

7.2 "Notturno con nebbia in Piazza San Marco" (Ippolito Caffi). Private collection.

well as the rebirth of Aquileja in a new guise. Although the dramatic action subsequently moves to the environs of Rome, Attila's ultimate demise from Odabella's sword-thrust signaled the realization of Foresto's prophecy. This, then, was an image of renewal and vigor that contrasted sharply with the portrayal of Venetian decadence in other operas of the period.

It was also an image that was disseminated more widely beyond Venice, even if its full significance might not have been apprehended. Thomas Kaufman lists productions in thirty-one other Italian theaters up to the end of 1849.[54] Few reviews, however, mention the dawn scene as a design exhibit. F. Tarchetti, writing for *Teatri, arti e letteratura* about a production at the Teatro La Pergola in Florence, described the instrumental effects of the sunrise as "marvelous and new" but made no mention of the scenography.[55] Luigi Casamorata's more equivocal response to that same production was published nine months later by the *Gazzetta musicale di Milano* on 17 January 1847, after the opera had opened on 26 December 1846 at La Scala—the review had been withheld, the editor wrote, until readers had first experienced the opera for themselves. Casamorata complained that the scene was simply a diversion from the narrative: the spectator is "dragged to witness the founding of Venice as the work of refugees from Aquileja" for no other reason than "to flatter the Venetians, for whom the work was written."[56] But the staging at La Scala was hardly equipped to counter such a negative response. Emanuele Muzio, writing to Verdi's father-in-law, Antonio Barezzi, the day after the first performance, described the success of the singers and the opera as a whole, but also the failure of La Scala to give sufficient attention to the dawn scene: "The *mise en scène* was wretched. The sun rose before the music indicated the sunrise. The sea, instead of being stormy and tempestuous, was calm and without a ripple. There were hermits without any huts; there were priests but no altar . . . and when the storm came the sky remained serene and limpid as on the most beautiful spring day."[57] Verdi even wondered grimly to Ricordi whether it was possible to stage the opera "worse than this"; his subsequent reluctance to have his works produced at La Scala lasted until well after Unification.[58]

La Scala had the resources, if not the will, to do justice to the dawn scene; some smaller theaters had the will but not the resources. An attempt to remedy matters came when Francesco Lucca, the publisher of *Attila*, reproduced Bertoja's design in his periodical *L'Italia musicale* in October 1847, complete with detailed instructions regarding the lighting effects (see fig. 7.3). The illustration was also accompanied by a long article on Attila written by Aurelio Bianchi-Giovini (a liberal and political

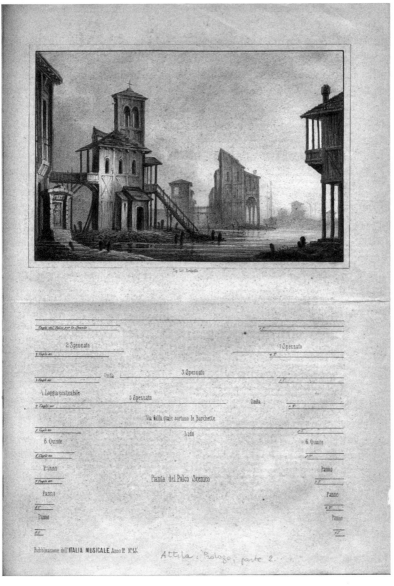

7.3 "Rio-Alto nelle Lagune Adriatiche" (Giuseppe Bertoja), set design for *Attila* as published by *L'Italia musicale*, 13 October 1847. Archivio Storico Ricordi.

activist) as a means of providing historical grounding. The influence of Bertoja's design on some later Italian productions can be seen, for example, in Romolo Liverani's almost identical sketch for a theater in Fano in 1850.[59]

By then, of course, the optimism of the sunrise in *Attila* had proved to be an inaccurate presage of the city's immediate future. Within two years Venice found the order of those opening scenes—devastation followed by rebirth—sharply reversed, as her once tranquil streets crumbled beneath the barrage of the Austro-Hungarian siege that, after seventeen long months, would in 1849 finally end the city's renewed claim to independence. Caffi's study of light and shade was no longer an essay in the picturesque but a terrifying reflection of life meeting death, as in his "Bombardamento notturno a Marghera 25 Maggio 1849"—events surely witnessed by Piave, serving with Manin's National Guard. Even when much of Italy found liberation in the second war of independence in 1859, Venice continued under Habsburg domination until 1866, and the Teatro La Fenice remaining obstinately closed during those years as a mark of protest. By the time it reopened, *Attila* and its "sublime" modeling of the cultural and physical geography of its environs was merely a faded memory of Verdi's early style (*prima maniera*) and that brief, exhilarating moment of defiance in the history of the Teatro La Fenice.[60]

EIGHT

Between the Frontier and the French Quarter: Operatic Travel Writing and Nineteenth-Century New Orleans

CHARLOTTE BENTLEY

Published in Paris in 1861, the short story "Le lac Catha-houla" opens with a mysterious nocturnal gathering: it is an August night in the late 1830s, and a group of travelers waits on the main square of the small town of St. Martin-ville, just over a hundred miles to the northwest of New Orleans. They are about to embark on a trip to the nearby Catahoula Lake for a few days of hunting, fishing, and exploring the wilderness;[1] the story follows their progress, documenting encounters with bears, wild bulls, and alligators in the densely wooded landscape, as their knowledge-able guide (a freed slave named Jean-Louis), his servant (an old French sailor named Lucien), and a slave named Harris ensure their safe passage. But these travelers are neither intrepid explorers nor run-of-the-mill tourists: they are opera singers from New Orleans's principal Francophone theater, the Théâtre d'Orléans, who have ventured into the depths of rural Louisiana on a summer tour.

If the appearance of opera singers in such an environment seems surprising to a modern reader, it might have been less so (although surely intriguing) to a nineteenth-century one. By the 1830s, when "Le lac Cathahoula" is set, New Orleans's operatic life was well established: founded in 1819 by John Davis, a Parisian-born entrepreneur who had come to New Orleans as a refugee from the slave uprisings on Saint-Domingue (modern-day Haiti), the Théâtre d'Orléans had cemented an operatic tradition in the city that dated back to the end of the eighteenth century.[2] At the heart of social life, the Francophone theater was renowned for the high quality of its performances (initially opéra comique, later Parisian grand opéra). Its troupe was recruited annually from Europe, a fact that New Orleans and Parisian newspapers alike would note. Each June, when the oppressive heat and constant threat of cholera and yellow fever in the crowded urban environment made theatrical performances unprofitable and downright dangerous, the troupe, like the performers in "Le lac Cathahoula," would leave the city. Sometimes, such as in the summers of 1828–33, 1843, and 1845, the theater management arranged tours of the northeast, during which the troupe introduced New York, Boston, Baltimore, and Philadelphia to the latest French operas from Paris.[3] But in many years, groups of performers undertook more local excursions of the sort described in "Le lac Cathahoula."

The author of the tale, Charles Jobey, would have known all this well: like his characters, Jobey had spent several years of his life as a performer at the Théâtre d'Orléans, as the principal bassoonist in the theater's celebrated orchestra. Born in Rouen in 1812 or 1813, Jobey arrived in New Orleans in 1834 and was contracted to the Théâtre d'Orléans for the next six seasons.[4] In 1840 he returned to France, where he turned his hand to writing, contributing regularly to periodicals such as *Le monde illustré* and publishing several novels and short-story collections, as well as two nonfiction works on hunting and fishing. Jobey seems to have had a less-than-harmonious relationship with the New Orleans theater management.[5] His experiences in Louisiana nonetheless provided ample material for a number of the works he went on to publish some twenty years after his return to Paris.

While none of Jobey's published stories solely concerned opera, the Théâtre d'Orléans, its performances and performers featured prominently on various occasions. Indeed, in the closing pages of "Le lac Cathahoula," when all the more adventurous pursuits have been completed, opera itself makes an appearance in the context of a series of impromptu musical performances, which serve as after-dinner entertainment on the final night of the trip. As the group sits beside the lake, Harris, the slave, improvises an eighteen-verse song in Louisiana Creole, to the delight of all gathered; Lucien, the servant, then performs a bawdy sailor's song, which fails to impress. For a while afterward the assembled company lapses into silence, listening meditatively to the plentiful, unfamiliar sounds—the "grand symphony of nature"—surrounding them. Vallière, the theater's oboist, eventually "dares to mix his voice" with the natural world, and performs the "Air du sommeil" from Auber's *La muette de Portici*.[6]

Finally, as the sun sinks behind the tree tops, the troupe comes together to sing the prayer from the third act of the same opera: "And the vaults of the virgin forest resounded, for the first and most likely for the last time, with this beautiful prayer from la *Muette*, of which the singing, so simple, so broad, begins with a pianissimo, resembling a breath of a breeze, and finishes with an energetic fermata, the echo of which returned to us like distant thunder."[7] As the sounds die away, the singers turn to see Harris, Lucien, and Jean-Louis on their knees in the sand at the lake's edge, moved to prayer by the performance. The scene takes on a special poignancy, as it turns out to have been an unwitting swan song for some of them: in a brief final paragraph, we learn that the following day the group set off back to St. Martinville, where three of them, Vallière included, soon died from yellow fever.

As an author—and one keen to find a niche in the increasingly saturated travel-literature market—Jobey always stressed that his works contained authentic local detail "of the greatest exactitude" and that "the names of many people who figure are real"; in the case of "Le lac Catahoula," which is written in the first person, we might even assume Jobey himself to be the narrator.[8] He stopped short, however, of claiming that any of his stories were actually true: instead, they seem to occupy a liminal realm, part way between travel writing (with its ostensible desire for authentic detail and reflection of authorial experience) and exotic fiction.

While Jobey's works were hardly unique in this respect, their in-between status opens up intriguing possibilities for opera's role in the stories. Drawing on James Duncan and Derek Gregory's positioning of travel writing more broadly as the "translation of one place into the cultural idiom of another" and Jennifer Yee's discussions of exotic fiction along similar lines, I want to explore the idea that writing about opera—and the specific (in the case of "Le lac Catahoula," highly unusual) geographical contexts of its performance—is one means through which Jobey began to "translate" New Orleans.[9] It became, in other words, a lens through which he attempted to understand his transatlantic experience and encouraged his readers to do the same.

While Jobey was hardly a literary giant, I suggest that his triple position as musician, traveler, and author granted him a very different perspective from that of many authors writing about their visits to New Orleans, and also from travel writers generally in this period. He was not traveling in the name of "exploration" or his own pleasure, but to work.[10] Instead of simply observing the city and its people, through his participation in operatic performances, Jobey played an active role in creating that society, in both the real world and in literary terms, and his writings are, as such, worthy of a closer look. Focusing on "Le lac Cathahoula" and *L'amour d'un nègre*, a novel Jobey published in Paris in 1860, I argue that the scenes featuring opera in his works open up new perspectives on European visions of the United States and offer insight into the nature of transatlantic cultural interaction in the mid-nineteenth century.

Opera in the Wilderness

If we return briefly to "Le lac Cathahoula" and to the series of sonic encounters in the wilderness with which it concludes, it seems at first that opera's role is quite a familiar one: a colonizing force in a non-European environment. As Rogério Budasz has suggested in the case of Rio de Janeiro, the building of an opera house in a colonized locale was a display of imperial might; in this case, it is the performance of opera that plays the colonizing role.[11] Indeed, Harris's song and the sounds of the natural world initially appear to lead up to the climactic performances of opera, and opera itself is presented as an art that, if not directly civilizing, has far more emotional power than any of the other sounds heard. Its impact on the story's three non-operatic Others (the freed slave, the slave, and the lower-class white man who fall to their knees during the performance), is, needless to say in this context, especially profound.[12]

Initially, the whole operatic episode seems set up to emphasize difference: between the improvised, "natural" characteristics of Harris's song and the sophisticated, prescribed operatic excerpt, between human music and the sounds of the natural world, and, of course, between opera's accustomed urban location and the rural wilderness. But on closer inspection, the narrative becomes less clear-cut, as these differences themselves seemingly enable operatic performance to reach a state of perfection. When music and nature meet, music gains its full emotional power, and, in true Romantic style, the sublimity of the landscape reflects the sublimity of the art performed, thus intensifying the experience of listening.

Even the opera troupe's "natural" harmony with the landscape could, of course, be read in terms of operatic power relations, with Europeans positioning opera against the magnificent environmental backdrop in order to bring the non-Europeans to their knees. But the natural world is more than simply a backdrop here, or something that can be tamed at will by humankind: it is an active participant in the story, which adds its varied voices—the "rustling of the foliage," the "plaintive note of the mockingbird," the "softened voice of the Ocelot," and the "sonorous cries of the caimans, tormented with amorous ardour"—to the exchange.[13]

Indeed, the natural world makes its presence felt in a way that complicates opera's relationship with its already unusual surroundings. When the troupe sings together, the physical features of the landscape resonate in response to them: the "virgin forest resounded" to their voices. In Jobey's description, this response is unique to operatic performance: there is no suggestion of such interaction in the case of Harris's or Lucien's songs. Furthermore, the sounds of opera are not only echoed back, but are adapted by the landscape, taking on qualities of the natural world as they resemble first the breath of a breeze and then distant thunder. The landscape, then, has the last word, as it returns the final notes of the troupe's performance in more sublime form.[14] Opera, a product of the urban environment, can stun other human beings into silence, but it cannot silence the natural world.

In "Le lac Cathahoula," Jobey complicates the narrative of operatic colonialism by refiguring the geographical setting—from the inside (of the opera house) to the outside, from the urban to the rural—and, therefore, the balance of power between the cultural product and its surroundings. Indeed, the tale is not so much about opera's power to tame or to civilize in ostensibly exotic environments, but rather about how opera has been absorbed and altered by those environments: its echoes return to the European performers with an air that is slightly disquieting, like the distant thunder simulated in Jobey's description. Rather than proclaiming French cultural domination in a quasi-imperial manner, "Le lac Cathahoula" seems to speak of letting French culture go, to be assimilated into an environment that makes it its own.

In this way, the episode reveals much about French (and, indeed, more broadly European) perspectives on the United States up to the mid-nineteenth century, perspectives that conflict in many respects. On the one hand, it buys into ubiquitous European narratives that saw the United States as a land without culture, while hinting at the role the Théâtre d'Orléans troupe played in remedying that lack.[15] Indeed, the troupe was known for the large-scale summer tours mentioned at the start of this chapter, during which it introduced the major cities of the northeast to French opera. The operatic performance in "Le lac Catha-houla" could then be seen as a wry comment on the troupe's role as missionaries in the cultural wilderness.

But at the same time, the wilderness is portrayed not as something that must be tamed, but as something that actually enhances opera's impact, acting as a sounding board for the troupe's performance: in this way, the scene presents a European fantasy of the Americas, of the kind that had been epitomized in the work of Chateaubriand.[16] In its focus on the natural world, "Le lac Cathahoula" presents a dream of European travelers' interactions with an untamed landscape that is fascinating, sublime, and, as the operatic performance shows, responsive to their influence all at the same time.[17] The American wilderness seems placed to fulfill the desire for "authentic" experience that was common to so much travel writing of the time. Jobey himself was at pains in his preface to stress that, unlike the numerous contemporary authors "who have never gone beyond the city walls of Paris but write most amusing stories, in which the action takes place some 4000 leagues from here," his experience was authentic.[18]

Inside and Outside the Theater: Opera and the Exotic

Another novel by Jobey, *L'amour d'un nègre*, published in Paris in 1860, provides a rather more conventional background for opera: the story is set firmly in the city. But this urban geography, it soon becomes clear, is in many ways just as alien and makes opera's position outside of Europe no more straightforward. *L'amour d'un nègre* follows a young Parisian named Charles Roger, who goes to New Orleans in 1834 to settle his late father's estate; a disastrous love affair, however, results in Roger's transformation from an eligible bachelor and wealthy heir into a hunted and broken man. As Jobey sketches out Roger's new surroundings, the theater emerges as a recurring if minor figure, until, at the very heart of the novel, there is a lengthy scene set in the Théâtre d'Orléans.[19] It is here, during a performance of Auber's one-act opéra comique *Le concert à la cour*, that Roger's fortunes take a sudden and dramatic turn for the worse.

Roger has gone to the theater at the instigation of his capricious new fiancée, Camillia. She wants him to report back on the evening's events: Mademoiselle Dupuis, a young soprano, and Madame Saint-Clair, the established prima donna, have been vying for position for weeks and earning passionate partisans within the Théâtre d'Orléans audience; that night is expected to be the moment of reckoning. At the theater Roger finds that the assembled throng is loudly discussing the matter, and disagreements are already breaking out. Not long after the start of the performance, Dupuis is drowned out by a barrage of whistles from her opponents, while a man clambers onto a bench in the parterre, shouting, "Mademoiselle Dupuis! The role of Adèle is not yours, and we will not let you play it!"[20] The ensuing melee is nothing if not operatic:

> At these words a tempest bursts forth, everyone is on their feet, in the orchestra, the parterre, in the boxes; they exchange insults, threats. The actress, cause of all this noise, takes the opportunity to faint on the stage. The disorder is at its height, personal provocations are exchanged; blows, bellows, fists rain down on all sides. A dagger is drawn, and in an instant twenty, fifty, one hundred, two hundred daggers, knives, dart-sticks, pistols, shine in the room.[21]

The rioting mob spills out into the street, where two people are killed, many are injured, and dozens more arrange duels to settle personal scores. As Roger and his companions go to leave their box, they are confronted by a young man named Simpson, Roger's bitter rival for Camillia's love. In the midst of the chaos, Simpson challenges Roger to a duel with pistols. When they meet the following day, Roger kills Simpson and is forced to flee New Orleans, destined to wander the United States in the hope of one day being reunited with his beloved Camillia.

What might we make of such an overly dramatic scene, and of opera's role within it? At first, it might seem as if, in his avowed quest for authentic detail, Jobey were drawing upon a memory of a real event: the singers in question here—Mme. St-Clair and Mlle. Dupuis—were indeed real-life contemporaries of Jobey's at the Théâtre d'Orléans, and the theater had played host, like so many theaters in the nineteenth century, to its fair share of "diva wars."[22] But there is little evidence to suggest that an altercation of such magnitude actually took place. Contextualizing Jobey's scene within long-standing literary traditions, on the other hand, opens up a very different set of interpretative possibilities. The *soirée à l'Opéra* was an oft-employed device in eighteenth- and nineteenth-century Parisian novels, to the extent that Cormac Newark has observed that the novels of Balzac's *La comédie humaine* "at times seem mainly populated by characters who like nothing better than discussing productions at the Opéra and its back-stage ins and outs."[23] Newspapers and magazines in New Orleans, too, were full of short stories and serialized works of fiction about operagoing, such as in the case of *La revue louisianaise* in the 1840s. Jobey's evening at the opera in *L'amour d'un nègre*, then, had an established literary precedent and a well-developed generic context on both sides of the Atlantic.

But Jobey's scene does more than simply capitalize on a literary vogue. As Newark has argued, opera became part of the frame of reference for French novels of the period because operagoing formed a regular part of the social experience of their bourgeois and aristocratic readership: the opera house came to function as a microcosm of society at large.[24] That society, of course, was almost exclusively Parisian, with occasional provincial detours, such as the evening at the opera house in Rouen that becomes the catalyst for Emma Bovary's second bout of adultery in Flaubert's *Madame Bovary* (1854). Jobey's microcosm of society, however, is explicitly not Parisian, nor from the French provinces, nor even from Europe. Instead, his *soirée à l'Opéra* attempts to provide the French reader with a distillation of a society some five thousand miles away.

Sure enough, through most of the book, Jobey paints an image of New Orleans as exotic and intriguingly (sometimes dangerously) unfamiliar: in this sense, he bought into an ever-growing vogue in France for exotic fiction.[25] Roger's bumbling Parisian uncle, Monsieur Potard, accompanies him for the first part of the novel, and it is through his eyes that Jobey introduces his reader to the unfamiliar world of New Orleans. Potard is unable to cope with either the climate or the behavior of his black landlady, and he is horrified when confronted with the local cuisine. In the few chapters before his return to France, he serves as a comic foil to his infinitely more open-minded and likable young nephew, as Jobey draws on long-standing tradition to satirize the small-mindedness of the European traveler abroad.[26] Although Roger, by contrast, embraces the unfamiliarity of New Orleans, he still finds himself marveling at the exotic. Indeed, when it came to New Orleans, articles in Parisian journals of the period, for example, focused on everything from the dangerous wildlife to the appearance of local women to unfamiliar burial customs, building the sense of peril found as early as L'abbé Prévost's *Manon Lescaut* (1731).[27]

But Jobey's use of the theater within the novel complicates the picture. Indeed, in contrast with the more straightforwardly foreign outside world, Jobey presents the experience of operagoing to his Parisian readers, at least initially, as entirely recognizable: Roger is listening to an opéra comique by Auber, the performers' egos are as inflated and fragile as those of any number of Parisian prima donnas, and they have the same power to inspire partisan support. Even the audience reminds

Roger of what he has left behind in France. Arriving at the theater early, he observes the rapidly filling auditorium: the Louisiana belles (and, of course, men of social status) in the first tier, the second tier occupied by free people of color, and the third tier filled with slaves.[28] The parterre (though not mentioned by Roger) would have been filled with less affluent white men; it is here, of course, that the unrest breaks out.

Although the social makeup differed strongly from that of the Parisian custom, while Roger watches the slaves, in "white dresses, yellow or red tignons," and eating their pecans, figs, and oranges, all he can think of is how like the "*titis* of the boulevards" they are, up there in New Orleans's equivalent of the *paradis*.[29] While such a description might be read as a quasi-satirical comment on the colorful social composition of Parisian theater audiences, Roger is in no way cynical: the initial experience of operagoing in New Orleans turns out to be oddly familiar, even uncannily Parisian. For one brief moment before the performance begins and chaos breaks out, the five thousand miles between New Orleans and Paris are reduced to almost nothing, as Roger sits in the theater and thinks of home.

In these terms, it is tempting to read Jobey's Théâtre d'Orléans as offering a utopian bastion of French culture, uniting New Orleans's disparate social factions—black and white, Francophone and Anglophone—in their shared desire for opera, even as they occupy distinct areas of the theater. Indeed, the theater is characterized as a place of relative equality; Roger is there because Camillia's slave told her it would be a night of great importance. All sections of society are equally keen to patronize the theater: the parterre and orchestra are so full that the audience overflows into the corridors, and the first tier is filled three deep; meanwhile, the second tier is completely full, with "only half of those who presented themselves at the box office," while in the third tier the slaves are "packed in like sardines."[30] Not only are all strata of society present in the theater, but they are actively participating in the debates: loud conversations about the forthcoming performance "take place in all parts of the auditorium."[31] The theater is therefore characterized as a space of relative privilege, where everyone from Creole elites to slaves can have an opinion on the performers in the troupe.

In other respects, too, the theater is not depicted as only a place of civilization, but as a familiar point of orientation for the European abroad. It seems to function within the novel as a kind of North Star, the guiding light by which Jobey helps the visiting Roger and his Parisian readers alike to navigate in the city. On several occasions, for example, meetings between characters take place in a restaurant that is described as being "opposite the French theatre," the location providing a counterbalance to the unfamiliar food the restaurant serves.[32] The geographical prominence of the theater at the heart of the French Quarter gave it a significance for the European visitor as much as for locals.

But if in all these respects the Théâtre d'Orléans appears to be a reassuring symbol of unity and civilization, Roger is soon confronted with the divergence of image and reality, with opera simultaneously a catalyst for social disintegration. The scene at the opera elaborates on and intensifies many aspects of New Orleans's unfamiliar and divided urban politics, sketched out in the earlier part of the novel. And, unlike the fictional violent events of the novel, these political skirmishes were real. In particular, this operatic moment evokes tensions that had been mounting in the city between French- and English-speakers since the early nineteenth century and which were reaching breaking point during the 1830s, as it sees the culmination of the love rivalry between Roger (a Frenchman) and Simpson (whose name reveals him to be Anglo-American) leading ultimately to the duel and Simpson's death.

The Théâtre d'Orléans was a particularly fitting setting for this moment of confrontation, as it occupied a highly charged symbolic position within New Orleans during this period. As New Orleans's Anglo-American population grew dramatically through the 1820s and early 1830s, French cultural hegemony came to be significantly threatened.[33] Specifically, the opening of James Caldwell's American Theatre on Camp Street in 1824 posed a direct challenge to the Théâtre d'Orléans's monopoly, although its usual repertoire (abridged and adapted English-language plays) did not overlap with that of the French theater, and its musical performances were generally agreed to be of lesser quality.[34] The founding of the St. Charles Theatre (another Caldwell enterprise) in 1836 created further competition for audiences, as it staged fashionable Italian-language works.[35]

In the early 1830s the French theater's operatic performances became the locus for heated debate concerning mounting tensions between the city's Francophone and Anglophone communities.[36] Indeed, the theater's significant physical location matched its important role in social life: located on Orleans Street, a stone's throw from the back of St. Louis Cathedral, and at the midpoint between Rampart Street and the Mississippi, Canal Street and Esplanade Avenue, it was almost exactly at the center of the city's French Quarter. Even as the city grew dramatically through the 1830s and '40s, developing an Anglophone "American Sector" to the southwest and the Faubourg Marigny (which was home to much of the city's free black population) to the northeast, the Théâtre

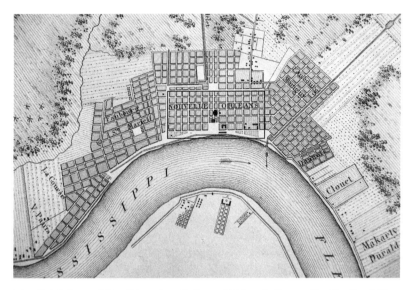

8.1 Detail of map of New Orleans from Guillaume-Tell Poussin, *Travaux d'améliorations intérieures projetés ou exécutés par le gouvernement général des États-Unis d'Amérique de 1824 à 1831* (Paris, 1834). Bibliothèque nationale de France. The central rectangle shows the French Quarter, while the large dot shows the position of the Théâtre d'Orléans.

d'Orléans remained roughly at the geographical center not just of the French Quarter, but of the expanding city as a whole for a number of years (see fig. 8.1).[37] Just as the Opéra formed the centerpiece of Haussmann's Paris, the Théâtre d'Orléans remained the focal point of New Orleans, albeit more by chance than by design.[38]

Any challenge to the Théâtre d'Orléans was read as a challenge to the Francophone community at large. And challenges there were: one particularly memorable incident, in the spring of 1835, saw the American Theatre preparing to stage Meyerbeer's *Robert le diable*. The Francophone press struggled to articulate their shock and dismay at the violation of their previously exclusive connection with the old country; meanwhile, the Anglophone press positioned the occasion as the Anglo-Americans' entry onto a global cultural stage.[39] Although the assault on the Théâtre d'Orléans's dominance was only temporary—the American Theatre soon returned to its regular repertoire—the idea that opera could be a focus of social disquiet remained important in New Orleans.

But, even if the events at the opera in Jobey's novel go some way toward undermining its image of cosmopolitan civilization, and the multitude of knives, daggers, and dart sticks appear to verge on savagery, in

CHARLOTTE BENTLEY

many ways the riot precipitated by the performance might have struck Parisian readers as familiar. On the one hand, the episode is simply a typical piece of novelistic drama, but on the other it hints at well-known stories of incidents both on and off the operatic stage. Descriptions of the scene's almost theatrical violence would surely have brought to mind the onstage revolutions of French grand opera, while the audience fracas might have recalled the fabled riot at a performance of Auber's *La muette de Portici* in Brussels that supposedly led to the Belgian Revolution of 1830.[40] After all, it is another Auber opera that is being performed in *L'amour d'un nègre*. Whether in Europe or New Orleans, then, opera's civilized image is shadowed by the potential for violence.

Jobey, therefore, uses the scene at the opera to facilitate a very particular kind of literary tourism for his readers. In his hands a familiar novelistic device, operatic performance, becomes the means through which an "exotic" society at large is translated and reduced into a recognisable form. But, as in "Le lac Cathahoula," the result is not a straightforward domestication of the exotic. Instead, while the world outside the theater remains largely Other, the opera becomes a way of presenting Parisian readers, such as the Parisian protagonist Roger, with an uncanny experience, as they are forced to recognize themselves in that Other. As a former colony—albeit one that had not been officially French since 1803 and, in reality, had not been governed by France properly for over forty years before that—there was doubtless much about New Orleans that must have seemed uncanny.[41] After all, here was a society that spoke French among its elites (and the French-based Creole in some other sections) and which continued to value French culture as its own, yet in many ways was so very different from the European experience.

The tensions between the recognisability of *L'amour d'un nègre*'s scene in the opera house and the more overtly exotic local features, and between the European Self and the non-European but uncomfortably recognizable Other, would surely have generated a frisson of excitement for Parisian readers, as much as a sense of discomfort.[42] The significance of Jobey's co-opted *soirée à l'Opéra* for *L'amour d'un nègre* lies in its recognizability: opera's dual position as a common novelistic trope and a complex sociopolitical phenomenon made it a revealing lens through which Jobey could negotiate issues of difference and Otherness in his writing.

There are not many texts more marginal than those considered in this chapter: as travel writing, they already sit at the very fringes of the nineteenth-century canon; their little-known author and apparently tiny print runs make them more peripheral still. Perhaps related to this marginal status, within these texts we begin to see hints of subversion; or, in this case, at least new possibilities for understanding both opera and travel writing. In Jobey's novels, operatic performance and its physical locations become mutually influencing, as opera shapes its surroundings and in turn finds itself shaped by the environments in which it is performed, in both literal and more broadly metaphorical ways. It seems, then, to borrow Jennifer Yee's formulation, that, while contributing to a thirst for imaginative journeys through (and thus imaginative control of) far-flung lands, these peripheral texts have the "potential to disturb a culture's monologue with itself, to remind it that it is not absolute."[43] For Parisian readers, the positioning of the Other in "Le lac Cathahoula" and *L'amour d'un nègre* becomes a way of reflecting back on French cultural dominance on a global scale: while it does not refute French claims to cultural influence in the period, it subtly complicates the picture by providing a glimpse (albeit through the medium of a European's travel writing) of another voice speaking back to the European subject.[44] As such, while neither text is overtly or sustainedly satirical, both can be seen as a nineteenth-century outgrowth of the eighteenth-century vogue—as exemplified in Montesquieu's *Lettres persanes* (1721) and Swift's *Gulliver's Travels* (1726), to give but two examples—for using foreign travel as a mode of reflection upon one's own society.[45]

Both, in a way, are examples of what Syrine Hout calls "the cultural dialogue" in travel writing, where the voice of the Other is not simply bypassed as the European writer soliloquizes, but is actively engaged (if ventriloquized) in the formation of the text.[46] It is interesting to note in this light that the Parisian publication of "Le lac Cathahoula" in 1861 (within a collection of short stories entitled *L'amour d'une blanche*) was not the first time it had appeared in print. Some five years earlier, in November and December 1856, it was featured as a serialized story in the short-lived arts journal *La loge de l'opéra*, published in New Orleans. The New Orleans version is almost word-for-word the same as the Parisian one, except for its title, "Souvenirs de la Louisiane," suggesting a more local Otherness designed for the sophisticated operagoers of New Orleans themselves.

The two strands of Jobey's career—as a musician and a writer—offer complementary insights into the global concerns of nineteenth-century Europeans. Travel writing and the international movement of opera (complete with the establishment of opera houses in distant locations) in many senses reflect a European desire to consume the rest of the world, in material and also imaginative forms. If the justification for both practices relied on mantras of civilization and progress, Jobey's writings suggest that the experience of travel and of operatic performance in foreign locales was rather less straightforward. In his stories (as in his career), the opera house becomes a site for the simultaneous affirmation of Eurocentric ideas of cultural consumption and the undermining of those ideas. Opera becomes a means through which aspects of the European self are centered and decentered, repeatedly destabilized in more or less unfamiliar settings, to the point where, in "Le lac Cathahoula," the opera house itself, with its veneer of civilization, disappears altogether. Ultimately, Jobey's stories show that international operatic production, like travel writing, can expose the European subject to alternative and sometimes troubling perceptions of even the most apparently incontrovertible features of their "civilized" cultural identity.

NINE

L'italiana in Calcutta

BENJAMIN WALTON

In an entry in the H.M.S. *Beagle* diary of June 1832, Charles Darwin records going for a walk into the country just beyond Rio de Janeiro in pursuit of biological specimens.[1] He enters a forest and at around five or six hundred feet reaches a viewpoint:

At this elevation the landscape has attained its most brilliant tint.—I do not know what epithet such scenery deserves: beautiful is much too tame; every form, every colour is such a complete exaggeration of what one has ever beheld before.—If it may be so compared, it is like one of the gayest scenes in the Opera House or Theatre.[2]

No surprise, perhaps. By the first half of the nineteenth century, the link between opera and depictions of the exotic was, after all, hardly unfamiliar. Yet Darwin's implication that opera might offer a portal out of cultivated Europe is worth setting against a letter he wrote two years later from Valparaíso in Chile, as he contemplated his return to England. It was addressed to Charles Whitley, a friend from undergraduate days, and juxtaposed yearnings for home's familiar pleasures with further eulogies on the tropical scenery, in its "glory & luxuriance," which, Darwin now suggests, "exceeds even the language of Humboldt to describe," adding that "a Persian writer alone could do justice to it." The letter's last two paragraphs play out the juxtaposition still more directly: Darwin recalls his first glimpse of one of the aboriginal inhabitants of Tierra del Fuego as

119

the sight that had astonished him most on his entire trip: "standing on a rock he uttered tones & made gesticulations than which, the crys of domestic animals are far more intelligible." Without a break, Darwin then switches the subject: "When I return to England, you must take me in hand with respect to the fine arts. I yet recollect there was a man called Raffaelle Sanctus. How delightful it will be once again to see in the FitzWilliam, Titian's Venus; how much more than delightful to go to some good concert or fine opera."[3] Here, too, is opera in familiar guise: a beacon of civilization, or else part of a comfortingly sturdy bulwark, including in its mass the canvases of Raphael and Titian, together blocking out the unintelligible cries and gesticulations of the naked Fuegian, "his face besmeared with paint."

I do not want to suggest that these two invocations of opera are incompatible; ultimately quite the opposite. But for the time being I would like to hold them apart, to explore the possibility that, taken with their differences intact, they might frame some of the questions we could ask about opera's actual and imagined trajectories in the first half of the nineteenth century. How, for instance, the real journeying of operatic troupes beyond Europe during this period intersected with European ideas about opera—such as Darwin's—as a way to understand the wider world. And to what extent the imaginative mobility enabled by these ideas was conditioned by metropolitan resources and technologies: when Darwin attended a performance of Rossini's *La Cenerentola* in Montevideo a few months after his Rio epiphany, did his mind return to that luxuriant hillside; was opera so far from home more or less exotic, or nostalgic, or just disappointing?[4]

My focus, then, is on opera's role in what Felix Driver and Luciana Martins term "the mediated nature of imaginative geographies."[5] And given that novels, paintings, travel narratives, plays, poems, and panoramas were all also vehicles for such mediation, it seems useful to start by considering whether opera works in a different way. For Darwin, possibly not: it is not clear in the citations above that there is a difference between the tropical scenery of an opera and that of a spoken drama, nor whether the viewing of a painting by Raphael would be more or less effective than an opera at shutting out the gesticulating Fuegian.

Another visitor to Rio in the 1830s, Emily Eden, provides a productive comparison. Eden too wrote in a letter home that the scenery "far exceeds all the amount of praise that has been lavished on it" but added only that "you can read an account of it elsewhere, in any book of voyages."[6] Then, traveling onward, Eden painted a panorama of Rio's Guanabara Bay, which was "the admiration of the ship."[7] And it was only when she reached Calcutta, in her official role as the sister of Lord Auckland, governor-general of India, that opera came to mind. On her disembarkation, late at night, she accompanied her brother to a dinner, hosted by the previous governor-veneral, Charles Metcalfe, and attended by eighty guests:

All the halls were lighted up; the steps of the portico leading to them were covered with all the turbaned attendants in their white muslin dresses, the native guards galloping before us, and this enormous building looking more like a real palace, a palace in the "Arabian Nights," than anything I have been able to dream on the subject. It is something like what I expected, and yet not the least, at present, as far as externals go: it seems to me that we are acting a long opera.[8]

Once again, there is plenty in this description that might appear commonplace in such a context: India seen through the filter of the *Arabian Nights* (in place of Darwin's Persian stylist), the substitution of one scenic staple—the untamed forest—by the magical palace. And for both Darwin and Eden, opera emerges as a reference point for the otherwise indescribable, thereby granting it a particular place within the imaginative resources of foreign description. Yet the contrasts with Darwin's Rio are also clear enough: whereas the appeal of the jungle lies in the absence of other people, in India the population are always present, to be co-opted to act as extras within the landscape. And while Darwin is a spectator, Eden becomes a performer, even a prima donna, within the gradually unfolding operatic world of colonial Indian life.

At first sight, these contrasts directly replicate the most striking differences between Robert Burford's grand panoramas of Guanabara Bay (*Bay of Rio Janeiro*, fig. 9.1) and of Calcutta (fig. 9.2), exhibited in Leicester Square a few years earlier. For the Rio view, which opened in 1827, there are ships in the foreground but few sailors, while the city itself—the largest on the continent at the time—is all but swallowed up in the folds of those verdant hills. Burford's Calcutta (displayed in 1830), on the other hand, was painted from drawings provided by Captain Robert Smith of His Majesty's Forty-fourth Regiment and showed the city from the huge park, the Esplanade, surrounding Fort William. This allowed a clear vantage of the proverbially palatial skyline of the area inhabited by Europeans, known as the White Town—a view, in the words of Burford's explanatory essay, that "appears equal in splendor to any [city] in Europe," and that also served to conceal "the bamboo huts and mud dwellings of the more northern part, or Black Town."[9]

Fundamental as this division was to the European mental geography of Calcutta, this act of concealment is hardly the most notable feature

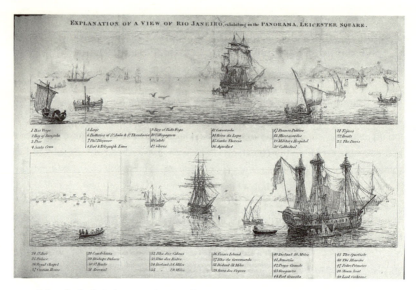

9.1　Robert Burford, *Explanation of a View of Rio Janeiro, exhibiting in the Panorama, Leicester Square* (London, 1828).

9.2　Robert Burford, *Explanation of a View of the City of Calcutta, exhibiting at the Panorama, Leicester Square* (London, 1830).

of the panorama.[10] This comes instead in the foreground, where, in contrast with other Calcuttan images of the period, the palaces of this proverbial "city of palaces" are themselves almost hidden from view by what one reviewer described as an "animating scene which presents itself on all sides."[11] It is a spectacle, according to Burford, such as might be encountered during a festival like the Hindu Durga Puja:

coaches, phaetons, and buggies, of Europeans, intermixed with the palanquins, hackeries, tonjons, and other carriages of the natives . . . military in their rich uniforms, Mahomedans in graceful and flowing costume, resplendent in all the gorgeous hues of China, and princely-looking Persians clothed in silk and brocade . . . processions in honor of the Deity, attended by bands of native musicians, playing on singularly formed instruments . . . jugglers performing extraordinary feats of skill . . . in fact, the whole male population appear, on these occasions, to leave their dwellings with one consent to mix in the busy throng.[12]

In Burford's words, it was "a scene difficult to describe, and certainly not to be equalled in the world." And where Rio's operatic sublime evoked for Darwin a painted backdrop, here Burford gave his spectators a whole opera—an opera such as, say, Rossini's *L'italiana in Algeri*, with its plot about Europeans interacting within a non-European setting.

Or perhaps not. Burford does mention the European carriages within the larger spectacle, and reviews at the time celebrated the possibility the panorama offered its London audience to be "present, as it were, in all parts of the world," immersing them within the scene in a fashion interrogated by many later panoramic historians.[13] Emily Eden's sense of dramatic centrality, however, is nowhere to be found. As a result, where the cultural hierarchies in *L'italiana* are rarely in doubt, Burford's image works as a fantasy that contains an awkward reality, providing all the local color any London Panorama-goer could desire, yet also offering a vision of life unfolding beyond colonial control. As Daniel E. Whyite has argued, the spectator's sense of autonomy and sovereignty was limited by the image's "combination of power, detail and authority at every turn."[14] How fortunate, then, that it at least remained static and seems to have included no accompanying soundtrack, thereby sparing the audience any risk of exposure to either gesticulations or unintelligible tones.

L'italiana was popular during this time in both Rio de Janeiro and Calcutta. It was first performed in Rio in 1821, and during the second half of 1827 it experienced a vogue thanks to the performance by the recently arrived soprano Elisa Barbieri in the role of the "Italian girl," Isabella. In January 1834 it opened the first Italian opera season in Calcutta, performed by a troupe of six singers and a violinist who had disembarked the previous month on the last leg of what, for some of them, had turned into a world tour, including performances in Rio, Montevideo, and Valparaíso just a few months before Darwin's own arrival in those towns.

For the opera historian sensitized to the plot's dynamics by over two decades of scholarship on nineteenth-century operatic orientalism, and conscious that much of it has taken the metropolitan locus of performance for granted, the question of how such a piece might have resonated within the slave-owning populace of Rio or the colonial society of Calcutta is hard to resist.[15] But newspaper reviews in Rio were more preoccupied by the rivalry between Barbieri and the previous Isabella, Maria-Theresa Fasciotti, while another visitor who stopped off in Rio en route to Calcutta, the French naturalist and member of Stendhal's circle Victor Jacquemont attended a performance in 1828 only to comment on the "detestable company," the "still more execrable orchestra," and the higher ranks of the audience, who yawned "as they do at Paris, as a sign of good taste."[16]

Just a few days after describing her sense of living in an opera, Emily Eden offered no comment at all on the performance of *L'italiana* that she attended in March 1836 at Calcutta's Chowringhee Theatre, toward the southern edge of the White Town.[17] But she soon gave her opinion on the Isabella in that production, Teresa Schieroni, after a subsequent concert at the Town Hall: "immensely fat, with a cracked voice—she is their Pasta," and added of the whole ensemble that "they all ask twenty guineas a night as if they really were *prima donnas*."[18]

For Jacquemont and Eden, then, the presence of opera wherever in the world one's ship might dock is perceived as little more than an extension of the existing ambit of substandard provincial imitations. As Jacquemont put it, the ballet in Rio was better than the opera, but only equivalent to that at Brest or Draguignan, and such imitative proliferation was not necessarily a good thing. And even before getting to South America, Jacquemont's excitement on arriving at Tenerife—"for it is a Spanish country, and I had never seen one"—and then seeing camels while on a trip into the mountains gave way to despondency at finding the men back in town dressed "in the newest fashions of London and Paris," while the women danced "French contre-danses set to Rossini's most popular airs." "Farewell to the stamp of locality!" he railed, in anticipation of one of the most popular themes of later globalization critique; "the whole world is tending to assume the same appearance, stupid, rather melancholy, and very vulgar."[19]

Such responses might collectively count as a negative variant of Darwin's Fuegian alternative: opera represents civilization even when badly performed, tedious, or unwelcome. And in the Calcuttan context, if Burford's description offered a sight of the reality that constantly undermined the supposed segregation of Calcutta's urban plan, then performances of opera could lay down the principles and boundaries of civilization by way of reinforcement.

This, anyway, was how several contributors to Calcutta's hyperactive press depicted the city's first performances of Italian opera in the weeks after the troupe's 1834 arrival. A preview in the *Bengal Hurkaru*, for instance, observed that "there is no society which piques itself more on its refinement than the enlightened community of Calcutta, and they will hardly allow it to be said that they do not appreciate the most *recherché* recreation of the fashionable world at home."[20] The *India Gazette*, meanwhile, hoped for "a permanent and beneficial support for those who professionally come forward to make their appeal to the taste and spirit of the community."[21]

The slight irony of the first report and the slight desperation of the second both indicate that the symbolic value of Italian opera could work only to a limited extent among members of an expatriate community who already considered themselves thoroughly civilized. And why should they not? After all, alongside local news, the same papers regularly reprinted in their entirety long and detailed accounts of operas, plays, and performances in London, headed simply "Drury Lane," or "Spiritual Concerts," thereby effacing the distance between the locations. As a result, *not* to attend the opera in Calcutta could be treated as of no more note than not attending in London, since in both cases one could always read the reviews instead. Yet in another way, the ever-dominating presence of London could also serve to underline the distance and deep differences between the two locations. Emma Roberts, for example, described in the *Asiatic Journal* how new arrivals in Calcutta from London would be expected to offer news of theaters, books, music, and pictures to prove their social credentials, but by doing so also risked being branded as bores, finding themselves, Roberts writes, "in much the same predicament as the narrators of tiger-hunts at home."[22]

Such an allergy to other people's travel stories translates across the centuries successfully enough, but the choice of imagery here is nonetheless revealing: the normalcy of the tiger hunt in India is exotic in London, yet still boring; likewise, metropolitan art potentially loses its appeal in the colony, both because it fails to fit with local concerns and because, as an established form of cultural currency, it is already in oversupply.

Meanwhile, the Italian singers themselves came not from London, but—most recently—from Macao, where they had spent a humid summer entertaining the European traders and their wives, followed by a brief stopover in Singapore and Penang. Reprints in Calcutta papers from the *Canton Register* and the *Singapore Chronicle* praising their performances accordingly served not only to puff them, but also to establish them with a pedigree as exciting yet respectable.[23]

It was an important balance to strike, not least because a lack of interest in the new arrivals from some—visible in relatively low ticket sales and in parties thrown to clash with opera nights—balanced a sort of fevered faddism from others. As the *India Gazette* critic wrote, "It is a fault in the good people of this city that they run too eagerly after what is new, only to desert it altogether when the rage for novelty has subsided." Instead, he went on, "we wish to inculcate the necessity of being moderately enraptured."[24]

While this reviewer invokes "moderate rapture" with a view to opera's long-term naturalization in Calcutta, the phrase also articulates something more fundamental about the response to Italian opera as imported entertainment in locations from Buenos Aires to Madrid, and even, despite their longer traditions, Paris or London. In all these places, enthusiasm was welcomed as a sign of taste, yet the expression of excessive Italophile infatuation was cast as an unhealthy affectation, in a way that equal passion for symphonies or oratorios typically was not. And in Calcutta, where the tension between overwhelming governmental desire for control and the constant threat of loss of that control was always so close to the surface, the press's promulgation of Italian opera as neither irrelevant nor of only momentary note formed part of a project to reshape the city's identity toward the European splendor described but then concealed by Burford's panorama. Successful opera in Calcutta might begin to make Calcutta less operatic.

Reviews of the opening production of *L'italiana* in 1834 can be read in this light. After the premiere, on 15 January, a lengthy account in the *Bengal Hurkaru* praised the "powerful voice" of Schieroni and described her method as "of the very best modern Italian school, uniting expression and brilliancy under the government of strikingly correct taste." The voice of the seconda donna, Margarita Caravaglia, had a "remarkably rich and sonorous quality," with "graceful and expressive" style; the first bass, Domenico Pizzoni, had "a fine bass voice of great extension and power," while the second bass, Gioachino Bettali, lacked Pizzoni's volume but nevertheless demonstrated a voice that was "full and mellow."[25] All, in other words, were granted the attributes required of them in their roles as representatives of operatic civilization. With each orotund sentence, it is as if the use of critical boilerplate might itself secure the operatic bulwark in place. The piece was "one of the best and most popular of compositions of that unrivalled master [Rossini], whose fertile and versatile genius equally shines in the gay, the tender, the sparkling, the pathetic, the comic and the heroic, and whose blemishes are but as the spots on the sun"; the music is graceful, "deviating as occasion requires into intricacies of Buffo, and into lofty aspirations of the heroic," and even grotesque elements in the score never threaten the unity or beauty of the whole. The plot, though, is passed over as "sad trash," a "peg on which to hang the music"; operatic relevance remains narrowly defined, and no parallel between stage and street, or between one set of Europeans abroad and another, could be countenanced.[26] By the time *L'italiana* was produced again in 1836, when Emily Eden attended, it was already a Calcuttan classic, with music sufficiently simple in its charms to please even the admirers of "those respectable periwig-pated personages Pacini [*sic*], Paesiello [*sic*] and Cimarosa, whom we might without offence designate the ancients."[27]

Such assertions sought not only to separate opera from the immediate world outside, but also to counter well-rehearsed arguments that took on new life once imported from London about the foreignness and potential undesirability of Italian opera. Doubts over the acceptability of attending the opera, for instance, gained urgency for a missionary population keen on offering a strict moral framework for a high-living European society, and for the much larger Indian population resistant to conversion. A letter to the *Calcutta Christian Observer* in February 1836, which argued that theatrical visits could be sanctioned by Christians provided the spectacle was itself moral, received a stern riposte from the journal's editors, who argued that "almost every Christian writer, ancient and modern, concurs in pronouncing [such] amusements worldly, seductive, and inconsistent with sober religious principles."[28] As a result, they fell into the same broad category as attendance at Hindu festivals, such as Durga Puja, whose long-standing attraction for "some respectable Europeans" was decried a few months later in the same paper as "an evil of no small magnitude."[29]

Aware that such language might seem off-puttingly strong, the editors turned to the Baptist missionary William Ward for support, and to his account of attending the worship of the goddess Durga at the palatial residence of Raja Raj Krishna Deb in the north of Calcutta in 1806:

The whole scene produced on my mind sensations of the greatest horror. The dress of the singers—their indecent gestures—the abominable nature of the songs—the horrid din of their miserable drum—the lateness of the hour—the darkness of the place—with

BENJAMIN WALTON

the reflection that I was standing in an idol temple, and that this immense multitude of rational and immortal creatures, capable of superior joys, were, in the very act of worship, perpetrating a crime of high treason against the God of heaven, while they themselves believed they were performing an act of merit—excited ideas and feelings which time can never obliterate.[30]

Old fears are being stirred here, and sure enough Ward adds a footnote: "The festivals of Bacchus and Cybele were equally noted for the indecencies practised by the worshippers." And Ward's concern over costumes, gestures, music, and performance in a darkened room might seem to cast even moderate rapture at the opera as suspect by association. It underlined the importance for its supporters of insisting on opera's separation from anything so unruly—no easy task, given the strong strand of Rossini's European reception that cast his works as unruly and cacophonous—with the famous din of bells, hammers, crows, and cannon that end act 1 of *L'italiana* as a locus classicus.[31] Meanwhile, explicit parallels between Indian and European musical entertainments were common enough: an account of the Durga Puja festivities in the autumn of 1833, for instance, noted the presence of "many respectable members of the Civil and military services," along with "native gentlemen of respectability" who had listened with pleasure to "the vocal performances of the Billingtons, the Catalanis, the Vestris, and the Pastas of the East."[32]

Equally, while Italians would frequently be grouped with the English as Europeans, their position in Calcutta was not straightforward. Despite a lively history of musical activity in the city, there was little tradition of visits by Italian musicians.[33] As a national group, they were considerably fewer in Calcutta than in London—too small a number, in fact, to register in the 1837 census at all.[34] Their potential exoticism was also underlined on the day after the troupe's arrival, when the *Bengal Hurkaru* chose to fill up column space with an extract from Frédéric Bourgeois de Mercey's recently published *Travels in Tyrol and North of Italy* (Paris, 1836) entitled "Italian Ladies," which mused in time-honored fashion (and in line with a central theme of *L'italiana*) on whether Italian women were "too affected or too natural" and commented on the general national atmosphere of voluptuousness and felicity.[35] When an (unnamed) English amateur took over Caravaglia's tenor roles, reviewers expressed astonishment at the feat, given that

no English organ, unless they have been modelled from early youth upwards in Italy, can ever attain the more delicate shades of expression or master the indescribable niceties which render the Italian language, from the mouth of a native, even in ordinary parlance, music, and which invest the music of that country with a charm which is sought in vain in the music of other countries and which perishes by translation.[36]

And so the hoary clichés of national stereotyping became refigured in a new environment. As a result, the amateur took on the role of Lindoro not only in the opera, but also in the troupe, as an outsider brought in to adapt to a foreign culture. Or, better, he became like one of the Indian students performing at the annual prize-giving ceremony of the Hindu College in 1836, also attended by Emily Eden, which took place days before the performance of *L'italiana*. As in previous years, this consisted of a series of recitations from the English classics: among others, Anund Kisen Bose offered a poem by Pope "in a tone of voice of great sweetness and with perfect propriety of accent," while Gopaul Kisen Dutt delivered Othello's address to the Senate with "truth, and accuracy of accent and feeling."[37]

It might seem a stretch to bring the performance of an English amateur in a Rossini opera into contact with such paradigmatic acts of colonial mimicry.[38] But beyond underlining the extent to which the Italians could be at once insiders and outsiders, it highlights Italian opera's unreliable role outside the European metropolis as a civilizing force. Successful mastery of Shakespeare was part of a well-established didactic program that actual performances, however bad, could not call into question. (On the contrary, amateur performances of Shakespeare's plays at the Chowringhee Theatre generated detailed and weighty articles analyzing the language and the drama and reminiscing over past performances.) For some, the arrival of opera promised the same: plugs for the performances' civilizing (and rationalizing) value appeared in selected Bengali-language newspapers, and a writer for the *United Services Journal* in 1837 hoped that opera's recent arrival might mark "the birth of a permanent and widely-diffused taste for musical entertainments in that remote quarter . . . in a ration commensurate at least with the sums so profusely lavished on other pursuits and recreations of a far less rational and refined nature," a hope supported by the observation that already

the natives . . . have been inoculated with a musical ardor. Orthodox Hindus and Mussulmans may be observed occupying an opera-box, and listening with doubtless unfeigned admiration to the beauties of "Il Barbiere," or "Semiramide," whilst one of the wealthiest and most intelligent of their body actually received lessons in singing from the *basso cantate* [*sic*] of the Italian company.[39]

BENJAMIN WALTON

It was not only Christian missionaries, however, who questioned opera's potential for edification. The extensive debates over "native education" during the early 1830s, culminating in Thomas Macaulay's notorious 1835 "Minute on Education," with its clarion call for "a class of persons Indian in blood and colour, but English in tastes, opinions, in morals and in intellect," never included music in the lists of necessary subjects of instruction.[40] And Macaulay's own opinion of 1830s Calcuttan social life was summed up in a letter of January 1836 as "vile acting, . . . viler opera-singing . . . and things which they call reunions."[41] When the Italian troupe performed Rossini's *Otello* in October 1834, meanwhile, the *Literary Gazette* supported the place of the opera in the city but disinterred Addison, Chesterfield, and Byron in order to deny that opera could communicate any truly useful purpose:

We should be sorry to be thought enemies to any refined amusement, and should blush to avow an indifference to the charms of music. As a musical entertainment, the Italian Opera is truly delightful, and we rejoice to see that it is beginning to be encouraged and appreciated in this city;—at the same time we do not hesitate to confess that its *dramatic* pretensions appear to us to be as despicable and ludicrous as its musical merits are transcendent and enchanting, and it provokes us to see them gravely defended by those who ought to guide and improve the public taste.[42]

Where does all this leave our Darwinian alternatives—the easy invocation of operatic simile outside Rio on the one hand, or, on the other, his speechlessness in Tierra del Fuego, when presented with a figure with a painted face standing above him, gesticulating and crying out in a foreign language, the recollection of which somehow turns his thoughts to the operatic stage? One final example might provide a clue, and a kind of conclusion. On the tour through India that Jacquemont undertook on his departure from Calcutta, he described being presented to the Great Mogul of Delhi. Jacquemont was brought to this specially convened audience "with tolerable pomp," accompanied by "a regiment of infantry, a strong escort of cavalry, an army of domestics and ushers, the whole completed by a troop of richly caparisoned elephants."[43] The emperor then presented him with a dress of honor, which he retired to put on, before being reintroduced so that some jeweled ornaments could be fixed to his hat. Struggling to keep a straight face throughout, Jacquemont records that he couldn't help feeling like Taddeo at the moment in the second act of *L'italiana*, when the Bey of Algiers, Mustafà, appoints him to the ceremonial role of Kaimakan—lieutenant—and makes him dress up accordingly, accompanied by a chorus singing *alla turca*.

130

For Jacquemont, once again, the experience of life through the prism of Rossinian comic opera leads only to ridicule. Everything, and everyone, can be comprehended; nothing threatens to fall beyond the range of artistic reference available to the cultured European traveler. As a result, Darwin's own two operatic positions no longer seem far apart, if ever they were. The portal offered by opera out of daily experience turns out to be an illusion, part of a closed system, defining the world from a single viewpoint: the jungle looks like a theater; Italian opera in Calcutta is safely trapped within the language of a stale set of aesthetic debates; and even Eden's magical palace was none other than Government House, the neoclassical center of British colonial power—no more a part of the Arabian Nights than the Palace of Westminster.

So the Kaimakan ceremony can stand as a symbol for European incomprehension at events beyond the ken of their own fantasies. But what of those Indian attendees at the opera? If nothing else, their presence offers a reminder of the rich possibilities—at least for a small elite of Calcuttan merchants—of a time, in Christopher Bayly's description, when "colonial encounters were not always about the construction of difference." This is what Partha Chatterjee has termed the "antiabsolutist early modern" moment in Indian history, which would quickly give way to the more familiar "authoritarian regime of colonial modernization."[44] That wealthy and intelligent figure mentioned above as taking singing lessons from one of the troupe, for instance, was Dwarkanath Tagore, whose great influence within 1830s Calcutta society can be demonstrated in many ways, not least in the fact that during the latter half of the troupe's stay he owned (and then merged) two of the newspapers from which I have quoted most extensively, the *Bengal Hurkaru* and the *India Gazette*. He also contributed impressively large sums to the donative fund set up to cover the Italian opera's costs in 1834, and in 1835 he bought the Chowringhee Theatre outright in order to keep it going.[45]

Yet, much as a more detailed exploration of such contexts would usefully break down some of the oppositions that haunt any study of opera in an unexpected location, at one level Tagore's enthusiasm, as well as his close friendship with many of those involved in the Calcuttan theater, just reinforced opera's traditional role as a pastime for the rich upper classes, harking back to eighteenth-century definitions of elite cosmopolitanism. So instead, let us return finally to the actual Italians in Calcutta in January 1834, who, after enacting the story of Europeans outwitting the natives overseas in *L'italiana*, followed it with *Il barbiere di Siviglia*. The critic in the *Hurkaru* celebrated in familiarly hermetic terms:

the most popular Opera of the most popular composer of this or any other Age, now for the first time spread forth in all richness and beauty in Calcutta. The music is familiar as household words to the ears of our community. Where indeed is it not? What corner of the civilised world resounds not with the joyous and delicious melodies which the genius of Rossini has supplied?[46]

How comforting, then, to read that Margarita Caravaglia *en travesti* stole the show in her depiction of the drunken Count, that the Peruvian bass Mayorga managed to make Basilio's irresistible Calumny aria fall flat, and that the orchestra frequently got so distracted by the action on stage they stopped playing altogether to watch. The normal stuff of touring opera, in other words, and hardly enough to present a serious challenge to the unyielding rhetoric of operatic civilizing. Yet each such detail, positive or negative, gives us purchase to prise open small, precious cracks in the thick rhetorical veneer holding in place opera's predetermined role within local and global cultures—whether we go in search of something other than moderate rapture, or hoping for a glimpse, through all the critical verbiage, of Darwin's speechless European, who, for a brief blessed moment, knows not how to express his feelings, and falls silent.

TEN

Thomas Quinlan (1881–1951) and His "All-Red" Opera Tours, 1912 and 1913

KERRY MURPHY

It is most refreshing to find a man who . . . has acted Imperially, and not merely parochially. ROBIN H. LEGGE, *DAILY TELEGRAPH [UK]*[1]

The English entrepreneur Thomas Quinlan's traveling opera companies made two extraordinary tours of "Greater Britain" in 1912 and 1913, singing "in English to English speaking peoples all the time, never leaving the red portions of the geographical map."[2] Through focusing on Quinlan's time in Australia, this essay explores the global cultural ramifications of these tours, suggesting that their imperialist mission was motivated by an idealistic desire to share a European cultural commodity demonstrating the highest artistic standards. In those far-flung—yet British—corners of the globe, audiences could feel as though they were transported to Covent Garden. As this essay will show, although Quinlan's educational mission seemed patronizing, Australian receptiveness to new works and audience enthusiasm also came to motivate Quinlan's enterprise.

Quinlan's first "all-red" tour (as he termed them) began in South Africa in 1912 and then moved to Melbourne and Sydney, where it was presented in conjunction with the Australian impresario J. C. Williamson.[3] On his second trip, in 1913, Quinlan also went to South Africa before traveling to Australia, but his visit there was curtailed by the violent Rand miners' strikes.[4] Miners' strikes also prevented him travelling to New Zealand, although he toured Canada on his way home (in two trains put at his disposal by the Canadian Pacific Railway); because of these disruptions, his grandiose plan of performing nine *Ring* cycles in six months (across the United Kingdom and its colonies in Canada, South Africa, Australia, and New Zealand) had to be abandoned. It was in Australia that he spent the most time, and documentation for these extremely successful tours is extensive.

White Australians were very familiar with touring opera companies by the early twentieth century. According to one 1944 study, after the gold rushes of the 1880s, "Melbourne, jingling its money, became a regular bidder in the world's opera market. Impresarios darted out with foreign companies and the latest scores . . . all the glittering operatic repertoire of mid-nineteenth century Europe—filled the antipodean air."[5] Yet desire for and experience of European operatic repertoire had been established rather earlier. It is worth explaining a little of this backstory in order to demonstrate both how lucrative Australia was seen to be as a market, and how readily waves of (sometimes competing) touring companies established a performing base for their successors: as Katherine Brisbane writes, it is a myth that Australia was isolated in the nineteenth century, since, with the discovery of gold, it became a financial center and quickly part of a world industry.[6] There was a desire for entertainment, and audiences were "cosmopolitan and discriminating."[7]

The most significant figure in colonial Australian opera was undoubtedly the Irishman William Saurin Lyster (1828–1880).[8] He first came to Australia from America in 1861 with soloists, a chorus, an orchestra, and a repertoire of French, English, and Italian opera. He stayed for the rest of his life, basing his touring enterprise in Melbourne, with a few trips away in the 1860s, including tours of New Zealand, and he leased theaters for short periods in the principal cities and larger provincial towns. The company sang (chiefly in English) repertoire including Rossini, Verdi, Bellini, Donizetti, Auber, Meyerbeer, Offenbach *opéra bouffe*, and the English operas of Michael Balfe and William Vincent Wallace.

In 1869 Lyster recruited singers in Italy for a temporarily expanded company. He then joined forces in 1871 with the Italian Cagli-Pompei company (visiting from Calcutta), who stayed until around 1875, performing in Italian.[9] Lyster provided them with theaters, a chorus, an orchestra, and costumes. They had some exceptional singers and also a fine conductor, Alberto Zelman (1832–1907), who stayed on with Lyster when the Italian company left. Lyster now managed both an Italian company and an English-language *opéra bouffe* and English-opera company, with Offenbach, Balfe, and Wallace as staples.

Lyster clearly demonstrated that there was interest in opera in Australia that others were subsequently happy to exploit, particularly with an experienced base of Lyster performers available. The major companies that toured after Lyster shared several key features: the Montague-Turner Company, touring principal cities and country towns of eastern Australia and New Zealand in 1881–85 and 1891–94;[10] Martin Simonsen's Royal English and Italian Company, touring throughout the 1880s and early 1890s, and his New Royal Italian Opera Company in 1886;[11] and the Italian Cagli Company (later Cagli-Paoli), visiting again in 1882–83, all boasted their own (often European) singers,[12] but they also recruited some of the Italian singers who had remained in Australia after Lyster's demise, as well as talented locals.[13] Choruses and the orchestra were often local (from Melbourne or Sydney), the New Royal Italian Opera Company, for instance, employing an orchestra of twenty-eight and a chorus of thirty-six in Melbourne for a year in 1886.[14] Montague-Turner and Simonsen were also family companies, a wife-and-husband singing duo heading the former and Simonsen's wife and daughters the latter, with Simonsen conducting. Both Martin and Fanny Simonsen and the Cagli Company were already known to Australians, having been introduced by Lyster. Most companies toured the state-capital cities of Australia as well as many large regional towns, and sometimes also included New Zealand. And, perhaps most importantly, many performers stayed on: for instance, the New Royal Italian Opera Company's conductor, Roberto Hazon (1832–1907), remained and, like Zelman before him, became a significant figure in Australia. Subsequent tours also depended on preexisting connections: in 1901 George Musgrove (Lyster's nephew) toured with a Grand English Opera Company (artists chiefly from the English Carl Rosa company), and that same year the entrepreneur J. C. Williamson brought out another Italian Opera Company. In 1911, the year before Quinlan's first visit, Nellie Melba visited Sydney and Melbourne with the Melba-Williamson Opera Company and singers recruited from all the major opera houses in Europe and America.[15] At times companies overlapped, but this does not seem to have mattered: there was an enormous appetite for opera. Nonetheless, and surprisingly, given Melbourne's very strong German community, leaders in the musical milieu, it wasn't until 1907 that George Musgrove brought out a company to sing Wagner in German, enabled by a subscription from Melbourne's German music lovers.[16]

Whatever the interest in different national traditions, Australia did not have the class division that Katherine Preston describes for the same period in America, where elite audiences attended opera sung in Italian and the middle classes attended English-language opera.[17] Although these two languages also dominated Australian performances, audiences were not large enough to be segregated according to status or language, even though strong followings existed for different operatic traditions. In addition, as Michael Cannon comments, "The gold rush . . . defrosted much of the class system inherited from Britain," perhaps because there was, in any case, a very small "upper" class.[18] According to Harold Love, Lyster's audiences would "have been much more representative of society at large than those of the European opera houses," and opera's success was accordingly "a tribute to the . . . Australian public at large."[19] While officials and professionals occupied the best seats, there were also large, highly popular areas of basic seating at some of the lowest prices ever charged for opera.[20] Of course, not everyone went for the music: Lyster ensured that it was also a spectacle, a place to observe and be observed. After all, for most of the colonial period, opera was performed in the same theaters as (often similarly spectacular) drama, vaudeville, ballet, melodrama, and pantomime, and for the same audiences. Nonetheless, as the century advanced, things changed. Lyster and the newspaper critics stressed that opera should be a "civilizing force," and a growing bourgeoisie came to populate opera boxes and the circle.[21] Lyster got rid of prostitutes by declaring that only women accompanied by men were permitted entry.

The somewhat uneasy balance underpinning these tours between exploiting a ready market, on the one hand, and offering civilization and refinement from Europe, on the other, was well established by Thomas Quinlan's time. Quinlan's own career demonstrates the degree to which, in Europe as in the colonies, industriousness and an eye to the main chance could forge a career. Trained as an accountant, Quinlan was also a singer with a fine baritone voice and, from 1909, the manager of Thomas Beecham's orchestra in London. From 1910 he managed the Beecham Opera Comique Company, which toured the English provinces,[22] but in August 1910 Quinlan broke from Beecham, with whom he had a difficult relationship,[23] and formed his own touring English-language opera company. He also worked as an impresario and manager of the Quinlan International Concert Agency, managing the concert contracts in America (and elsewhere) of Caruso, Kreisler, and Sousa's band, amongst others.[24] Quinlan managed Sousa's band tour to Australia, New Zealand, and South Africa in 1911 and thus would have been familiar with the issues involved in an Empire circuit.[25]

In 1911 the thirty-two-year-old Quinlan announced his own ambitious proposal of a British Empire tour. He must have had substantial money behind him, because his touring party comprised singers, soloists, chorus, full orchestra, and all the costumes, properties, scenery, and other things required for a full season. He aimed high: Louis P. Verande (b. 1870), a well-known Covent Garden director, accompanied him as the stage manager;[26] costumes were designed by Dorothy Carleton Smyth (1880–1933), an authority on historical theatrical costumes; Quinlan claimed he used the largest scenic studio in London to research his costumes; and the sets for all the operas were designed by the modernist designer Oliver Percy Bernard (1881–1939), who also accompanied the party.[27] Quinlan had to pay salaries—including for a four-month rehearsal period—ocean fares, and copyright fees. It was a colossal undertaking. Different reports give slightly different figures, but in 1912 he brought out roughly 163 singers, 60 chorus members, and an orchestra of 55. Unsurprisingly, he was keen to publicize his Herculean endeavor. According to one Australian review, "with a repertoire of 25 operas, the music library of the company comprises 56 cases, each weighing 1 cwt. [approx. 50 kilos today], and including 900 band parts and 700 vocal scores. The wardrobe department has charge of 700 wigs and 3300 costumes, whilst the stage staff has control of over two miles of scenery."[28] To the *Sydney Morning Herald*, Quinlan claimed to be touring with "475 tons" of scenery and wardrobe, and he estimated that he would spend over "£100,000; . . . salaries will be £30,000; steamer fares, £16,000."[29] Unquestionably, Quinlan's touring company was the largest to visit Australia; the logistics of such an entourage must have been extraordinary.

As already mentioned, Quinlan's visit to Australia was managed by J. C. Williamson, who also managed Melba's tour in 1911. A "Memorandum of Agreement" between Quinlan and Williamson outlines that Quinlan was to provide and pay for an "efficient [and] capable orchestra," a competent conductor, and a stage manager to be sent ahead, "to paragraph and in other ways boom [promote] the Company's season in Australia."[30] Quinlan chose as his stage manager the expatriate Australian journalist Agnes Murphy, who had been Melba's social secretary on her 1909 visit to Australia. Williamson was to provide the theaters in Sydney and Melbourne, the front- and back-of-house staff, lighting, advertisements, and fifteen "chorus ladies," as well as access to the scenery used on Melba's tour (both companies performed *Faust* and *Rigoletto*). It is unlikely, however, that Quinlan used either scenery or "chorus ladies," as both Quinlan and the critics made constant references in the press to the company's being "self-contained."[31] Quinlan was to receive 66⅔ percent of gross receipts (after shipping costs and other initial outlay), and Williamson 33⅓ percent. Williamson must have been confident of the enterprise: when the company arrived, after the South African leg, it was in severe financial trouble, forcing Quinlan to obtain an advance of £2000 from Williamson. The impresario easily made his money back: the company's treasurer, E. J. Gravestock, later recounted that takings in Melbourne were astronomical, far higher than anything they had received before. Quinlan apparently took to purchasing corona cigars by the boxful.[32]

Quinlan's boasts about the scale and self-sufficiency of his operation created tension, however. As already observed, previous touring companies had often recruited orchestras and choruses from within Australia, so there were complaints about Quinlan's failure to use any local performers. But there was also some understanding, as one critic commented: "To bring chorus-singers and orchestral players to Australia sounds very much like taking coals to Newcastle; but anyone who has had experience . . . of training and rehearsing a number of strange and unfamiliar works with strange and unfamiliar players and singers, will understand readily why Mr. Quinlan should have decided as he has done."[33] Melba did not escape so lightly some ten years later when she again toured Australia with Williamson's agency and an all-Italian company: the Australian Theatrical Alliance asked the government to prohibit the Italian choristers from landing, under the provisions of the Immigration Act.[34]

Quinlan's stated rationale for the comprehensiveness of his touring party was that they worked as a cohesive team, with a sense of collegiality and high performing standards, and with no superstars. Before embarking on the "all Red tour," they had rehearsed for four months in London and also toured the English provinces. Knowing each other and the repertoire well was essential, given the schedule to which Quinlan was planning to submit them. The company also had great respect for Quinlan, making a public presentation to him in Sydney before their return home in 1912:

Before returning to the homeland, we, the members of the Quinlan Opera Company, desire to place on record our tribute to your great courage and ability in undertaking and carrying to a brilliantly successful issue an operatic enterprise of an unprecedented character . . . the necessarily strenuous activities of this great tour have been considerably ameliorated by the artists' knowledge that they had in you not only a resourceful director, but a truly sympathetic comrade.[35]

Perhaps Quinlan sought to highlight a contrast with Melba's tour the previous year: although it had started with rapturous enthusiasm, when Melba became sick and stopped performing audiences dropped off, and comments were made in the press about under-rehearsal and lack of cohesion.[36]

Quinlan's 1912 season lasted only five weeks each in Melbourne and Sydney, during which time they performed fourteen operas, with four Australian premieres in eight days. Quinlan did not venture farther afield, despite Adelaide music lovers' pleas; he told them a subscription list would be needed as guarantee, but they could not raise sufficient funds.[37] Given the size of his party and the scale of their ambition, his caution was understandable. The premieres were Offenbach's *The Tales of Hoffmann* (1881), Puccini's *The Girl of the Golden West* (1910), Wagner's *Tristan and Isolde* (1865) and Debussy's *scène lyrique, The Prodigal Son* (1884). *Tannhäuser*, in the Paris version, was also presented for the first time in Australia. The other operas performed were *Rigoletto, Aida, La boheme, Carmen, Lohengrin, Madame Butterfly, Faust, Hansel and Gretel*, and *La Traviata*. In advertisements before his arrival, Quinlan reassured audiences that all his singers were also actors and that they would be seeing the equivalent of Covent Garden productions, but at much cheaper prices. There was great excitement in the press and pedagogical preparation for the company's arrival; for instance, Ernest Truman, the Sydney Town Hall organist, performed arrangements of all the advertised operas, meticulously divided into acts and with annotated programs provided.[38] A letter from a Melbourne Conservatorium student in the *Australian Musical News* reveals that normal classes were abandoned in favor of sessions on forthcoming operas, Wagner in particular.[39]

The company was a huge success with audiences and critics; Quinlan himself maintained a very public profile, and members of his company were also happy to be interviewed and photographed. A demonstration either of canny selection or of an Australian enthusiasm for novelty, the four Australian premieres were well received. *The Tales of Hoffmann*, which opened the season, was extremely popular. Success with ordinary citizens was, indeed, what one critic looked for: "There was not a musical man in Melbourne or Sydney whose face did not kindle with delight when [he] recalled the *Tales of Hoffmann*. I heard the boys whistling the 'Barcarolle' on the surf bathing beaches."[40] Australian audiences were familiar with Debussy's piano music, *L'après-midi d'un faune*, and the op. 10 String Quartet, which had been played by the Austral String Quartet in 1910.[41] Some were initially fearful of encountering "excessive modernity" in *The Prodigal Son* but found themselves surprised by its melodiousness, as it showed none of the composer's "'cranky' harmonic effects."[42] It had an exotic staging with swaying palms and dancing girls, and the critics found the music pleasantly exotic as well. Quinlan had first performed Puccini's *Girl of the Golden West* in English in Liverpool in October 1911, a year after its New York premiere, with Puccini in the audience. Australian critics were slightly taken aback by the opera's realism. The *Sydney Morning Herald* critic referred to "the Californian miner's life" as "naked and unashamed—rude, primitive, and almost savage in the violence of its emotions . . . [with its] dark and vast background of primitive characters and untrammelled nature."[43] Another claimed that Puccini had set the "wild and lawless times" to "wild and lawless music,"[44] yet reports described enthusiastic houses and almost uniform appreciation of Puccini's orchestration. The premiere that had the most impact, however, was *Tristan and Isolde*, in which, as one critic commented, "a terrific display of passion . . . sweeps through three wonderful acts like a whirlwind."[45] "Whirlwind" is not a word that springs to mind in relation to *Tristan*, and there were also rumblings about slowness and length, along with suggestions that it should have been cut to three hours.[46] Nevertheless, such was the audience interest that houses in Melbourne and Sydney were sold out and extra seats had to be put in the pit. A Melbourne critic remarked that "the house realized that it was, musically, the greatest work ever heard on the operatic stage in Melbourne."[47] Australians, in particular Melbourne audiences, were rabid Wagnerites, and perhaps all the more so because of cultural isolation, as a critic summed up in 1900: "We live far enough from the art centers of the old world; but even here the man who should think it incumbent upon him to reconcile us by argument to Wagner's theories or to 'convert' us to his views, would waste his breath."[48]

Quinlan was evidently happy to take advantage of the enthusiasm for Wagner. Before leaving Australia, he posted a letter in major newspapers stating that he would be back the following year and was willing to put on the *Ring* in English for the centenary of Wagner's birth, if "1000 (one thousand) subscribers, one guinea for each of the four performances," could be found to provide an advance subsidy.[49] He asked less for the stalls and gallery, but would only issue tickets for the complete cycle,[50] grandly proclaiming: "The Ring, which is the supremest expression of music drama, and which should be of incalculable service to the advancement of Australian musical art has to be done on a scale of splendid completeness or not at all."[51] His ambition extended more broadly: to do nine *Ring* cycles in six months across "Greater Britain"—a union of the United Kingdom and its settler colonies in Australia, Canada, New

Zealand, and South Africa, with subscriptions also sought elsewhere.[52] Quinlan obtained his subsidy easily.

By 1913 the members of his company knew each other even better and could thus be set an even more punishing schedule, as can be seen from the following program for Melbourne, as advertised in the *Age* (with Anglicized names): opening night on the Saturday was *The Mastersingers*, Monday *Rigoletto*, Tuesday *The Rhinegold*, Wednesday matinee *Tales of Hoffmann* and evening *Tosca*, Thursday *Samson and Delilah*, Friday *The Valkyrie*, Saturday *Faust*, Sunday free, Monday *Siegfried*, Tuesday *Aida*, Wednesday matinee *Tales of Hoffmann* and evening *Tannhauser*, Thursday Charpentier's *Louise*, and Friday *Twilight of the Gods*—fourteen different operas in fourteen days.

Apart from the *Ring*, the other Australian premieres of the 1913 tour were Charpentier's *Louise* (1900), Wagner's *The Mastersingers* (1867), and Massenet's *Manon* (1884). Other operas in the tour included Puccini's *La boheme, Tosca, Madame Butterfly,* and *The Girl of the Golden West*; Verdi's *Il trovatore* and *La Traviata*; Mascagni's *Cavalleria rusticana* and Leoncavallo's *Pagliacci; The Barber of Seville; The Marriage of Figaro*; and an assortment of French operas: *The Prodigal Son* and *The Tales of Hoffmann* once again, and also *Samson and Delilah, Carmen,* and *Faust*.

Wagner's *Ring* was obviously the star attraction, and leading up to Quinlan's return there had again been a flurry of activity to prepare audiences. The organist Ernest Truman returned with arrangements of the *Ring* on the Sydney Town Hall organ.[53] The musician and critic Arundel Orchard gave public lectures. Critics quoted extensively from the English press on Wagner, in particular the writings of Ernest Newman. Demand for information on Wagner was high; one young man was driven to steal a book on the *Ring* from the State Library of Victoria. A State librarian caught the hapless thief a few days later, engrossed in the stolen tome while queuing for tickets for the *Ring*. He was imprisoned for a day![54]

The scenery for the *Ring* had been, according to the program note and information in the press, especially made in Berlin by the artist and set designer Leo Impekoven (1873–1943).[55] New singers joined the company for the *Ring*, including the Englishwoman Maud Percival Allen (1880–1955), who had sung Brunhilde in Richter's English-language production of the *Ring* in London in 1908, and the handsome young German heldentenor Franz Costa; both the chorus and the orchestra were enlarged, the former from sixty to seventy and the latter, from fifty-five to sixty-five. The main conductor, Richard Eckhold (1856–?), was also new, and although he had a German Wagnerian pedigree (he had played the violin at Bayreuth), he had conducted Wagner mainly in English in Edinburgh and at the Metropolitan Opera in New York.[56] Eckhold, like Quinlan, was a popular personality, customarily greeted by cheers in the theater. Of his (and Wagner's) reception he said in one end-of-tour interview: "I have conducted Wagnerian music in Germany, England, America and South Africa, never have I encountered a more enthusiastic and sympathetic audience than in Sydney . . . Australians . . . are simply overwhelming in their unaffected enthusiasm and I don't believe they fully realise how they inspire the conductor and soloist to do their very best."[57] If such comments hinted at flattery, other reports confirm audiences' genuine enthusiasm.[58]

The *Ring*'s success was such that a petition, headed by the well-known patrons Elise Wiedermann and Thomas Pinchoff, was presented to Quinlan requesting a repeat performance—which they were granted, despite the demands it placed on the performers.[59] The press often stated that the petition came from the German-speaking community (Pinchoff was the honorary Austro-Hungarian consul, and his wife, Elise, was a Viennese soprano teaching at the Melbourne Conservatorium); in fact, there were a number of well-known non-Germans as well, such as the Italian conductor Alberto Zelman and the English musician Fritz Bennicke Hart.

The other operas in the season were also highly appreciated, even if the subject matters of *Louise* and *Manon* were a little challenging for some. Probably the works least appreciated by critics were *Traviata* and *Rigoletto*; one critic expressed a widely shared sentiment: "Why do these two operas still keep a place—lagging, superfluous veterans—on the stage?"[60] The operas' popularity with audiences explains why they remained a safe choice for Quinlan, whose second tour was even more rapturously received than the first. On the last night in Melbourne, members of the gallery paid a very Australian homage, throwing down a carefully constructed boomerang on pulleys, decorated with flowers, with a silver plate bearing the good wishes of the gallery patrons.[61]

Challenges of Australia

The size and scope of Quinlan's tour was particularly herculean given the infrastructural constraints of the period. The company traveled to Australia on the *Orsova* steam ship, a two-funneled passenger liner that first travelled from London via Suez to Australia in 1909, a route that reduced the length of the journey to Australia from several months to thirty-five or forty days.[62] All-day rehearsals started on board ship: "The orchestra would be rehearsing on a hatchway on the foredeck, the chorus in the deck aft, and the soloists in the music room. As we occupied nearly all the accommodation on the ship, the few remaining passengers had a great time."[63] Travel within Australia was complicated too (which may explain some of Quinlan's reluctance to visit Adelaide). The journey by train from Melbourne to Sydney was long and slow; the company appear to have caught the train in the evening about three hours after a performance. Until 1962 the railway gauges in Victoria and New South Wales were of different sizes, so passengers and goods (including, in the company's case, sets and costumes) were obliged to change trains at the state border, usually in the middle of the night.[64]

The company encountered practical difficulties in their performances, at Her Majesty's Theatres in Sydney and Melbourne (both controlled by J. C. Williamson). Her Majesty's in Sydney had "the latest technology," including "an intricate pulley system on a grid above the stage to facilitate set changes."[65] Her Majesty's in Melbourne did not have the latest technology and had (and still has today) an inadequate pit. Often the orchestra had to overflow into the stall boxes on either side of the stage.[66] Quinlan faced challenges in fitting his sets onto the two stages. He traveled with his own stagehands, but, unexpectedly, they were prevented from working in Australia (a highly unionized country at this time) by the Australian Federated Stage Employees Union, and instead had to twiddle their thumbs in the wings while the Australians, inexperienced with the sets, did their best.[67] For the performances of the *Ring* the Australian crew were obliged to lower the curtain between scene changes, which was seen as unprofessional.[68]

Where practical difficulties impinged on the musical performance, however, they were not likely to be treated leniently. A small press debate was created on the issue of gaps in the woodwind and brass sections and the presence of only one harp (the only female performer in the orchestra).[69] A critic reviewing *Tristan and Isolde* rebuked Quinlan for the incompleteness of the woodwind section and the absence of the second oboe and third bassoon.[70] Quinlan replied in a letter to the paper stating that "not only have we a first and second oboe, but a bass clarinet; unfortunately the second oboe ha[d] to play the Cor Anglais in this opera, as we could not procure a player in Melbourne to take the second oboe's place . . . but the music didn't suffer, as the first oboe played all the important second oboe cues, and in addition, one of the first violins played the second oboe part throughout the opera."[71] Ill will for not using a local orchestra may have led to reluctance to assist the company.

The third bassoon part was reported by Quinlan as being played "by one of the cellos." The critic replied robustly that "the question of the second oboe and third bassoon is purely one of orchestral colour, and the fact that these parts are 'cued-in' for strings does not compensate for the loss of characteristic tone."[72] Clearly, the critic felt he had to make Quinlan aware that there were people in Australia who were familiar with the score and that he should not think he would get away with shortcuts without notice; generally, however, the orchestra was thought to be extremely good.[73]

Many Australian critics were already familiar with the *Ring* from European trips; for instance, the *Argus* critic T. E. Guennet had visited Bayreuth in 1888, and G. W. L. Marshall-Hall had attended Richter's London Wagner concerts in 1889, and in 1913 was sending back articles to the *Argus* from London, reviewing Nikisch's *Ring* at Covent Garden.[74] Others, however, who were not familiar with it, reacted defensively, seeking to bring it down in some way. The *Bulletin* reviews, in particular, aimed to deflate pretension while endeavoring to be humorous: "Heaven knows what all these silly prehistoric people want with the jingling metal . . . A god incapable of sterilising a gnome's curse, or stopping his wife's tongue is not much of a person to write a four-volume opera about."[75] Another critic stressed the awkward silliness of the libretto by a bathetic comparison: "Learning of the wonder of a beautiful woman surrounded by a wall of flame, . . . [Siegfried] rushes through the fiery barrier with as little concern as a Melbourne citizen through a dust storm, awakens the sleeping beauty with a kiss, and claims her as his own."[76] Achieving bathos through local comparison was common. One Sydney critic wrote, "Rhinemaidens circl[e] . . . round quite in the style of the Bondi surf ladies . . . [although] far more picturesquely attired,"[77] while another observed that "Sydney bathers could find no name for . . . [the] swimming strokes [of the Rhine maidens]."[78] These self-conscious similes were typical of the period, with deliberate irreverence, use of vernacular, and avoidance of pomposity.[79] This approach was part of what Dirk den Hartok has called the "democratic identification of [Australian] writers . . . with a 'levelling' egalitarianism rather than with . . . noble conceptions."[80]

As this analysis might suggest, there was some tension between those who (in multiple press articles) advocated doing some homework before going to the *Ring* and those who felt one could "over study."[81] As one critic commented, "One can become so absorbed in the literature of Wagner and the meaning of the *Ring* motives that the thought of . . . actual diversion is dispelled altogether. . . . Yet it was not, as it had been feared by some, heavy and difficult to understand," so others claimed it was possible "to enjoy the music without knowing anything."[82]

The Rhetoric of Heroism and Civilization

Both Quinlan's and the critics' rhetoric surrounding the tours revolved around two of the core beliefs of imperial patriotism—heroism and militarism.[83] Quinlan's tours were on a heroic scale: he led a small army; he was a "general" with "horse, foot, and artillery at his disposal."[84] "The man behind the gun [is] more of a Bismarck than a Barnum";[85] he was the "Napoleon of opera, . . . seeking new worlds to conquer," or the "Alexander of the operatic world."[86] Similarly, in a letter to the Williamson management, after the riots in Johannesburg, Quinlan said, "I [felt] like Napoleon in his retreat from Moscow—only I [had] my army with me."[87] The company also used the rhetoric of heroism, speaking of Quinlan's "great courage . . . in undertaking . . . an operatic enterprise of an unprecedented character," "under conditions which are epoch-making."[88]

Quinlan was on a civilizing mission, reanimating connections between cultural (and racial) cousins. For him, as for many, the main cities in Australia were extensions of the English provinces in which his tours had started. William Lyster saw Melbourne in the 1860s as a "prosperous English provincial city suddenly hurled to the other side of the world."[89] Quinlan's desire to advance the cause of opera among the world's English-speakers included the "American cousins," and he was proud of "having proved that whenever English is spoken there is a vast field for opera in that language," convinced that this would both convert Anglophones to opera and encourage English-speaking composers to write more operas.[90]

Before Quinlan's arrival there had been discussion in the Australian press about the use of English. There was a general feeling that it might attract audiences, and, when the performances began, some critics expressed relief at finally understanding what was happening on stage (although others still complained about comprehensibility).[91] Press desire for excellent translations and everyday language was met by Quinlan's use of the popular Frederick Jameson translation of the *Ring*.[92] In press interviews he listed new translations among his costs and claimed he commissioned translations from writers such as Ernest Newman.[93] However, some translations were poor, such as that for *Traviata*, as one critic remarked: "If the elder Germont must introduce himself to Violetta with the words 'I am the father of the incautious stripling you to ruin are luring,' it is much better he should do so in knee-breeches."[94]

KERRY MURPHY

Quinlan was, as we have seen, in no way bringing opera to the "great unwashed"; at the same time, however, his finely balanced program had an "educational" and aspirational veneer that reflected the period's burgeoning music-appreciation movement:

Our All Red tour . . . is largely educational. I hope to find Australian audiences even more receptive of new music . . . than those of Europe. . . . Mind you, I don't pretend to go in for advanced art and novelties all the time. Salaries must be paid, and the public has every right to its little preferences. . . . moderation in all things. My aim is to please my public and lure them forward at the same time.[95]

However, Quinlan's emphasis on "novelties" and didacticism was neither purely a marketing ploy nor simply patronizing. He, like others in his company, repeatedly spoke of finding Australians "exceptionally emotional audiences" who were receptive to new works: "On that account the conviction has come to me that (as proved by our Melbourne season) they will rapidly assimilate novelties."[96]

The sense of debt to Quinlan, voiced by the critics but also in letters to the editors and actions such as the Gallery's "boomerang" was obvious. He was seen as "doing an immense work in stimulating the growth of the musical atmosphere so sadly needed in this part of the world," and was considered as "deserving of a Victoria Cross for valour in undertaking such a tremendous enterprise . . . [he] certainly will rank as a hero amongst the impresarios of the world."[97] Quinlan's visit was also seen by some as consolidating Empire. The businessman and politician Sir William McMillan, at a luncheon given in honor of artists from the Quinlan opera company who were alumni of the Royal Academy or Royal College of Music, London, stated that their visit "could not but be beneficial to imperial interests as a whole." They were "missionaries of Empire, who would return with a good word for the Commonwealth, and would report that they recognized in it those principles of Justice, honour, and courage which animated all who looked for inspiration to the enlightened Constitution of England, and for guidance to the British flag. (Cheers.)"[98] No other touring company received the accolades Quinlan's company did, even though reports of loan defaults suggest that his finances were not so healthy.[99] Above all, he reinforced the need for Australians to establish their own national opera company. As G. W. L. Marshall Hall wrote in a letter to the editor of the *Argus* in 1912, Quinlan's success had "demonstrated that Australia is now ripe for an annual, or at least a biennial, opera season. . . . Lead us from the Sahara . . . to the promised land flowing with musical milk and honey."[100]

It is telling that Marshall-Hall hoped Australians themselves might lead the way to the promised land of a permanent opera company. That land was not reached until the mid-1950s and, sadly, was one full of those "lagging superfluous veterans" of which earlier critics had complained.[101] Quinlan's vision and anticipation of an Australian audience more receptive to new works than the British was not to be fulfilled.

ELEVEN

Empires in Rivalry: Opera Concerts and Foreign Territoriality in Shanghai, 1930–1945

YVONNE LIAO

On 9 December 1941 the (British-biased) *Shanghai Times* reported on the Japanese invasion of the International Settlement:

Rudely awakened long before the break of dawn by the rumbling of naval artillery and the chatter of machine guns as Japanese forces swung into action against the British gunboat Petrel, Shanghailanders [British and other foreign settlers] later in the morning watched silently as units of the Japanese Special Naval Landing Party swiftly and efficiently took control of the International Settlement.

The occupation of the so-called International Settlement—an area associated with British interests in treaty port Shanghai—was of obvious military significance following Imperial Japan's declaration of war on the British Empire and the United States.[1] But such encroachment was played out as much in cultural terms as militarily, and not merely during the Second World War. Indeed, the imperial scrambling for possession of the International Settlement, Shanghai's political and financial hub, reveals a hitherto unexplored relationship between opera and foreign territoriality in the

148

1930s and 1940s.[2] The present essay aims to analyze this relationship by discussing opera concerts of the Shanghai Municipal Orchestra (上海工部局樂隊) and its successor, the Shanghai Philharmonic (上海愛樂協會愛樂樂團), in the context of British territoriality in the 1930s and Japanese territoriality in the first half of the 1940s.[3] In examining opera and territoriality across this period, the essay brings to light not only the workings but also the changing conditions of imperial rivalry in Shanghai.

The idea of territoriality assumes particular force given the highly prized and contested nature of the port of Shanghai in the first half of the twentieth century. Territoriality, according to Robert Sack, is "the attempt by an individual or group to affect, influence, or control people, phenomena, and relationships, by delimiting and asserting control over a geographic area."[4] Sack's definition highlights not only the cultural instability of a geographic area due to human intervention, but also the act of laying claim to such an area. Similarly, and more recently, Yvonne Whelan has stated: "Unlike the more benign concept that is 'place,' with its connotations of space made meaningful by human habitation, territory and territoriality are suggestive of more malign forces at work."[5] Although it may seem extreme to equate territoriality with malevolence, its expression as the acquisition and/or preservation of (geo)political power entails adopting strategies through which such power is exerted.

Territory and commerce became intertwined issues in Shanghai and other ports in China in the second half of the nineteenth century. Following China's military defeat in the Opium War and the Treaty of Nanjing in 1842—the first of the so-called Unequal Treaties between the Qing Dynasty and such signatories as Britain and France—Shanghai was designated as a port in which foreigners could trade, reside, and lease land at fixed and favorable rates.[6] Settler communities began to take shape in Shanghai and other locations deemed strategic along China's coast, waterways, and railways. In terms of their legalese, the treaties were exploitative but also labyrinthine agreements ratified with no fewer than a dozen signatories; in the case of Shanghai, they included Italy, the Netherlands, Portugal, Spain, and the United States in addition to Britain and France. Treaty ports were thus multi-colonial settings—or colonial formations—and differed from centrally managed colonies due to their various foreign-leased areas, and to their (initial) focus on commercial exploitation rather than systematic colonization.[7] Of the peculiar and fragmented nature of colonialism in China, Bryna Goodman and David Goodman rightly observe that although Shanghai and Hong Kong were both "locations of British colonial presence," they were "structurally dissimilar and have [therefore] engendered different historical narratives."[8]

To their remarks can be added another important point: because the treaties brought rival empires into physical proximity in designated ports such as Shanghai, major signatories and their subjects, with their own prerogatives, had competing visions of how leased areas should be planned and administered.[9] The British and Americans, originally with separate Concessions, merged in 1863 and became the so-called International Settlement. The French, wary of the Anglo-American alliance, established their municipality, the French Concession, in the 1860s, which was presided over by the French consul-general in Shanghai. The International Settlement became synonymous over time with British (and allied) interests; the Shanghai Municipal Council, the Settlement's de facto governing body, served to safeguard those interests. The International Settlement was effectively Shanghai's political and financial hub, while the French Concession, south of the Settlement and about half its size, developed into an upscale residential district. According to Tess Johnston, "The conventional wisdom of the day was that in the International Settlement the British would teach you how to do business, but in the French Concession the French would teach you how to live."[10] Johnston's dichotomous view of Shanghai's foreign-leased areas was captured in similar fashion in maps drawn by Chinese cartographers in the 1930s and 1940s, offering distinctive bird's-eye snapshots of the French Concession and the International Settlement (see fig. 11.1). The Concession was characterized by comfortably spaced private residences and green plots; the Settlement, by contrast, was densely packed with banks, department stores, and hotels.[11] There was also Greater Shanghai, a municipality administered by the Chinese Nationalists—namely areas outside the Settlement and Concession, notably the Chinese City (Old City). Shanghai of the 1930s was thus a "city of cities" composed of foreign and local municipalities. This landscape shifted, however, with the Japanese occupation of Greater Shanghai in 1937 and the International Settlement in 1941.[12]

Sack's definition of territoriality—the attempt by an individual or group to delimit and assert control over a geographic area—is especially apposite in view of Shanghai's treaty-port history and various municipal administrations. Foreign territoriality in 1930s and 1940s Shanghai can be defined in terms of British, French, and Japanese forms of imperial expansionism and attendant displays of political—specifically municipal—power. Multiple uncoordinated administrations led to rivalry, which was manifested, for example, in attempts by the Shanghai

11.1 Map of Shanghai, 1934, with the International Settlement (A), French Concession (B), and the Chinese City (C). Reproduced with kind permission of the Bodleian Library, Oxford.

Municipal Council and the French Municipal Council to stamp their marks on the International Settlement and the French Concession respectively, and attempts by the Japanese, following occupation, to restructure treaty-port institutions to their own ends. Such institutions included not just those bound up with the practicalities of government and administration, but also those whose province was the expression of culture, notably the Shanghai Municipal Orchestra.

The foreign monopoly of musical institutions was especially symbolic with regard to opera, which was still the jewel in the crown of imperial emblems. As Michael McClellan has shown in his study of opera in colonial Hanoi, the French government of Indochina used the construction of theaters in Hanoi, Haiphong, and Saigon as a means to project their command over Southeast Asia and to showcase their cultural superiority.[13] Opera thus "constitute[d] a basic tool of empire," with the opera houses "contribut[ing] to the reorganization of the colony's physical space," while the operas themselves "reified the lessons of colonial order" by means of dramatic narrative and theatrical spectacle.[14] In late nineteenth-century Shanghai—not a colony as such—operatic performance did not reinforce a particular colonial order in the same manner; nonetheless it enabled British settlers' notions of (European) high culture to hold sway.[15] Opera stagings at the Lyceum, a British-constructed venue, rendered evident settlers' influence over the emergent, colonial "theatrical environment" and the engagement of performers, be they touring companies or the Shanghai Public Band.[16] While staged opera in the 1930s saw ambitious productions by émigré-run groups such as the elusive, Russian-oriented Shanghai Opera,[17] opera's associations with empire not only intensified, but also became more complex. This could best be observed in another type of event—opera concerts—presented in the 1930s by the British-run Shanghai Municipal Orchestra (formerly the Shanghai Public Band), and in the 1940s by the Japanese-affiliated Shanghai Philharmonic Society. Such concerts were potent symbols of foreign territoriality and imperial expansionism. Just as opera in Hanoi can be analyzed in terms of the colonists' (attempted) glorification of Third Republic France, opera concerts in 1930s and 1940s Shanghai can be examined as manifestations of British and Japanese territoriality, and, at a deeper level, as (shifting) discourses of municipal power. Indeed, evidence relating to such concerts, which favors settlers' and occupiers' perspectives, points to attempts by the British and the Japanese to steer matters civic and cultural.[18] Given the multitude of foreign powers in Shanghai and the strategically important International Settlement, opera can be seen not only as an imperial emblem, but also as a form of imperial rivalry writ large.

British Territoriality in Shanghai: Opera Concerts in the 1930s

The Shanghai Municipal Orchestra (SMO) of the 1930s—an institution with imperialist undertones, based in the International Settlement— provides fascinating insight into the nature of British territoriality.[19] Despite the name accorded the area, which was leased to a host of foreign powers under the Unequal Treaties, affairs of the Settlement, according to Robert Bickers, "served the interests until the late 1930s of British settlers."[20] Key posts in the Shanghai Municipal Council (SMC) were occupied by British or British-affiliated administrators. The image of the Settlement as a British-run municipality also derived from the SMC's control, along the Huangpu River (黃浦江), of the upper stretch of the Bund—a (British-constructed) waterfront symbolizing Shanghai's global financial status, with the headquarters of the Hongkong and Shanghai Banking Corporation, Jardine Matheson and Company, and other major British (as well as non-British) enterprises. By contrast, the lower stretch of the Bund, which belonged to the French Concession, consisted mainly of wharves. Although different foreign subjects resided and conducted business in the Settlement, the British presence that had developed over nearly a hundred years (since the establishment of the first treaty ports) was doubtless pervasive.

Founded by the settlers as the Shanghai Public Band in the late 1870s, the SMO was initially organized along the lines of a military band, for performances of bandstand music in Europeanized civic spaces such as the Public Garden. The Spanish musician Melchior Vela was originally tasked with the recruitment and training of musicians, many of them Filipino (a regular source, as the Philippines was an established colony), though always with European stiffening. Following the introduction of "art music" into the repertoire and the band's gradual development under successive conductors into a fuller orchestra, the Italian Mario Paci was appointed music director in 1919, and the ensemble was re-organized and renamed the Municipal Orchestra and Band in 1922. Its funding came from the Settlement's ratepayers. The Orchestra, with its growing bulwark of European players, aspired to be Shanghai's most high-profile Western-music ensemble, operating within and beyond the International Settlement—for example, also giving concerts in Shanghai's French Concession, such as on Bastille Day.[21] Orchestra members doubled as brass-band members in the summer, giving performances in Shanghai's public parks.[22] Yet, despite its activities in the French municipality, and the inclusion of Chinese, Filipinos, Italians, Japanese, and Russians among the audience members and players, the Orchestra was essentially a British-controlled institution.

During the winter, the Orchestra gave concerts at the Grand Theater in the International Settlement and, from 1934, at the Lyceum Theater in the French Concession. Such concerts designedly revolved around nineteenth-century European "art" repertoire (a priority since the days of Rudolf Buck, Paci's German predecessor), for example "a special program of the most popular works by Franz Liszt in memoriam of the fiftieth anniversary of the Master's death."[23] The Lyceum also hosted performances by other groups, notably touring companies and the English-speaking Amateur Dramatic Club (ADC), and showed the latest motion pictures, including Hollywood releases (which of course required accompaniment). Home to the ADC, the Lyceum (named after the theater in London) was lauded in 1867 by the (British-biased) *North-China Daily News* as "one of the best-planned and most commodious houses in the East."[24] When the Lyceum was rebuilt in 1874 after a fire, it was "constructed in the western manner, with external brick walls, and roof covered with tiles."[25] Its design included a thirty-foot roadway allocation that created "convenient harbourage" for chairs and vehicles, attesting to the theater's role as a (foreigners') social hub.[26] The

ADC acquired new premises in the French Concession in the late 1920s and rebuilt the theater again. Yet, despite that location, the Lyceum as a performance venue had by then become synonymous with British-affiliated groups such as the ADC and the SMO.

The Orchestra's programs in the 1930s highlight the prominence at the Lyceum of opera concerts presented by the SMC, which centered on selections rather than staged complete works. This approach, which might seem wanting by modern standards, was deemed to hold utilitarian virtue, according to the column "From the S.M.C. Orchestra" in the English-language *China Press*. For example, on Wagner's *Der Ring des Nibelungen*:

> A most attractive method of presenting [his] music has been devised for presentation at the second symphony concert of the season at the Lyceum Theater tomorrow. Those musical amateurs and others to whom Wagner is either anathema or unintelligible cannot but delight in the short musical biography of Siegfried [a series of orchestra interludes] which has been selected from the *Nibelung* tetralogy.[27]

The SMO's opera concerts certainly made an impact: they were monthly events during the winter season, publicized in the press (including Chinese newspapers), and attended by foreign and Chinese patrons alike. The international mix of the audience was evidently important to the SMC: officials kept records of attendance by nationality.[28] Just as internationally diverse were guest singers brought from opera houses in Europe (La Scala and the Théâtre de la Monnaie, for example) to Shanghai and other cities in China by such agents as the Latvian-born Avray Strok, who was originally responsible for recruiting Mario Paci (as a touring pianist) in 1918 and was described by Floria Paci Zaharoff, Paci's daughter, as "the Far Eastern impresario."[29] Other singers performing in the SMO's opera concerts included émigré Russian musicians who fled after the Russian Revolution and established themselves initially in such places as Harbin and Tianjin but moved further south during the 1920s, partly due to increasing Japanese military presence in northeastern China.[30] Vladimir Shushlin, a bass active in 1930s and 1940s Shanghai, was one such: he relocated from Harbin to Shanghai in 1929 and became a professor at the National Conservatory of Music.[31] The "stateless" status of Russian émigrés (following the cessation of diplomatic relations between the Republic of China and the Soviet Union in the 1920s) afforded some measure of neutrality, and with it, job opportunities regardless of imperial rivalry and its attendant politics; as such, these émigrés could provide continuity between different regimes.

154

Programs from the 1930s bring to light a rich and wide-ranging repertoire, as the following samples from 1934–35 show. The SMO's opera concert in February 1934 was a three-part extravaganza with German, Russian, and Italian operatic excerpts. Part 1 comprised excerpts from Wagner's *Parsifal*, *Tannhäuser*, and *Die Walküre*. Part 2 was made up of excerpts from Russian operas: Mussorgsky's *Khovanshchina*, Rimsky-Korsakov's *The Tsar's Bride* and *The Snow Maiden*, and Glinka's *A Life for the Tsar*. Part 3 consisted of excerpts from Ponchielli's *La Gioconda*. The concert in March 1934 featured a "Grand Italian Operatic Program" and the soprano Elisabetta Silinskaia, "from the opera houses of Milan and Turin." Silinskaia performed popular arias such as "Signore ascolta" from Puccini's *Turandot* and "Si, mi chiamano Mimì" from *La bohème*. Interspersed among the arias were the Intermezzo Sinfonico from Puccini's *Manon Lescaut* and the Intermezzo from Mascagni's *L'amico Fritz*. The opera concert in May 1935 featured arias from Tchaikovsky's *The Enchantress* and Puccini's *Tosca*, which were supplemented by the Sinfonia from Mascagni's *Le maschere*.

Judging from these and other surviving programs, the SMO's opera concerts demonstrate a broad repertoire extending from Gluck and Rossini to Saint-Saëns, Puccini, Tchaikovsky, and Wagner. Such variety can probably partly be attributed to Mario Paci, who remained music director until May 1942, when the Japanese transferred the Orchestra to private auspices.[32] Paci seems to have traded on his European credentials, emphasizing his Milanese training and support from legendary operatic figures in the program notes for the Orchestra's farewell concert (as a municipal unit) on 31 May 1942. Of one encounter at a soirée hosted by a La Scala patron in 1896, he wrote: "Luck assisted me. . . . I played [Beethoven's] 'Appassionata,' and who should come and compliment me? Puccini!"[33] According to Paci, that encounter proved fruitful: "During my four years at the Conservatory, thanks to Maestro Puccini's [help] again, my apprenticeship as a conductor was under great conductors such as Ferrari, Mugnone, Mascheroni, and the [greatest] of all: Toscanini."[34] If Paci's apparent exposure to such figures as Puccini and Toscanini in 1890s Milan helped to create a vibrant municipal opera scene in 1930s Shanghai, his (purportedly extensive) connections with Milan must also have added a sense of authenticity to the presentation of the Orchestra's opera concerts—always a concern for the municipality.[35]

Yet, the social function and success of such concerts depended on more than artistic qualities alone, for they were also contingent on the approval of the SMC and the local British community, and as such were plagued by civic and financial politics around their expense and choice of programs.[36] For example, the Council's Orchestra and Band Committee issued the directive that "the Conductor [Paci] should submit [proposed programs] to the Committee weekly for information prior to publication."[37] At a deeper level, the SMO's opera concerts—though brought to audiences by Paci and soloists of diverse nationalities—were inextricably linked to the display of British municipal power, and to settlers' cultural values and sense of superiority. A local British report in the 1930s boldly touted Shanghai's International Settlement as "an unprecedented chapter in the history of the world's municipalities."[38] Clearly, British administrators sought not only to celebrate what they deemed a unique political achievement, but also to distinguish themselves from Shanghai's French and Chinese administrations. And since signatories of the Unequal Treaties and their subjects had, from the outset, different vested interests, over time the British settlers' exercise of municipal power in Shanghai's International Settlement became a matter of both hegemony and self-distinction.

Seen in this light, the SMO's activities were a specific manifestation of British territoriality. Its opera concerts served to steer and define a civic culture, as the settlers saw it, and, more broadly, reflected the SMC's pride (and control) as Shanghai's "British" municipal administration. With regard to a proposed bandstand in Jessfield Park, for example, the Council emphasized that "a cheap type of construction is unsuitable and incompatible with [its] dignity."[39] The imperative of maintaining the Orchestra as a "British" institution also pervaded print media such as the *North-China Daily News*; in a letter to the editor, a reader opined: "When the Municipal Council accepted a small donation from the Italian Government it surely was not the intention that in return the Municipal Orchestra should give us an absolute surfeit of Italian music at the weekly concerts, and advertise Italian composers on every possible occasion."[40] So, while Italian repertoire was in many ways the mainstay of the SMO's opera concerts, it was also subject to appraisal by the local British community. The "international" feature of the Settlement— and political muscle flexing there, such as the Italian government's presumed attempt to influence matters of programming through sponsorship—were materially conditioned by British territoriality in Shanghai. Even then, not all had the same aspirations to uphold the SMO's function as the bastion of (British-defined) high culture; it is salutary to note that in 1934, the SMC found itself obliged to promise "more good music of a lighter kind" in order to ensure continued ratepayer support.[41]

Japanese Territoriality in Shanghai: Opera Concerts in the 1940s

Following Imperial Japan's declaration of war on the British Empire and the United States in December 1941, and the subsequent occupation of the International Settlement, the SMO gave its final concert as a municipal ensemble on 31 May 1942—with the program for which Paci's autobiographical reminiscences were penned. The Orchestra was then placed under the auspices of the newly established and privately funded Shanghai Philharmonic Society (or the Society), overseen by Japanese secretaries general. Reporting on the transfer, the *Shanghai Times* quoted the Japanese embassy spokesman as saying: "Technically all the members of former Shanghai Municipal Orchestra will remain [if] the members apply for inclusion in the new band [of] the Shanghai Philharmonic Society."[42] Paci did not apply, possibly because he was not in favor of the arrangement and did not see eye to eye with the Japanese authorities. Other regulars were employed instead: Takashi Asahina (朝比奈隆), a Japanese conductor; Arrigo Foa, an Italian Jew and previously concertmaster and deputy conductor of the Orchestra; and Alexander Sloutsky, a Russian émigré and previously assistant conductor of the Orchestra. Foa's and Sloutsky's increased roles could be attributed to the fact that changing regimes affected Shanghai's musical life in unexpected ways, including through the expanded presence of Russian and European Jewish performers. There were (apparently) two contributing factors: the Soviet-Japanese Neutrality Pact of April 1941, which had the effect of separating Russian subjects from British, American, and other Allied subjects; and the Japanese internment of perceived enemy nationals (including members of the SMC), which provided new sources of employment for "stateless" musicians such as Russians and European Jews, long-standing proponents of opera and operetta in Shanghai.[43]

At first glance, the Philharmonic's opera concerts, which were presented by the Society, did not differ greatly from those of the Municipal Orchestra. The Philharmonic continued to perform a rich and wide-ranging repertoire. Paci guest-conducted the "Grand Italian Operatic Concert" at Koukaza Park in August 1942, which featured excerpts from Rossini's *La scala di seta*, Ponchielli's *La Gioconda*, Mascagni's *Cavalleria rusticana*, and Verdi's *Aida*. In the same park in July 1943, "An Evening of Opera Music" was conducted by Foa, and featured selections of French and Italian operas, including Gounod's *Faust*, Bizet's *Carmen*,

Donizetti's *Don Pasquale*, and Catalini's *La Wally*. There followed "An Evening of Russian Opera" the next month, which was conducted by Sloutsky. Sloutsky also conducted a complete performance of Leoncavallo's *Pagliacci* at the Koukaza Park in September 1944, billed as a "Special Opera Performance under the Stars." Like Sloutsky, the soloists and chorus members were all Russian émigrés.

Yet, operatic programming was also different: staged and complete performances became a prominent feature of the Philharmonic's opera concerts in the 1940s, according to copies of programs donated in 1989 to the Shanghai Symphony Archive by Yoshito Kusakari (草刈義人), who had been secretary general of the Shanghai Philharmonic Society during the occupation. They indicate that a succession of full-scale opera concerts took place in the months prior to Japan's surrender in August 1945. In March 1945 there was a "Grand Opera Performance" at the Lyceum of *Cavalleria rusticana*, which was conducted by Sloutsky and produced by Kusakari. May saw a production at the same premises of Lehár's *Count of Luxembourg*, whose personnel included the Japanese choreographer Masahide Komaki (小牧正英), who was also a principal dancer at the Shanghai Ballets Russes. Performed in the same month and again in June was Offenbach's *Tales of Hoffmann*; produced, again, by Kusakari, it featured the Chinese soprano Gao Lanzhi (高蘭芝), who had studied with the Russian bass Vladimir Shushlin. June saw Kusakari's production of *La traviata* at the Lyceum, with Sloutsky on the podium, Gao as Violetta, and the émigré Russian tenor Peter Markov as Germont.

The programming of the concerts might suggest that Shanghai's opera scene was largely undisturbed by bloodshed and hostilities, and that it even became more vibrant thanks to staged and complete performances with a readily assembled "all-star" cast.[44] Yet, these concerts cannot merely be viewed as social events; they were part of Japan's assertion of its credentials as an empire. Its declaration of war points to changing conditions of imperial rivalry, specifically a propagandized East-West competition distinct from that among predominantly Western powers in the International Settlement in the 1930s. Below is an excerpted version of the declaration of war:

More than four years have passed since China, failing to comprehend the true intentions of Our Empire, and recklessly courting trouble, disturbed peace in East Asia and compelled our Empire to take up arms. Eager for realization of their inordinate ambition to dominate the Orient, both America and Great Britain, giving support to the Chungking [Chinese Nationalist] regime, have aggravated disturbances in East Asia. We only desire to do away with the tyranny of America and Britain and to restore East Asia to its proper and undefiled state of existence. The rise and fall of our Empire and progress or decline of East Asia depend upon the present war.[45]

In juxtaposing the "tyranny" of the West with "progress" in the East, Imperial Japan sought to project itself as a liberator: the alleged oppression of the United States and the British Empire (particularly in the treaty ports) served to justify Japan's military presence in China and promulgation of a "Greater East Asia." As John Stephan notes, "[During the war] 'Greater East Asia' reverberated through radio broadcasts, newspapers, magazines, academic monographs, Diet [Japanese parliament] speeches, classrooms, and barracks."[46] Its "us versus them" strategy was designed to encourage unity among East Asians and support for Japanese endeavors, papering over Japanese atrocities in China and elsewhere. But, as Barak Kushner observes, the message was nonetheless "that Japan was the most modern country and race in Asia and that it alone could lead Asians through the twentieth century."[47]

Imperial Japan's posturing—notably its antagonism toward the British Empire—gives context to the occupation of the British-run International Settlement and the restructuring of the SMO. Indeed, the founding of the Shanghai Philharmonic Society produced a different state of imperial rivalry: Japanese municipal power, a symbol of expansionism in the name of a "Greater East Asia," effectively displaced British municipal power, a symbol of expansionism in the name of increased trade. That the Society was privately funded did not mean that it was not politically controlled, for ultimately it was loyal to the Japanese military cause, presenting nationalistic concerts such as the "Grand Symphony Concert in commemoration of the Third Anniversary of the Greater East Asia War."[48] Restructuring the Orchestra was said to "greatly improve the musical activities of the group," but this policy—civic culture on Japanese terms—served to extol Japanese virtues and diminish the British influence.[49] At the same time, the attempt to transform a "British" orchestra into a "Japanese" one was, according to Tang Yating, an inherent contradiction, because the orchestra itself (as a concept and in its repertoire) was a Western import.[50]

Nonetheless, the inauguration of the Society a mere six months after military occupation illustrates how the occupiers staked out territory both politically and culturally in the International Settlement. Opera concerts presented by the Society were a specific manifestation of Japanese territoriality, reflecting the occupiers' ambition to outdo a Western enemy power and to stamp their mark through "new," regular offerings, notably staged productions of complete operas. Territoriality *in* the Settlement was particularly significant: Japanese occupation of the French Concession did not matter to the same degree, as the Concession was (nominally) under the purview of Vichy France (a regime collaborating with Nazi Germany, Imperial Japan's ally). Thus, the Society's opera concerts were characterized not only by a continually varied repertoire, staged performances, and an abundant supply of "stateless" musicians, but also—and above all—by the occupiers' attempt to assert political control over the International Settlement. This control was rendered especially apparent in the Japanese Proclamation in Shanghai of December 1941, which attempted to cast occupation in a positive light—as aspiring to "the peace and order and preservation of the prosperity of the Settlement," with a readiness "to respect the life and property of the general public"—but which also warned of affiliations with the "enemy."[51] So, just as the Orchestra in the 1930s was for the Shanghai Municipal Council a matter of articulating and maintaining British influence, the Philharmonic in the 1940s was inextricably linked to the occupiers' reputation and justification of their invasion of Shanghai.

Operatic Geography In and Of Shanghai

This emergent geography points inevitably to the influence of empire(s), but at a deeper level, it also complicates opera as a *distinct* imperial emblem. Michael McClellan's study is worth revisiting here. Highlighting the flaws of "French imperial performance" in colonial Hanoi, he notes that "the theater was a byproduct of [the colonial] community's attempt to overcome its separation from and nostalgia for what it had left behind, but opera, which was expected to mediate between the metropole and the colony and draw them closer, ultimately reinforced Hanoi's peripheral status."[52] He explains: "The size, location, function, and opulence of the building heralded the power of empire, but its irregular use and mediocre performances in combination with a lack of Vietnamese interest discredited it as an emblem of authority."[53] In other words, rehearsing a French imperial ideology had the opposite effect, tempering that ideology. "Opera," supposedly a symbol of the metropole in the colony, was instead locally and spatially negotiated and hence became a compromised discourse. If the politics of empire in Hanoi were fraught with contradiction, the situation in 1930s and 1940s Shanghai was doubly ambiguous. In a setting that was not anyone's colony, but was contested by various foreign powers, one is compelled to ask: whose imperial ideologies and values mattered, and, ultimately, were they able to take root? If the efforts of both British and Japanese administrators rendered opera concerts exemplars of a "model" civic culture, multiple forms of expansionism also meant that, unlike the (attempted) glorification in Hanoi of Third Republic France, "empire" was neither clearly defined nor universally understood. There was no single colonial presence and, further, no single metropole to which Shanghai's foreign administrators professed loyalty.

Seen in this light, operatic geography in and of Shanghai is shaped not only by imperialisms in the plural, but also by different manifestations of foreign territoriality. Such particularities also prompt some other historical reflections, specifically regarding the ingrained epistemological distinction between imperialism and colonialism. In *Postcolonialism*, for example, Robert Young offers the view that imperialism is "typically driven by ideology from the metropolitan center and concerned with the assertion and expansion of state power."[54] Colonialism, by contrast, is principally "economically driven by migrant settler communities, speculators or trading companies and [is] concerned with more ad hoc, localized matters of territorial and economic administration."[55] In *Colonialism/Postcolonialism*, Ania Loomba initially appears to theorize imperialism and colonialism in terms of cause and effect; she is quick to point out, however, that the distinction between them is "defined differently depending on their historical mutations."[56] Opera concerts in 1930s and 1940s Shanghai certainly provide one such example of mutation, highlighting (geopolitically) the entwining of imperialism and colonialism. Imperialism and colonialism both operate as forms of intrusion, leveraging resources on forcibly claimed territory; as such, they are equally contingent upon the degree to which power is maintained on the ground. Indeed, neither treaty-port nor wartime Shanghai can be characterized simply as extensions of metropoles or colonial formations; rather, they constitute political arenas in which existing and emergent empires attempt to safeguard their interests, and are thus marked by power relations produced and negotiated within those arenas. Opera concerts in the context of multiple Shanghais—albeit an intriguing story of "Western" music-making in the East—bring to the fore these local spatialities of power, presenting a complex tale of encroachment and occupation.

TWELVE

"Come to the Mirror!": Phantoms of the Opera— Staging the City

PETER FRANKLIN

The theatre . . . assumed so important a role in public life as hardly was possible in any other city. For the Imperial theatre, the Burgtheater, was for the Viennese and for the Austrian more than a stage upon which actors enacted parts: it was the microcosm that mirrored the macrocosm, the brightly colored reflection in which the city saw itself. STEFAN ZWEIG, *THE WORLD OF YESTERDAY*, 1943[1]

Stefan Zweig's description of the peculiarly urban nature and apparently popular function of theater in late nineteenth-century Viennese society inevitably went on to include opera. One had not to walk far along the Ringstrasse from the Burgtheater to find the Austrian capital's celebrated showplace for top-quality musical theater, old and new, in August Siccard von Siccardsburg's and Eduard van der Nüll's enormous 1869 Hofoperntheater (Court Opera Theatre). But was the role of theater in public Viennese life at that time really particular to that city, as Zweig suggested, or was it exemplary of wider European cultural practice? Such a question could be addressed in terms of physical geography by mapping theater locations and the routes and routines of performers and entrepreneurs moving between them; here my focus will rather be on the sociocultural geography of the imagination, examining what was experienced by audiences around the turn of the cen-

162

tury, and specifically opera audiences.[2] Might opera across the Continent, often constructed as an irrelevant and expensive indulgence of the privileged, also have been turning into a more modern form of urban entertainment, mediating the real world for a European high culture that often sought to escape from it?

Theodor Adorno, who grew up in this period (b. 1903), could serve as a guide in exploring the tensions underlying this process. In his valuable, characteristically labyrinthine 1955 essay, "Bourgeois Opera" (revised for inclusion in his 1959 collection *Klangfiguren*), Adorno began by confirming retrospectively that opera was, in his opinion, of little direct relevance to any broader consideration of "contemporary theatre" (implicitly, therefore, of contemporary culture).[3] His reasons for doing so were illuminating. Reinforcing the essentially critical, modernist perspective from which he viewed early twentieth-century opera, Adorno considered Alban Berg to have been "the only first-rate opera composer in the new century."[4] Subscribing to the Weimar-period *Opernkrise* view of then-current popular opera as a decadent and debased form of musical drama, Adorno nevertheless believed that the medium as such, for all its apparently "peripheral and indifferent" position in contemporary culture, remained a key ("prototypical") theatrical form.[5] He located its essence in magical operas such as Mozart's *The Magic Flute* (1791) and Weber's *Der Freischütz* (1821); these he associated with later works that included Wagner's *Lohengrin* (1850) and (implicitly) *Parsifal* (1882). He could have expected his German and Austrian contemporaries to have made another comparison with one of the most popular new operas of the 1880s, Humperdinck's *Hänsel und Gretel* (1883, after the Grimms' tale)—often programmed as a "children's" fairy-tale opera, with its pantomime scene in the woods where the sleeping children are protected by fourteen angels who descend on a ladder that "reaches down from Heaven." Such works, Adorno felt, might now embarrass us, although he identified with interest the childlike qualities that resided in their marriage of music and scenic appearance. As with many other late Romantic operas, pictorial language was a key to their character, for all that we "no longer understand it"; we may ridicule the swan in *Lohengrin*, he suggests, but without it the opera becomes "pointless."[6]

While preparing the way for a skeptical reading of modern opera as a supremely urban and bourgeois form (its warhorses, such as *Il trovatore*, had "sunk into the living-room treasure-chests of the petite bourgeoisie"),[7] Adorno nevertheless tried to locate an "anti-romantic" aspect in an operatic type that preserved magic "in a world bereft of magic," thus embodying bourgeois disillusionment.[8] A more nuanced and flexible

PETER FRANKLIN

critical perspective than that typically offered on popular opera (including by Adorno himself) thus seemed still in prospect. However, frequent and intentionally derogatory references to the historical and practical connections between late Romantic opera and film led him to suspect that their comparably sudden historical invention as cultural forms—albeit at different times—and their aim to present "common knowledge to the masses" (presumably as ideology), emphasized more recent operas' increasing "similarity to the modern culture industry." After all, Adorno felt, their libretti (in their "manifold silliness") demonstrated their "affinity with the marketplace, [and] the commodity character which in fact made them placeholders for the as-yet-unborn cinema."[9] Typically, Adorno became formulaic in his Marxist conclusions, but the subtlety and historical richness of his critical perceptions and evocation of his own lived experience make him a powerful and articulate historical witness. The underlying Romantic idealism of his characterization of "the marketplace" and theatrical works of art as "commodities" is frequently surpassed by more complicatedly nuanced recollections of his immersion in the "old" nineteenth-century European cultural milieu Zweig evoked.

Adorno, Modernism, and "Popular" Opera

From a viewpoint less constrained by his ethical mythologizing of high modernism, Adorno's contradictory and even conflicted account of bourgeois opera might reinforce the relevance to Zweig's urban theatrical experience of "popular" opera—the form that Adorno understood and rejected. The pedigree of this kind of opera may be traceable less to the then still emerging bourgeois "great tradition" of high-art musical theater than to the realm of popular boulevard melodramas and operettas, and even of those eighteenth-century *intermezzi* in which "the bourgeoisie also forced its way onto the musical stage."[10] That account, alongside Adorno's linked observation that "feudal lords always allowed art to be created by members of the bourgeoisie," who typically "prefer to address their equals rather than their patrons,"[11] provides a coherent working model of what we might call popular bourgeois opera (c. 1880–1930) of the kind usually subsumed under such vaguely derogatory labels as "late Romanticism," "post-Wagnerism," or (for the Italians) verismo. The overarching category includes many works still in the repertoire and a good many once-successful operas that have now fallen out of it. Relevant composers would be Puccini, Mascagni, and Leoncavallo in

"COME TO THE MIRROR!"

Italy; and Humperdinck, Strauss, Schreker, Korngold, Pfitzner, and Schillings in Germany and Austria. Still more names might be added from France, Poland, Russia, and beyond.

The problem with late Romantic opera, derogatorily deemed both modern and "popular" in the richest period of its production—roughly 1880–1920—is thus (perhaps unintentionally) both revealed and explained by Adorno from his explicitly partisan modernist viewpoint. From Adorno's perspective, the only fully acceptable, widely performed German opera of the 1920s was Berg's *Wozzeck*, whose evident political radicalism was bolstered by its composer's left-wing connections and high-modernist credentials as Arnold Schoenberg's pupil. By contrast, the late Romantic genre (of which *Wozzeck* was in reality an extreme, tonally dissonant product) was generally associated in highbrow aesthetics with notions of debasement and a resultant lowering of "progressive" operatic taste and manners. In this way, European modernism oddly reified and relied upon these associations between late Romantic opera and supposedly debased taste as much as it sought practically to challenge and perhaps frustrate them. Popular operas' debasing attributes were typically deemed to be sensationalism, "kitsch" effects, sentimentality—all characteristics of the unchallenging, consumer-driven taste of middle- to low-brow audiences. This perspective has continued unhistorically to dominate a critical discourse aiming to erase this repertoire, or at least to restrict our understanding of it to one mode of its discursive construction in the period. The operatically late Romantic has accordingly been condemned to a cautionary or transitional role in the grand narrative of the bourgeois-shocking march of modernism.

In more recent scholarly attempts to restore (for decent historical reasons) what has been erased, the danger of course remains that we reinscribe the very values that constructed the critical binary in the first place. A model diagnosis of how the reclamation of socially functional popular culture can perpetuate such binaries is found in Simon Frith's essay "Towards an Aesthetic of Popular Music" in Richard Leppert and Susan McClary's *Music and Society* (1987), the book that effectively marked the arrival in the United Kingdom of "new" musicology. Accepting that the discipline had opened its doors to the popular musical Other, Frith noted the ideological implications of the fact that, nevertheless, the popular was still commonly subjected to contextual, "sociological" reading, while the classical was covertly celebrated by ostensibly objective "aesthetic" analysis.[12]

The cultural work done by the latter is now better understood, while the musical sociologist Peter Martin has highlighted the former's

PETER FRANKLIN

problematics.[13] Seen in this light, the sociological treatment of late Romantic opera has often relied upon a rather too conventional, and arguably unhistorical, insistence upon access to opera's being contingent upon money and "class." We might bypass the complex issue of assessing the comparative aesthetic quality of such musical theater by adopting Frith's implied desideratum that widely consumed music might best be approached as experienced (primarily by audiences of the time). The experiment might be to look at late Romantic opera without inherited presuppositions, as if it really were something like the popular, even childlike genre many of its detractors (Adorno and numerous older and more reactionary critics in Germany and Austria) often accused it of being.

Rather than cataloguing the genre's high-modernist shortcomings or fixating on its old-fashioned Marxist characterization as a quasi-medical symptom of some larger social and cultural malaise, we could start from the popular bourgeois character of late Romantic operas and seek to understand better how they were valued and utilized as "products" (even "commodities") reflecting modern metropolitan society and culture. As interpreted by audiences eager for both escape and arrival, they might in their turn (recalling Adorno's assessment of romantic opera's "anti-romantic" impulse) be seen to incite critical reflection. The "reality"they mirrored constructively negotiated the social, cultural, and ideological fears and desires of the audience and the prevailing historical circumstances in which such operas were produced. There need be no unwitting "victims" or scheming "victimizers" in this more nuanced, participatory reading.

Phantoms and Fears (Beneath and Beyond the City)

The operas with which I am concerned here are set in and to some extent thematically address the modern city, not simply as a place in which to live and work, but as one that both enables and polices activity on behalf of groups of inhabitants who felt either oppressed or disenfranchised. To adapt Hamlet's subversive notion, opera might be the thing wherein we catch the conscience of a culture, or (to take Hamlet a step further) the conscience of an historical audience that might be as suspicious of kings and their avarice for power as of those who would seek to overthrow them and disturb a broadly acceptable status quo. The efforts of those with power to secure or justify holding onto it had, of course, long provided the subject matter and plots of stage dramas, musical and otherwise. However, in focus here is the immediate post-

Wagnerian period, when court opera houses were turning into, or giving way to, city or state institutions, as they had already in France; they did so gradually in Austria and Germany, although in Vienna it was only after the First World War that court gave way to state and the Catholic-censored Hofoper became the Staatsoper. In these theaters the gradually diversifying audience was frequently reminded, overtly or covertly, of its potential political kinship with the shape-shifting operatic chorus: variously a conservative rural crowd or a threatening urban mob, alternatively instigators of control (those ever-processing or marching soldiers) or inflamed revolutionaries. Was it not also with the chorus that the bourgeoisie en masse, in Adorno's formulation, "forced its way onto the musical stage"?

Their doings, and those of their protagonist representatives (whether heroes or villains), were never more interesting than when they were represented as citizens—dwellers in modernity's metropolitan centers. The politics of the late nineteenth-century "modernizing" of major cities (Berlin, Vienna, and Paris are all revealing examples) have interested such historians and scholars as Richard Sennett, Michel de Certeau, and Michel Foucault. All have drawn attention to the need and ability of citizens to negotiate the double purpose of such modernizations as the banished encircling walls and fortifications, replaced by the liberating scenic Ringstrasse of supposedly "liberal" Vienna, its width gauged to accommodate the troops that might be needed to quell the mob's enthusiasms. And then there were the healthy, healing parks where burghers could stroll in simulacra of ordered nature, but where by night strategems, treasons, and Dionysian debauchery might be found and even relished by those same burghers, whom we encounter there, in not a few operas of the period, in compromising situations and activities.

This essay's title deliberately alludes to one specific operatic entertainment that was not really an "opera" at all, but the phantom of one: *The Phantom of the Opera*, indeed, whose 1925 film version with Lon Chaney—long before Andrew Lloyd Webber arrived on the scene—excited audiences both in its original "silent" form and its later, 1930 version, partly dubbed with sound (a version now lost). That adaptation could have been the one my own mother recalled seeing in London (or perhaps just being told about), with the spooky invitation, playfully rehearsed by her teasing brothers, of the "unseen" Phantom to his beloved singer: "Come to the Mirror!" That mirror was also a forerunner of the one in Jean Cocteau's 1949 *Orphée*, which was similarly a portal into another world. Here it was the Phantom's world, below the grandiose Paris Opéra, with its subterranean lake, passageways, and chambers, that seemed part of a compressed version of the city. Or perhaps it was an expanded version of the bourgeois home, one of the period's favored metaphorical images of the modern psyche: beneath the comfortable and elegant upper rooms lay a dark, subterranean basement of libidinous desires and dreams.[14] In the Phantom's case such desires and dreams were articulated by music, played on his private pipe organ, which represents a demonic, Lisztian inversion of that instrument's normally "sacred" role—one that would, indeed, become institutionalized in the great Wurlitzer organs that rose out of and sank back into the depths beneath the stage of major movie theaters by the 1920s.

Few fin-de-siècle texts better exemplify the relationship between what went on in dark, subterranean spaces and changing perceptions of both the psyche and the city than a central passage in Hugo von Hofmannsthal's proto-expressionist ventriloquism of the inner crisis that would lead to his abandonment of lyric poetry, some years before he began collaborating, as opera librettist, with Richard Strauss. This is the short piece of 1902 simply called "Ein Brief," but usually known as the "Letter of Lord Chandos" (by whom it was ostensibly addressed to Francis Bacon in 1603). Beneath the scholarly, historical English mask, Hofmannsthal explored a crisis in his ability to rank and differentiate fields of experience in ways that usually facilitated the distinction between self and other and enabled him to make secure intellectual and descriptive evaluations. The world had now, however, become "one great unity": "The mental world did not seem to me to be opposed to the physical, likewise the courtly and the bestial, art and barbarism, solitude and society. . . . And in all of nature I felt myself."[15] Before expressing his lack of an available language in which to communicate his new vision, he gives examples of the moral dilemma into which he finds himself pitched, the most memorable of which relate directly to his experience of, and in, natural and architectural spaces, including that of the milk cellar of one of his estate farms, where he had had rat poison spread. He becomes obsessed with his mental image of the imprisoned rats' violent death struggles—a vision he compares with Livy's account of the residents of Alba Longa as they await the destruction of their city:

The cool and musty cellar air, full of the sharp, sweetish smell of the poison, and the shrilling of the death cries echoing against mildewed walls. Those convulsed clumps of powerlessness, those desperations colliding with one another in confusion. The frantic search for ways to get out. The cold glares of fury when two meet at a blocked crevice. But why am I searching again for words, which I have sworn off! My friend, do you remember Livy's wonderful description of the hours before the destruction of Alba Longa? The people wandering through the streets that they will never see again . . .[16]

Here, like the cellar full of dying rats, the city explicitly becomes an extreme prophetic site of mass destruction.

In many dramatic narratives and operas of the period, the wider city is revealed as an extended but bounded site of shared experience that may indeed prove dystopian and terrifying, as in the exotic, unnamed Japanese city in which Mascagni's Iris, in the eponymous 1898 opera, would be abducted as a sex slave, to escape and be denounced by her blind father (he assumes she had chosen her path) before dying in a sewer. Puccini's one-act *Il tabarro* (1918) would be set on a barge moored at a working quay in Paris. Stevedores, match sellers, and organ grinders animate the background to a melodrama in which the barge owner Michele murders his much younger wife's lover; she yearns only for the utopian Parisian suburbs in which she had grown up. In fact, the utopian city, in a period dominated by verismo and popular melodrama's "low-life" characters, was much more rarely encountered outside Viennese operetta, in spite of the model provided by Wagner's 1868 Romantic-nationalist *Die Meistersinger von Nürnberg*. More commonly, the dystopian city might shrink to the Symbolist confines of a gloomy castle, like that of Maeterlinck's "Allemonde" in Debussy's 1902 *Pelléas et Mélisande*, or Bartók's *Bluebeard's Castle*, behind whose doors ("forbidden" to the titular character's new wife, Judith) lie treasure and torture chambers, landscapes and lakes of tears. These represent the wider domain of Bluebeard's power, typically signifying submission and/or death to his wives and those other women who are, in every sense, the frequent subjects of these dramas.

It will be clear from this account that operatic music and subject matter of the decades around 1900 responded to changing constructions of the city in ways later relished by the sound cinema. Often the stylistic range and location of noises and music "naturalistically" evoke the experience of passing through or around urban spaces, confines and structures, like the organ grinder's out-of-tune waltz and the ship's siren heard in *Il tabarro*. Still, it is the psyche of the modern subject that is most often represented, and repressed, by the city in these operas. In Korngold's *Die tote Stadt* of 1920, Bruges by night is explicitly a dream world beyond the nostalgic and claustrophobic private space of Paul's apartment. This dream city is not only filled with operatic music, but also inhabited phantasmagorically by operatic characters and singers; in it the protagonist will commit a murder (or appear to, since the whole opera is framed as a dream).

Viewed as a proto-popular genre, processing and revealing contemporary anxieties and aspirations, opera has a theatrical richness that embraces not only the technical devices that anticipated the cinema, but also the ideological complexities of its penchant for revelatory spectacle (reduced by Adorno to childlike "magic" tricks). Among these devices might be numbered Wagner's elaborately engineered "cross-fade" scene changes in *Das Rheingold*, or the scrolling scenery to suggest movement (with accompanying underscore) in *Parsifal*, where it leads the protagonist, like a novice operagoer, toward the site of ritualized revelation— the kind of thing in which the grandest of more conventional "grand operas" might once have been assumed to deal. The Hall of Monsalvat is also, of course, the location of temporal power, musically manifested. These are, after all, arms-bearing and purposefully marching Grail knights, for all their quasi-monastic celibacy. Not for nothing has Wagner been invoked as a source of musical and dramatic features for George Lucas's 1977 *Star Wars*, with its imperial storm troopers and even more numerous revolutionaries gathered in the great temple of fame at the triumphant, musically resounding end.

Other early twentieth-century operas more obviously accommodated the darker shadows of Wagner's National Socialism–inflected mid-twentieth-century reception, even in Italy. Such works moved from the display of exotically theatricalized societies (often nightmarishly dystopian—as in the two and a half acts of Puccini's posthumous 1926 *Turandot*, up to Liù's death) to an ostensibly more "modern" type in which we encounter a discursive problematization of urban space as both ceremonial stage and labyrinth of routes and places in which power operates, as in Puccini's much earlier *Tosca* (1900). La Tosca herself, a singer playing a singer, may have dreamed, as would Giorgetta in *Il tabarro*, of her secluded villa in a garden suburb with Cavaradossi, but Scarpia and his police need those interconnecting central urban spaces and private places from and in which to gain intelligence, exercise influence, and satisfy their desires—often simultaneously.

Still other operas functioned more specifically as the settings for renewed idealistic and Romantic desires to escape the city's bounds into nature's parks, woods, and mountains. That kind of escape is richly thematized and problematized in the relative distancing of Alviano Salvago's fantasy island, "Elysium," from the city of Genoa in Franz Schreker's lavish 1918 Renaissance spectacle *Die Gezeichneten*. The opera's Frankfurt premiere was attended by the youthful Theodor Adorno, filled with media-hyped anticipation (in journalistic essays, interviews with the composer, and so on).[17] Schreker's own libretto presented the ancient city unequivocally as a site of class tension (between a decadent aristocracy and the citizens represented by a burgher Podestà, or mayor). This tension was melodramatically heightened by young aristocrats' abduction of and sexual predation on girls from Genoa, utilizing, unbeknownst to its crippled patron and creator Duke Alviano Salvago, the grottoes of the island "Elysium." His dissolute young friends (aristocratic thugs) fail to understand his theme-park creation's vision and intention, as do the townsfolk to whom Alviano gifts it. The preludial business to the long third act is a fine example of Schreker's proto-cinematic art, with Genoa being distantly visible from Elysium across a stretch of water dotted with small, lighted boats, ferrying across the invited audience of townsfolk who are to be the variously reluctant and bemused participant-spectators in Alviano's intended aesthetic utopia. Here the chorus represents "us," the audience, as a motley group of locals, admitted to Elysium's artistic theme park by aristocratic benefaction—but without the kind of educational program that might have prepared them to see "Art" in place of near pornographic display, as in the impressively staged but misleading festival procession. Looking like one of Hans Makart's Viennese spectaculars, this presented a classical allegory of "Culture" as a site where the self-elected bourgeois Artist may hitch a lift in Apollo's chariot and evade the questionably affiliated Dionysian rioters and celebrant Maenads. Beneath their feet, in underground grottoes and chambers that might recall Hofmannthal's cellar with its dying rats, Alviano's out-of-control young friends are indulging in orgies that will cause the death of his beloved Carlotta and his own consequent derangement as the curtain falls. The police and many of the locals had been led down into the grottoes by Alviano to see for themselves the shabby and shocking mess of the outwardly repressed hedonistic indulgence that lies, quite literally, beneath the surface of "art."

Operatic Points of View

Late Romantic operas construct their audiences in often complex ways. They can also be as open to critical reflection upon their own problematics as any of modernism's officially canonized masterpieces. Turning Adorno's observations about the link between Romantic opera and the cinema to additional advantage, we might adopt an approach similar to that used in discussion of the cinematic "point of view": instances where acts of looking and seeing may be highlighted as part of a film's narrative technique. Such an approach is relevant not least to the modern trope of

PETER FRANKLIN

the urban landscape as something experienced by a solitary watcher or someone walking through a great city's streets—often an outsider, someone from elsewhere, and frequently lacking social or economic power.[18]

It might therefore seem odd to turn here to Strauss's *Der Rosenkavalier* (1911), often regarded as a sort of Mozartean pastiche served up with lashings of late Romantic whipped cream and conservative nostalgia and infrequently marked by critical self-awareness. This viewpoint sorely underestimates the subtlety of Hofmannsthal and Count Harry Kessler's libretto, nostalgic and conservative as both men undoubtedly were. The subtlety is readily apparent in the second act, set in the home of the nouveau riche and recently ennobled widower Faninal, whose overawed daughter, Sophie, is shortly to be the object of an invented aristocratic ritual of courtship: the presentation of a silver rose by the representative and herald of her suitor, the vulgarly dreadful old Baron Ochs. Neither vulgar nor dreadful, however, is the handsome Octavian—the young Count Rofrano—who is the travesty-role rose-bearer. Specifically relevant here is the way in which the Faninal home, grand in the nouveau riche's somewhat over-the-top style, is located in its urban setting, as the focus of a mode of social theater which includes the public space of the street outside. We don't see the external theatricals' beginnings, but they *are* seen and described excitedly by Sophie's duenna, Marianne, employed partly to stand in for her deceased mother. She is glued to the window intended to be visible to the right of a central door and gives an excited running commentary as Octavian and his retinue arrive outside. Her excitement is inflamed by the fact the she can see that *others are also seeing it*, both in the street and from other windows—a grand and, as it were, licensed version of suburban curtain twitching. This piece of urban social theater is meant to be public: it starts as an ostentatious procession to the house, accruing considerable cultural capital for the family who will, with us, become its proud and privileged "private" audience. We are thus perhaps willingly manipulated into Sophie's point of view and are ready to join with her in exclaiming "It's so *beautiful!*" as the visual and musical richness of the presentation scene unfolds, following Octavian's glitteringly spectacular entry. This, at the very least, is grand opera thematically presenting and defining itself in terms of urban spectacle and privilege of a kind that really might look "heavenly" ("ein Gruss vom Himmel") to those not quite of the class to have experienced such things themselves.

The elements of a thematized critical discourse about the nature of art and specifically opera in the urban environment, are, as I have suggested, found in other late Romantic works of the period normally excluded from the category of modernism. We need only return to *Tosca* in Brian Large's 1992 version, filmed in the "actual settings" in Rome and at the appropriate times of the day, which was, in its way, rather wonderful and emphasized how here, too, the city was as much protagonist as mere setting—or rather that the acts' urban locations needed to be understood as linking and intersecting the city's and the drama's private and public realms.[19] Floria Tosca's performances as a singer are sponsored primarily as adornments to the intrigues of power. At their center the decidedly powerful police chief, Baron Scarpia, acts as both entrepreneur and audience of the grand operatic drama of his desires—of which his own death at Tosca's hand is an entirely unintended consequence.

One of the locations of Scarpia's explicitly "operatic" pieces of staged crowd and news management is striking. In act 1 we see a great church, Sant' Andrea della Valle: a public space whose more secluded corners and recesses may be utilized for varied purposes. At first the purpose is art: the area near the Attavanti Chapel is where Cavaradossi is completing his painting of the Virgin, whose assumed model provokes Tosca's jealous suspicion. Before Cavaradossi enters to take up his work, the locked chapel itself is seen as a hiding place for the fugitive rebel Angelotti, whose sister has hidden a key so that he might enter and conceal himself. Meanwhile a large church's daily business passes around and through the space, the central nave becoming, in the last scene of the act, the stage for a great religious procession. This initiates the public celebratory service, with the Te Deum, that Scarpia had instigated: in reality it is pure propaganda (there had been no victory over Napoleon at all). The grand-operatic religious spectacle relies upon what in cinematic terms we would call the diegetic music of the onstage processional and Te Deum, with its grandiose obbligato punctuation by offstage cannons. To all this Scarpia provides a "private" foregrounded monologue, which reveals the duplicity of it all as he plans to recapture Angelotti and achieve his conquest of Tosca by torturing Cavaradossi. It is of course this action that provides the shock content of act 2 in this opera, which is otherwise neither "shabby" nor "little" (to recall the late Joseph Kerman).[20] Tosca's private performance for the murderous Scarpia takes place in his Farnese palace apartment, through an open window of which we have already heard her singing a "victory" cantata offstage at the banquet. Faced with Cavaradossi's torture and the threat of his death, Tosca's aria "Vissi d'arte" (I lived for art, I lived for love), sung for Scarpia alone (in whose plot we become momentarily complicit), expresses anguish that her art of sacred song must lament art's inoperancy, or what Adorno might have considered its inevitable failure.

Cities are, as already noted, places where music and other sounds are located spatially, inside or outside the stage setting. The tragedy of *Tosca*'s third act famously takes place in sight of the heart of the city, on the Castel Sant' Angelo's parapet, whence the bells of Rome's churches, calling the faithful to matins, are heard from near and far in elaborately naturalistic imitation, along with the song of a passing shepherd boy evoking the surrounding countryside.[21] That pastoral arcadia is one from which modern urban political prisoners are terminally banished. We should remember that at the opera's premiere *in* Rome in 1900, attended by Queen Margherita and members of the government, the theater was at the center of a real-life political drama in which a demonstration or what we would now call a terrorist attack was indeed feared (the actual assassination of King Umberto took place only weeks later).

The Roman locatedness of *Tosca* might appear exceptional in its political character and specificity, but as I have suggested, operas by Franz Schreker featured similarly sensational plots that made much of audience-like crowds of citizens being both manipulated by art, specifically music, and at times rebelling against its designs upon them. I have already referred to the Renaissance Genoese setting of *Die Gezeichneten*, but Schreker was fascinated by the necessarily, and prosaically, urban location of theaters able to present the "Romantic" art of such lofty idealists as Romantic artists were believed to be. This he explored perhaps most potently in his first staged opera, *Der ferne Klang* (The Distant Sound, 1912). It is often compared to Charpentier's 1898 *Louise* (and we should not forget Puccini's *La bohème* [1896], that other great Parisian "city" opera). But *Der ferne Klang* is in a class of its own with respect to the way its final act thematizes the location of opera, indeed *this* opera, in its urban space, creating a most elaborate effect of *mise en abyme* while also contemplating dramatically the effect of such opera on the notionally "ordinary," malleable audience (see fig. 12.1). Among its tightly packed and architecturally distanced members (far up there in the fourth gallery, say) might be found its most truly and, in their way, sophisticatedly sympathetic audience.

In light of Adorno's description of this highly successful opera as fit only for "maidservants," the opera's own focalized audience member is appropriately female: a prostitute, Grete.[22] She had been left behind by her aspiring composer boyfriend Fritz in act 1; she tried to follow him but was picked up by an old procuress who turns her into the main attraction of a rather classy bordello-*cum*-nightclub in the Venetian La-

12.1 Franz Schreker, *Der ferne Klang*, Frankfurt, 1912, act 3, scene 2, set design by Alfred Roller. Grete (Lisbeth Sellin) can be seen lying in the foreground (left), having been brought from the nearby opera house (visible in the background). By permission of the Franz Schreker Foundation.

goon (in some respects the precursor of Alviano Salvago's "Elysium"). Eventually corrupted by modern life and the city to which she had returned, the former simple country girl finds herself viewing her old flame's opera from what we take to be the cheap *Stehplätze* (standing places) of the Vienna Court Opera, whose bulky side elevation was alluded to in Alfred Roller's original set design for the Frankfurt premiere (although composed in Vienna, *Der ferne Klang* was in fact never performed there in Schreker's lifetime). Seeing the opera about her, the opera of her life—the opera we have been watching and of which we now hear recognizable snatches from the nearby theater—Grete has become emotionally involved and moved to the point where she has fainted and been brought out to the nearby *Theaterbeisl*, or street café, where we have witnessed the posh and sometimes scornful talk of singers, critics, and snobs discussing the new opera's apparent failings. The emblematically sentimental and overwrought female convalescent fits to a T the type of the modernist's or Brechtian's mocked "identifier" with the characters of bourgeois opera.[23] Yet Schreker does not mock his heroine, who is presented in a sense as the very person for whom the opera was written and who truly understands it in all its fallenness and self-deluding qualities. Even Fritz, its onstage composer, will confess to such delusions, and to the opera's failure, in the final scene, in which he and Grete are re-united but where he, the famous male author-composer, will literally die in the arms of this naturalistically realized descendant of a "redeeming" Wagnerian heroine. Grete is quite literally left holding the dead baby, in a conclusion of veristic bleakness.

The operatic subject as part of a crowd or mob is here also the powerless but intensely feeling individual who may be both spectator and victim of the modern city and the art forms, opera not least, in which she is figured. This is something rather different from the overdeterminedly privileged viewpoint of Thomas Mann's Siegmund and Sieglinde Aarenhold, whose attendance at a performance of *Die Walküre* is described in his powerful if richly problematic 1905 story "Blood of the Walsungs":

They were in the city's heart. Lights flew past behind the curtains. Their horses' hoofs rhythmically beat the ground, the carriage swayed noiselessly over the pavement, and round them roared and shrieked and thundered the machinery of urban life. Quite safe and shut away they sat among the wadded brown silk cushions, hand in hand. The carriage drew up and stopped. Wendelin was at the door to help them out. A little group of grey-faced shivering folk stood in the brilliance of the arc-lights and followed them with hostile glances as they passed through the lobby.[24]

Grete was more likely to have been found in that group of "grey-faced shivering folk," but her exclusion from the world of power and privilege did not in fact exclude her from the gallery, or the standing room. From there, too, the future Viennese music critic Max Graf would rejoice, in the Prague of his childhood, at Prague's first production of Wagner's *Ring* cycle in 1885, under the baton of the young Gustav Mahler and with the original Bayreuth sets and costumes:

Already by midday I was hanging around the old theatre where, that evening, the Rhine-maidens would be swimming, the gold gleaming amongst the waves, the storm-clouds gathering, the rainbow shining forth—for it was the scenic wonders that most strongly gripped our imagination and heightened the tension of our anticipation. How happy I was when the swimming machine of the Rhine-maidens chanced to be drawn past, clumsy and heavy, with its swings on iron poles. This would nevertheless bring to life our fantasy picture.[25]

His recollection of standing closely packed with his friends in the theater that evening was one of total immersion and captivation that "never faded from my memory, even after the passing of many years."[26]

Not even his sight of the much-derided 1876 technology uncertainly facilitating that immersion could generate a subsequently modish alienation from this experience, all aspects of which his parents' generation roundly mocked (his conclusion was that "we were already aroused by the opposition of the adults to become Wagnerians").[27]

We glimpse here what Adorno, in spite of himself, would perhaps have recognized as an echo of that "essential" operatic experience, with all its childlike pictorial fantasy, whose loss his 1955 essay entitled "Bourgeois Opera" concluded by discreetly if dialectically lamenting, without making more of his own suggestion that this kind of musical-theatrical experience would indeed be appropriated and developed by the less than wholly new medium of cinema.[28] Yet that experience was still available before the First World War, when recent or new operas continued to feature significantly in the repertoire of many larger urban opera houses.[29] If my own aim here is colored in part by a restorative impulse, it would not be to advocate Large's site- and time-of-day-specific *Tosca* or the famous 1998 production of *Turandot* in Beijing's Forbidden City.[30] But even to think about the possibility of occasional attempts at "historical," or naturalistic, productions would be to conceive that Lloyd Webber's *Phantom of the Opera* might not remain the only means to see anything approaching a convincing evocation of the visual, scenic character of a nineteenth-century opera—appropriate as the low- to middlebrow context might be (that musical has for long been a staple of contemporary London's theater land).

For the other trajectory taken by popular late Romantic operatic and symphonic music one would, of course, have to return to the cinema— and not least to films where the urban spaces explored by late Romantic opera become the backdrop to performances in high-culture theaters and concert halls. The music heard there is then carried out into *noir* nocturnal streets and railway stations and into the homes of middle-class heroines with radios and gramophones, where it forms the passionate underscore to the "real" urban life that seems so often hostile to such passion, its soundscapes crowded, perhaps, with the noise of traffic, advertising jingles, marching bands, or the tense private silences of jealous lovers.[31]

THIRTEEN

Open-Air Opera and Southern French Difference at the Turn of the Twentieth Century

KATHARINE ELLIS

Mention of opera performed in ancient outdoor spaces most likely conjures up images of the Verona arena or the Terme di Caracalla in Rome, which present summer opera festivals originally dating from 1913 and 1937 respectively. These enduring meeting points of culture, tourism, national memory, and myth creation form a belated Italian analogue to the French use of such sites as protagonists in expressions of republican universalism, *latinité*, and local identity or celebration from the 1860s onward. In this essay I examine open-air opera and other forms of theater music in southern France (Provence and the Languedoc). I probe the physical, visual, sonorous, and social specificity of performances of preexisting and new music in spaces filled with tens of thousands of spectators, and I explore the nature and implications of their success as site-specific experiences.

Open-air opera was unusual in France in this period because there was no system underpinning it. The phenomenon was catalyzed by a single instance of state subsidy (at the Théâtre Antique, Orange), but it emerged on the back of funding from wealthy individuals and interested munic-

178

ipalities, as free-market events run by local managers who had hitherto spent their careers promoting sport, or (from at least 1910) as bought-in packages from specialist providers.[1] Two artistic entities could have encouraged coalescence into a unified regionalist movement, bringing activity within an institutional framework: for theater, the Félibrige, promoters of Occitan languages and culture in the Midi; and for opera, the avowedly "regionalist" (but Paris-based) Schola Cantorum. Neither did. The phenomenon of open-air opera was not, then, the result of a concerted regionalist campaign. It was owned by everyone and no one, resulting in developments as unpredictable as their modes of local expression were varied and generating a whole range of regionalist tensions both locally and with Paris.[2] At the highest-profile venues (Orange and Béziers), new commissions upset the traditionally Paris-centered structures of French lyric theater, while the widespread reframing of standard operatic repertoire in outdoor environments brought new relationships between audience and work.

The *théâtre en plein air* Phenomenon

The roots of the *théâtre en plein air* phenomenon lie in the *félibre*-inspired but government-sponsored restoration of the ancient theater at Orange, starting with a *fête romaine* featuring Étienne Méhul's biblical *opéra comique Joseph* (1807) on 21 August 1869. Subsequent events at Orange in 1874 (Bellini, *Norma*; Adam, *Le châlet*; and Massé, *Galathée*), 1886 (plays by Alexis Mouzin and Molière), and 1888 (Rossini, *Moïse*) led to a near-annual festival of plays with incidental music from the mid-1890s. Regular opera—inaugurating what are still known as *chorégies*—started with Gluck's *Iphigénie en Tauride* in 1900. Had funding not proven problematic, they would have continued in 1901 with Saint-Saëns's *Les barbares*, a new commission with a plot set in that self-same Orange theater.

Further south, the Béziers wine merchant Fernand Castelbon de Beauxhostes contributed financially to, rented, and adapted the town's brand-new bullfighting arena for the first of a series of important commissions, *Déjanire*, on 28 August 1898, by the avowedly centralist Saint-Saëns. Through the Republican Beauxhostes, Béziers would become nationally renowned for its spectacular (though generically indeterminate) new works. Involving a double cast of actors (principals) and singers (secondary characters, choruses), it encompassed drama, melodrama, opera, oratorio, dance, and historical pageant, the whole within high-rise, walk-through sets with multiple receding horizons, and with occasional

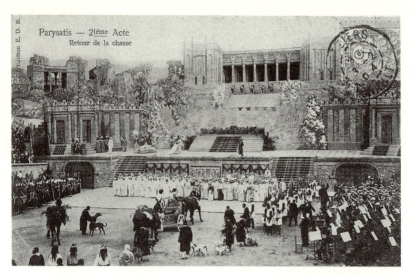

13.1 Postcard of the Arènes de Béziers during a performance of Dieulafoy/Saint-Saëns, *Parysatis* (1902): "The Return of the Hunt." Photographer unknown. Author's collection. This image shows a performance from the 1903 revival, in which greyhounds and pack dogs have been added to the horses we find in postcards illustrating this scene at the premiere.

use of the arena as processional gathering place for horse-drawn chariots, dogs, and sundry extras (see fig. 13.1). Moreover, at Béziers, a sense of what might later have been termed a "happening" was amplified both by avoidance of the repertoire concept (one work per year, with very limited revivals), and by Beauxhostes' reluctance to let "his" works travel—save to Paris. Béziers works, then, aroused huge regional and Parisian interest; but most of them lived, like butterflies, for a few days, and died where they were born.

Two main trajectories are evident: ancient classicism and localism. The ancient classicism of Roman ruins, or (at Béziers) the creation of a Roman-style space, and adherence to a shared sense of *latinité*, ran in parallel with localism rooted in the specifically French traditions and history of the Midi. Most of the time *latinité* dominated, creating a sense of place that was local, Greco-Latin, and French all at once. Classical plays in French translation (Euripides, Sophocles), classical French plays (Racine, Corneille), new plays on classical subjects, and operas based on mythological or Roman themes presented audiences of thousands with centerpieces for local summer festivals. Biblical subjects, especially from the Old Testament, were subsumed into the same category. However, as Béziers shows, the artistic register could vary widely. As at Orange,

Beauxhostes focused on ancient European or Near Eastern subjects, beginning with Greek myth (Gallet/Saint-Saëns, *Déjanire* [1898–99]; Hérold and Lorrain/Fauré, *Prométhée* [1900–1901]; and Dieulafoy/Saint-Saëns, *Parysatis* [1902–3]), occasionally resorting to historical repertoire (Gluck, *Armide* [1904], and Spontini, *La vestale* [1906]), and introducing overt regionalism into plot lines apparently only after local pressure (Hérold/Lévadé, *Les hérétiques* [1905] and Népoty/Rabaud, *Le premier glaive* [1908]).[3] But progressively he and his temporary successor Joseph Charry played to the gallery by yoking ancient classicism to sensational, gory plots about murderous queens (Magre/Gailhard, *La fille du soleil* [1909]), decadent emperors (Sicard/Séverac, *Héliogabale* [1910]), or tyrants and slave rebellions (1911 Payen/Kunc, *Les esclaves* [1910]).

In all its forms, open-air theater had a populist element; indeed, it swiftly became implicated in nationwide arguments about *théâtres populaires*. The Midi's audiences were of dizzying size: up to 25,000 (Nîmes). Here, it was claimed, was not just "the public," but "the people"—underscored by the fact that Nîmes, Arles, and Béziers had arenas big enough to accommodate around 33 percent of their population, while Orange could accommodate an astonishing 80 percent (based on 1901 census figures). Admittedly, top ticket prices at Béziers outstripped those at the Paris Opéra; but in addition to low-cost performances (*représentations populaires*), there were always thousands of cheap seats in the upper terraces, and the extension of local traditions from sports to opera opened the gates to the widest possible social mix. Was the popular being grafted onto opera, or was opera being grafted onto popular, venue-based entertainment? At Nîmes, for instance, the traditional free entry to witness the death of the last bull of the day was translated into free admission for the last act of operatic performances. The arena manager, Arthur Fayot, reportedly dared not do otherwise.[4]

As the open-air vogue spread in the early 1900s, the environments became more diverse: not just Roman monuments, but medieval castles, woodland clearings, valleys, river islands, and country parks. There were *théâtres de la nature* and *théâtres de verdure*, and a wag writing for *La cigale* in 1907 even suggested a new category of *théâtres de la mer*, implying a theater not beside the sea, but actually *in* it.[5] By the 1920s around fifty such venues ran regularly, spread across southern France and now, due to their immense popularity, stretching northward. Most did plays, including new works; many did opera (mostly standard repertoire dating from after 1850). The city of Paris was not well served. By 1907 there was a Théâtre Antique de la Nature at Champigny, twenty kilometers east of the city, and by 1909 the Pré Catelan in the Bois de Boulogne

13.2   The Arènes de Lutèce, Paris, in 1912. Bibliothèque nationale de France. The scene is notable for the tenement buildings, the undersized classical columns, and the rain. http://gallica.bnf.fr/ark:/12148/btv1b69182009?rk=64378;0.

had a small *théâtre de verdure*. But there was nothing remotely comparable to the great Provençal sites until the Arènes de Lutèce were fully excavated after World War I (see fig. 13.2). Even then, the Arènes, which are hedged around with tenement buildings in what became Paris's fifth arrondissement, approach neither the grandeur of Orange's theater nor the sheer scale of the Midi's Roman or Roman-inspired amphitheaters.

Resisting Centralism: Environmental Authenticity

What, then, were the ingredients that made open-air opera and theater music belong to the Midi? Separately the most common ingredients did not seem unique: Latin subject matter, a local backdrop, and an open sky. Classical subject matter was after all a double-edged sword. Because every regime from Louis XIV onward had appropriated it as symbolic of French universality, it was both "official" and "national"; but within Roman ruins the synergy of artwork and backdrop added an extra dimension of Latin authenticity and rootedness that only the Midi, with its unparalleled density of such ancient sites, could provide. The combination was intoxicatingly site-specific. At Orange works were selected to exploit the Théâtre Antique's famous acoustic Wall—the largest and best-preserved in France—enabling Provence to demonstrate, though performances collapsing the present into the ancient past, that it was the cradle of French civilization. Provence was not only a confluence of Greek and Roman influences, but the first part of France to be Christianized, which justified the selection of biblical and Near Eastern works with no obvious classical content. The *Iphigénie et Tauride* of 1900 seems (after a twelve-year operatic famine) to have convinced the *félibre* local organizer Paul Mariéton of the genre's merits as a festival centerpiece, specifically because it seemed to key opera to this unique and ancient environment. He called Gluck's work the "ideal spectacle" (*spectacle idéal*) for Orange, its music marvelously appropriate "to the proportions of the sublime Wall" (*aux proportions du Mur sublime*).[6]

But content mattered, too. A widespread critical demand for scenic realism (part of a long French tradition of *vérisimilitude*) meant that entire genres were quickly dismissed as inappropriate to the grandeur of Roman ruins: comedy, historical opera dealing with anything after the Dark Ages, and opéra comique.[7] Hence the procession of Gluck *tragédies-lyriques* at Orange, its *Troyens à Carthage* and *Norma*; and by extension, hence Gluck's *Armide* and Spontini's *Vestale* as the only "old" works done at the Roman-style Béziers arena. The Gothic would have been especially incongruous: just as *Samson et Dalila* seemed perverse at the fairy-tale medieval fortress of Carcassonne, so medieval costume seemed nonsensical at Orange.[8] As the example of *Les barbares* for Orange shows, the keying of plot to performance site could become intensely local. It also extended beyond ancient classical ruins: in 1911 Jean Poueigh favored doing Gounod's Provençal opera *Mireille* "in the middle of the Camargue" (*en plein Camargue*), where he thought the "vast horizons of the Rhône estuary" would make "marvelous backdrops" (*vastes horizons de l'estuaire du Rhône seront des décors merveilleux*).[9] Nevertheless, even the most authentic backdrop could have its impact undercut: if architectural remains were meager, authenticity also depended where one looked and what one filtered out. The Théâtre Antique at Arles—of which little is evident save for two columns—prompted the academic and *félibre* Émile Ripert to write in 1925 that "the essence of the set is formed by all the houses that surround it and which seem to await the appearance of Joanna of Naples or of King René [of Anjou], rather than those of the Furies or of Nero."[10] And to destroy audience suspension of disbelief one needed only to point to the unavoidable sonic intrusions of modern life, including the bells and whistles of public transport.[11]

Despite such precariousness, authenticity and rootedness remained non-negotiable because they were so closely intertwined with ideas of identity, especially in the Midi, where a love of outdoor life was touted, often poetically, as a collective sign of meridional spirit. In 1901, for example, Louis Reboul referred to the Béziers audience as "a Latin people, people of the open air, of life outdoors" (*race latine, race de plein air, de vie extérieure*), as if the outdoors provided their defining characteristic.[12] It is hardly surprising, then, that sky and weather loomed large in critiques of open-air stage music. Wind (the famous mistral) was an occupational hazard, but the sky was both roof and horizon, backdrop and protagonist, and paradoxically there are frequent references to it as a *velarium*—the awning that had kept ancient Roman elites cool and dry.[13] If it was not in the scene painter's job description to engineer a seamless transition from canvas backdrop to natural environment, critical responses suggest it was nevertheless expected. The backdrops and walk-through scenery at Béziers, especially, prompted gasps at the effects of trompe l'œil on several levels. Saint-Saëns recalled that locals in 1898 wondered why they had never before noticed the far-off mountains— part of Marcel Jambon's set design.[14] Two years later, for *Prométhée*, Jambon surpassed himself with stack upon stack of triangular rocks reaching the summit of the arena, where, in Gustave Larroumet's words, "they merge with those which form the real horizon, so that the eye cannot make out where the painted canvas ends and where the blue line of the Cévennes begins" (see fig. 13.3).[15] Larroumet, who had traveled from Paris as theater critic for *Le temps*, had seen nothing like it.

Almost invariably the sun conferred on the proceedings the limpidity of light that painters such as Cézanne in Aix or Van Gogh in Arles were intent to capture on canvas. Gray skies were unacceptable: if the sun misbehaved, as happened at the beginning of a festival *Mireille* in Arles (1899), it was "an irredeemable accident" (*un accident irréparable*) for the locals,[16] instantly removing their proud sense of meridional distance from the foggy north. Sun and sky, however, were more important as protagonists. While Orange and Nîmes tended to schedule evening performances under electric lights, Béziers retained an afternoon slot seemingly timed to allow sunset and dramatic peroration to fuse. Arles did likewise for *Mireille*, where the sunstroke that afflicted Mistral's heroine in the Crau desert claimed her young life at sunset. More symbolic was the denouement that Émile Sicard and Déodat de Séverac engineered for Béziers' *Héliogabale* (1910), where the decadent emperor's death was marked by a hymn to the setting sun. And beyond brute conjunctions of the end of life and diurnal rhythm was the way the changing quality

13.3   Hérold/Fauré, *Prométhée* (Béziers, 1900), act 3, "Pandore supplie Prométhée d'implorer Zeus, mais Prométhée reste inébranlable." Bibliothèque de l'Opéra, Bibliothèque nationale de France. This detail of the set differs from the classic postcard images of Béziers productions in being taken at 45 degrees to the stage, emphasizing the usable depth and height and showing clearly the effect of receding horizons in Jambon's design. It was probably taken in rehearsal: most of the harpists are standing with their backs to the conductor, watching the singers.

of light slowly transformed performance spaces. For Pierre Lalo, music critic for *Le temps*, nature conjured a suitable closing spotlight for *Prométhée* in 1900:

The drama nears its end. Prometheus is fixed on the highest rock and Pandora weeps at his feet. The sun sinks, and the shadows creep up slowly, like water, within the crucible of the arena. Arena and stage are now both fully submerged. Only the rock still rises above, tinted golden and fiery by the low rays of the day's end, and the group of the Titan and his lover flares in the tragic clarity of the sunset. In turn, this supreme clarity fades; the shadows have overtaken everything. And the spectacle is complete.[17]

The *Prométhée* of 1901 saw the skies behave even better, offering a "supernatural" experience to the usually witty *Figaro* critic, Alfred de Mortier. Not only did well-timed rumbles of thunder seem to reflect Zeus's displeasure at the theft of his fire, but a large bird of prey began circling high above the arena, at whose highest point lay the now defenseless

KATHARINE ELLIS

Prometheus. "These are fortuitous juxtapositions," he wrote, "but they are no less prodigious for that, and relegate to an inferior level all the artificial effects that a closed theater has ever achieved."[18]

The ne plus ultra of contextual realism in opera was the introduction of bullfighting into a Latin work on a distinctly modern subject: *Carmen*. As Sabine Teulon-Lardic has noted, this combination of popular gladiatorial sport and Parisian art was a jarring one. It was underpinned by a fierce localism in the form of protests against the French government's 1894 ban on Spanish bullfighting (*corrida de la muerte*).[19] The first attempt to bring the two together appears to have taken place in the Toulouse Arène des Amidonniers in the summer of 1899, just three years after the Spanish bullfighting ban was lifted in 1896. Thereafter it spread to Nîmes, where it became a fairly regular occurrence despite purist condemnation from musicians and the bullfighting fraternity alike.[20] In Toulouse, picadors and toreadors processed into the arena in parallel with the performers' onstage commentary on these same events (act 4), and opera and bullfight vied for the audience's attention. It is a moot point whether the audience/crowd was there primarily for the bullfight or the opera, and what proportion was equally attracted by both. The opera, however, seems to have been abandoned soon after "Escamillo" entered the arena. The singers went to change out of their costumes; chairs were brought out so the cast could watch in comfort from the stage; the orchestra looked on from the upper tiers where they had been safely placed. Carmen and Don José sang nothing of their final confrontation.[21] By contrast, it appears that in Nîmes they intended to keep opera and bullfight in play together;[22] the pictorial evidence, however, suggests that either they failed, or (more likely) that on the occasion in the photograph the bullfight simply continued beyond the opera's final quarter of an hour. Whatever the reality, the informality and disorder on stage and the lack of musicians in the pit (see fig. 13.4) rather indicates a situation as in Toulouse, where art has morphed into spectator sport.

Not least because Spanish bullfighting had only recently reached Toulouse, we might detect desperation in a claim made by the critic of the city's main paper, the *Dépêche*, to the effect that this was Toulouse's analog to the Wall at Orange.[23] Yet within a broader perspective the splicing of opera and *corrida* represented both a local form of defiance of Parisian centralization and a natural extension of pre-existing musical traditions for bullfights: by the 1890s in Toulouse the overture to *Carmen* had become the traditional entrance music for toreadors, and the decree banning Spanish bullfights, though swiftly overturned, meant that such an

13.4 *Carmen*, act 4, Arènes de Nîmes, 12 May 1901. Nîmes, Musée du Vieux Nimes © Ville de Nîmes. The scene is similar to the description of the Toulouse experiment of 1899: the cast is onstage, but the opera has either finished or been suspended. There is no sign of an orchestra in the high-walled pit in front of the stage.

attempt to cement sport and art rammed home a point about the preservation of local culture—even if that culture was imported from a near neighbor.

The local could also be reclaimed via new political works, in greener and more intimate outdoor settings that reinforced their message. In summer 1913 Émile Sicard's play *La fille de la terre* received its first performance with incidental music (by Déodat de Séverac) in the small Occitan town of Coursan. A *théâtre de verdure* featured a few trees, a stone well, a farmyard, and tools for working the land; it was done similarly the following September in French Catalonia, at Amélie-les-Bains (see fig. 13.5). At one level, Sicard's drama, about the murder of the "townsman" who has tempted an old farmer's daughter away from her wholesome peasant life, reprises a ruralist trope about the confected stench of bourgeois city life; at another, it provides a commentary on the very real problem of rural *dépaysement*—the exodus from the countryside to manufacturing and other semi-skilled jobs in French towns. An unsigned critic attending the Amélie-les-Bains performance wrote that Séverac

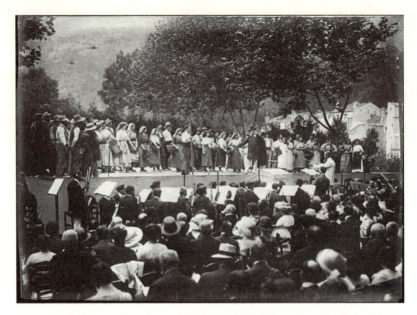

13.5   Sicard/Séverac, *La fille de la terre* (1913), at the Théâtre de verdure at Amélie-les-Bains, 14 September 1913. Bibliothèque nationale de France, http://gallica.bnf.fr/ark:/12148/btv1b6927134q?rk=21459;2. The staging and lighting are minimal, the seating temporary, and the whole flanked by trees, with the countryside visible beyond.

had composed "voluptuous and profound music, which seemed like the very emanation of the surrounding nature" (*une musique voluptueuse et profonde, qui a semblé l'émanation même de la nature ambiante*).[24] That same relationship emerges in a 1921 review, where François de Fortis compares Séverac with the French impressionists, painting rural landscapes outdoors: "Within this expansive fresco he discovered the original model, and, placing it in front of him, he worked *from nature*."[25]

Acoustics, and a New Genre

The question of scale presents a further aspect of open-air opera that helped define the regional difference of newly composed works. Paul Mariéton noted in 1900 that the selection of Méhul's *Joseph* for Orange in 1869 seemed to indicate a need for "slow opera" (*un opéra de manière lente*), adding that he awaited "a work specially adapted by Saint-Saëns to the theater's acoustic" (*un ouvrage adapté spécialement par Saint-Saëns à l'acoustique du théâtre*).[26] The latter piece was undoubtedly the 1901

Les barbares, a commission Saint-Saëns tried very hard to reject precisely because the acoustics at Orange, though excellent for spoken theater, were suboptimal for music. At the very least, in Saint-Saëns's view, stage depth needed to be reduced to help musical projection, and "a monster orchestra" (*un orchestre monstre*) assembled.[27] By contrast, he loved the acoustics at Béziers, which were, seemingly quite by accident, far superior for music to those of any other Midi arena.

Nevertheless, huge forces were still required, together with compositional adaptation to the sheer scale of the space. And in this respect the succession of premieres inaugurated by the Gallet/Saint-Saëns *Déjanire* at Béziers in 1898 seemed to offer a welcome challenge to Parisian (over-) refinement, potentially even creating a new genre in the process. The difference in practice that resulted from a difference of context struck home. Reviewing the vocal score of Fauré's *Prométhée* and not having seen the work, Paul Dukas suggested its genre might be that of a "poème lyrique" or an "ode dialogué," or even an "oratorio en action." He also (like others) noted Fauré's streamlined style—one in which melody and rhythm became simplified, while harmonic interest was retained through the richness of individual chords and from originality in their progression.[28] Émile Baumann said something similar of *Déjanire*, calling its musical style "strangely simplified" (*d'un simplisme étrange*), with choruses "sculpting" (*sculptant*) their texts in unison and benefiting from a stripped-down harmonic palette. The whole seemed related as much to ritual or liturgy as to theater.[29]

Meanwhile the term invariably used to describe Séverac's music for *Héliogabale* was "fresco"[30]—which counterpointed what Séverac now dubbed the "excessive pointillism" of his previous major opera, *Le cœur du moulin*.[31] *Héliogabale* answered decisively the question of whether Séverac could compose anything more than "picturesque watercolors," just as, for Dukas, *Prométhée* addressed Fauré's ability to cope with anything large-scale.[32] The implicit broader operatic question was articulated by the Parisian critic Émile Vuillermoz in 1911. Searching for an escape from decadent subtleties, Vuillermoz looked to two sources that appeared both robust and exotic: Russian primitivism and the Midi.[33] In an article provocatively entitled "Trial by Fire" ("L'épreuve du feu"), he described the summer round of open-air performances as a series of "lessons from the Arena" (*leçons de l'Arène*) that were both salutary and necessary for Parisians. They burned off the unhealthy detritus of the Paris winter season, providing a welcome alternative to the finesse and psychological interiority of a national operatic practice that had strayed too close to the lied—doubtless referencing Debussysme: "Our delicate and intimate style is killing us. We need a bigger, more expansive even, coarser, vision" (*Notre délicat intimisme nous tue. Nous avons besoin de voir plus grand, plus large et même plus gros*).[34] These were attributes posthumously conferred on Séverac, he being seen, in the words of Breton composer Louis Vuillemin, as "at once primitive and learned" (*primitif et savant à la fois*).[35]

But was the Béziers formula a long-term answer? Despite applauding *Héliogabale* from beginning to end, its ignoble sensationalism and fragmentation of the musical experience meant that the otherwise supportive Nantes critic Étienne Destranges thought not.[36] Vuillermoz reached the same conclusion: it was bad enough that the 1911 Béziers commission, Payen/Kunc's *Les esclaves*, seemed musically conventional, academic and Prix de Rome–like;[37] but worse was that the amphitheater's unique magic was wearing off, perhaps exacerbated by the death in 1908 of its illusionist set designer, Marcel Jambon. It now appeared to Vuillermoz as a "lowly suburban velodrome"—"a dreary fairground circus-ring whose roof has been ripped off by a gust of the mistral" (*pauvre vélodrome suburbain. . . . un morne cirque forain dont un coup de mistral aurait arraché le toit*).[38] None of that, however, prevented Beauxhostes resuming his annual festivals successfully in 1921, with two new commissions in 1922 and 1925–26.

Open-Air Opera as Site-Specific

Such shattered illusions notwithstanding, it was when transferred to the capital that Béziers's musical dramas became most gravely impoverished. Scaling down was unfortunate, but a generic mismatch of work and new environment was usually more serious. Upon *Déjanire*'s winter transfer to Paris in 1898, Saint-Saëns forestalled criticism by detailing, in *Le Figaro*, the new orchestration, reduced to fit the smaller, enclosed, and generic space of the Odéon theater. The outburst that follows is leavened by his comment on harps—large numbers of which were the stuff of Béziers legend—but he is only half joking:

Where is the great space of the stage at the Béziers arena? Where is the sun, which itself seemed to be part of the play, lighting with its rays Hercules' pyre, at the moment where he asks his father, Jupiter, to send him the fire of the heavens to put an end to his terrible torment? Where are the fifteen orchestral harps? At the Odéon we have only two: we are lucky they are excellent.[39]

OPEN-AIR OPERA AND SOUTHERN FRENCH DIFFERENCE

For Gustave Larroumet at *Le temps*, the problem was different: too much music for this venue—a national crucible of spoken theater—and too few words. He was blunt: "I was bored stiff" (*je m'étais ennuyé ferme*).[40] *Déjanire* in its natural habitat, however, was another matter, and in 1899 he described it as "a fine artistic experience" (*une grande impression d'art*).[41] Indeed, in 1899 his mention of the underwhelming experience in Paris was almost certainly calculated to set in relief the success of the authentic Béziers version, from which he had just returned.

Prométhée prompted similar reactions. When it was presented at the Paris Hippodrome in 1907, Larroumet's colleague Pierre Lalo offered an extended lament:

Unfortunately, Paris cannot offer it [*Prométhée*] everything that Béziers could. No longer do we have the open sky, the sonority, at once crystalline and shimmering, of the voices and instruments in the air of a beautiful summer's day, the play of light, the wind, the costumes that lift and fall with the breeze, the changing splendor of the sun, and the stage set merging with the mountainscape on the horizon. There is a circus ring, dreary, ugly, and badly proportioned; there are electric globe lamps that cast a hard and cold light from the ceiling; there is a terrible acoustic, which muddles instrumental and vocal timbres. . . . In this shift from one climate and venue to another, *Prométhée* has rather changed in appearance.[42]

As Moore, notes, after a much later Paris performance (1917), Lalo's view hardened—indeed, in line with that of Vuillermoz on Béziers more generally, it reversed: in hindsight *Prométhée* now appeared a simple generic error of hybridity, its effectiveness in Béziers "illusory" rather than magical.[43] Even a straightforward opera, Saint-Saëns's *Les barbares*, was not immune to such charges in Paris, because it started off in the wrong kind of venue—the Palais Garnier. Here, the prologue's oratorio-like narrator immediately rendered the work generically suspect, while the *farandole* concluding its huge ballet fell flat—a peasant dance, embarrassingly overdressed.[44]

Despite its variety of performance venues, then, Paris, had no "place" for open-air works—which accordingly retained the allure of a foreign culture. By the standards of the Midi there was nothing on the correct scale. In 1907 Gabriel Boissy, keen that the state should take up the educational and propagandistic opportunity that open-air theater afforded, and perhaps aware that, without a Parisian home, it was unlikely to have a mainstream future, offered a solution both enterprising and implausible. He suggested that the six-thousand-seat Salle des Fêtes of the

Trocadéro—a remnant of the 1878 Exposition Universelle, with dreadful acoustics—should have its elliptical hall fitted with terraced seating, upper-level galleries looking out onto the city, and, crucially, a retractable roof.[45] The plan would have amounted to a rebuild, in miniature, of the "fake" Béziers arena, with a roof as concession to the northern climate. It never happened.

Critiques of the Midi's open-air works "falling short" in Paris illustrate not so much the failure of the open-air phenomenon as a double bind: site specificity could never work within a context of artistic centralization where, if difference cannot be accommodated, it must be rejected. They also reveal the hope invested by the musical elite (Parisian and otherwise) in the new artistic currents they saw taking shape in the south. In their focus on art alone, however, the enthusiasts perhaps underestimated the social function of open-air opera. These events did not yield museum pieces that could be presented within any proscenium arch; they were part of local leisure as much as of burgeoning national and (even) international tourism.

Opera as "Total Spectacle"

Often attached to agricultural shows, wine festivals, or the celebration of local heroes, open-air opera was not just site-specific in the sense of performance environment, but in that of audience experience. It was another entertainment alongside processions, street dancing, and festive banquets. As such it was more of an outing than an event. The journey, lead-up, and return were as important as the event itself, with audiences frequently cast, or casting themselves, as protagonists. Festival performances of Gounod's *Mireille* in Arles saw women turn up in the black-and-white version of Arlésienne costume; and their attendant costumed "Bals Mireille" neatly inverted the usual Parisian snobbery about provincial fashion when visitors from the capital tried (and failed) to replicate the look.[46] Moreover, while actors and principal singers for the vast majority of open-air events were sourced from the national theaters of Paris and the local orchestras and conductors were often professionals, the choruses, bands, and onstage crowds were made up of local amateurs, which fundamentally changed the balance of art and leisure.

It could also form part of a ritual of belonging, as indicated by the groups of displaced meridionals, many of them from communities in Paris, who made pilgrimages to Orange to attend performances they considered expressions of their *félibre* identity.[47] Neither was the journey to an open-air event necessarily over when one disembarked from a long-distance train: at Saint-Rémy de Provence in 1913 the walk to the Vallon de Saint-Clerc to see Gounod's *Mireille* in a *théâtre de nature* took two hours, and one had to leave time for a picnic before the performance started. The return completed a full day's outing. The opportunity to purchase souvenirs added to the holiday feel and reached epidemic levels in Béziers, with its rapid production of postcards showing rehearsals and performances.

The performance experience was itself informal. These arenas, clearings, and valleys were not controlled environments: the mistral could, and did, make much of the opera inaudible, placing the emphasis on spectacle, stage business, and atmosphere bold enough to be intelligible from a distance of fifty meters. Moreover, if the Greek theaters at Orange and Arles had histories of decorous tragedy, the Roman arenas at Nîmes, Arles, Fréjus, Saintes, and elsewhere had far rowdier associations. Whatever the main event, audiences here constituted crowds. When Paul Mariéton referred to "slow opera," he doubtless thought of leisurely dramatic and musical tableaux as an aesthetic necessity; but such techniques also meant that massed audiences could easily pick up the thread if their concentration wandered—to food and drink, for example. Béziers audiences were infamously noisy, but even at upmarket Orange, the popping of lemonade corks acted as a barometer of audience (in)attention.[48] And depending on what happened in the bullfight, the noise levels during the latter part of act 4 of a "spliced" *Carmen* must have veered wildly from hushed to deafening.

Conclusion

Operatically, what remained, especially at Béziers and in the combining of *Carmen* with bullfighting (or bullfighting with *Carmen*), was a phenomenon that was inseparable from its environment. It refused to fold itself into any one of the Parisian prototypes of what Dan Rebellato has, in relation to the globalized musical, called "McTheatre";[49] even when it presented classical "universals" or standard repertoire, it insisted on its specificity of place, meaning that small rural *théâtres de verdure* could be as transformative, for audiences, as the major open-air venues of Orange or Béziers. However, the rootedness of the open-air phenomenon was both a strength and an impediment to the development and

KATHARINE ELLIS

consolidation of new repertoire, and when Béziers folded in 1926 an entire genre folded with it, while open-air opera using standard repertoire went from strength to strength.

And yet some aspects of the Béziers open-air legacy are perhaps to be found elsewhere, as a by-product of its experimentalism. Productions here pointed generically beyond the confines of the stage and toward the screen—where Saint-Saëns was himself a pioneer composer and to which Henri Rabaud, another Béziers composer (*Le premier glaive*, 1906), was a major contributor.[50] The Béziers musical dramas, with assorted animals, chariots, and massed extras in togas, prefigured the screen spectacular. As early as 1907 it was predicted that cinema would be the death knell of open-air theater because it could outstrip the latter's emphasis on the visual.[51] That prediction was as premature as more recent pronouncements about the death of the book, but it was prescient nonetheless. Two years before, in 1905, a critic signing himself as "Dorlange" wrote in *Lyon mondain et sportif* that open-air performance was part of a trajectory of rejection of the constraints of eighteenth-century theater. Performance had grown so big that "finally, with a last push it cracked open this restrictive strapping of sets in paper and canvas . . . revealing itself finally amid nature" (*enfin d'un dernier effort il fit craquer cette ceinture de décors en papier et en toile qui l'enserrait . . . se montrant enfin en plein nature*).[52] Generically, the next stage was to return the audience indoors while exposing them to an artwork with infinitely changeable landscapes and infinitely permeable borders.

194

FOURTEEN

Pastoral Retreats: Playing at Arcadia in Modern Britain

SUZANNE ASPDEN

Is not the very idea of opera, most elaborate and urbane of entertainments, in a field in Sussex preposterous?[1]

Frank Howes's question in the 1953 *Glyndebourne Festival Opera Programme Book* was posed rhetorically, but, even with twenty years of Glyndebourne successes behind them, it was not unreasonable. A glance at the history of British country-house opera shows it has largely been considered something of an oxymoron: while operas might meditate on the countryside, to stage one there was an unlikely undertaking. There is, after all, a logical basis to the connection between opera and the city, deriving from opera's inherent combination of the arts. As Allen J. Scott has observed, theorizing the urban environment's vitality, cities "emerge out of a need for proximity when large numbers of individuals are caught up [in] certain kinds of mutually interdependent activities."[2] While these activities have traditionally been those concerned with the production of material goods, urban industrial agglomeration has come to include what Adorno and Horkheimer coined the "culture industries."[3] And although modern forms—cinema, popular music, gaming, and information technology—have been the focus of recent scholarly interest, the agglomerative rationale of the "culture industry" might as readily apply to opera and other urban cultural enterprises born in earlier periods. Indeed, there was widespread "strategic investment" in culture and the arts in nineteenth- and (as Margaret Butler and Michael Burden remind us in this volume) even eighteenth-century cities as a means of economic advancement and civic expression.[4]

The urban environment's economic and practical advantages for opera are obvious, then; however, the countryside's corresponding disadvantages for staging opera, which gave rise to Howes's parodic skepticism, are not this essay's focus.[5] Instead, I want to explore why (given the problems) such ventures have come to thrive in twentieth-century Britain—as demonstrated, particularly, by the success of Glyndebourne, which celebrated its eightieth anniversary in 2014 and has gone from strength to strength. What cultural developments underpin this mode's success? Put another way: what ideological positions attend it?

Et in Arcadia ego

The origins of English country-house opera marked the mode as both deliberately distinct from city opera and at the same time oddly reflective of urban cultural imperatives of the past hundred years or so. One such imperative was a burgeoning nostalgia for an idealized form of country living associated (increasingly) with the country house, arising from the inevitable dissatisfactions of industrialized urban life.[6] Of course, we might trace both nostalgia and the emergence of a "countryside ideal" in Britain much further back, to an increasingly urbanized eighteenth-century society's cultivation of city pleasure gardens, or to nineteenth-century spa towns and seaside resorts (deemed healthful because apparently in touch with the "natural").[7] This long-standing idealization of the countryside and country living was both in tension with and (therefore) produced by the growth and press of urban life.[8] Thus Michael Bunce proposes that "urbanisation established four basic conditions for the nurturing of the countryside ideal" (social structures, a political economy, an intellectual and cultural climate, and the landscapes themselves), while Peter Borsay points out that the spas and resorts of the eighteenth and nineteenth centuries were designed to service particular urban markets.[9] W. J. T. Mitchell has suggested, more broadly, that "landscape" should be seen as a verb rather than a noun, "a process by which social and subjective identities are formed."[10] Certainly, the notion of the "countryside" was, from the beginning of its widespread usage in the early eighteenth century, also intrinsically tied to the articulation of power and control, in both ownership of the land so defined (enclosure and other processes meant that by 1700 in England around 70–75 percent of cultivated land was in the hands of major landowners), and in its aestheticization and definition by amenity value (as the site of recreational hunting, fishing, and shooting).[11]

Within the developing idealization of the countryside, the country house—long the object of domestic touristic fascination, and part of an appeal to an "old rural social order" in which everyone knew his or her place[12]—attained institutionalized significance at the turn of the twentieth century. At that time, nostalgia for a better (and quintessentially English) past found formalized legitimation and a tangible focus in organizations such as the Society for the Protection of Ancient Buildings, founded (by the designer and socialist William Morris) in 1878, and the National Trust, founded in 1895. Any idea of social reform was leavened in mid-century by the National Trust's shift of focus from preservation of open spaces to preservation of country houses, a response not only to general anxiety about Britain's cultural heritage, but also to a postwar economic environment in which landed gentry argued that they could no longer afford to maintain their estates. Evelyn Waugh's *Brideshead Revisited* (1945) struck a chord in its consuming concern about the national patrimony as represented in the stately home; the middle-class protagonist Charles Ryder, who takes a kind of ownership of the aristocratic estate he loves by painting "portraits of houses that were soon to be deserted or debased" (as if painting a person), believes that "we possess nothing certainly except the past."[13] The aristocratic past that Waugh and others sought so anxiously to possess had been epitomized by the country house since at least the seventeenth century, when Ben Johnson's *To Penshurst* (1616) made such houses an acceptable theme of panegyric.[14] And aristocratic nostalgia lives on: television costume dramas such as *Downton Abbey* continue to promulgate the country house as the epitome of an English life seemingly all the more ideal because unobtainable.

For twentieth-century country-house opera, too, rural nostalgia and an idealized Englishness have been ever present, with aspirations to recapture an idyllic past often established by appeal to natural surroundings. Vita Sackville-West's 1958 contemplation of the connection between Glyndebourne's gardens and music—"The graciousness of civilization here surely touches a peak, where the arts of music, architecture, and gardening combine"—is exemplary in this regard.[15] The most significant early attempt at opera in the countryside in Britain, by Reginald Buckley in Stratford and Rutland Boughton (along with Buckley) in Glastonbury, also invoked nostalgic contact with a mythical, originary past, achieved through authenticity of location, as their 1913–14 "National Appeal" for funds for an opera on Arthurian themes made clear:

At last the opportunity has come for producing an English Music Drama upon the Arthurian legend as told by Malory. Glastonbury, the ancient Isle of Avalon, affords the most ideal and appropriate setting for this venture. Glastonbury is, according to well-founded tradition, the site of the first Church in Britain, built by Joseph of Arimathea and his companions. Chalice Hill, Glastonbury, derives its name from the belief that the Holy Grail was buried there. To Glastonbury the mourning queens brought Arthur himself after his last great battle to be healed of his wound, and here he and Guinevere were buried.[16]

Their appeal to a genius loci is worth seeing in context: Rolf Gardiner, a contemporary proponent of a return to rural roots (and of an outlook with fascist tendencies) suggested that "a place takes on the memory of noble doings and adds their spell to itself."[17] And from 1905, town and village pageants linking locale, history, tourism, and community (most involved thousands of local people, as well as "the great and the good") enjoyed a brief efflorescence, to the extent that "Pageantitis" was deemed to have struck the country.[18] Similarly, theater and concerts in rural locations burgeoned at this time. The *théâtres de nature* described by Katharine Ellis in this volume were also attempted in early twentieth-century Britain, in ways that suggest a link with later country-house productions: in 1920 at Glastonbury "Britain's First Orchard Theatre" was proclaimed, in which an "open-air playhouse" was built "amid beautiful surroundings suitable for pageant work and allegorical work of all descriptions" (though, in view of Britain's rather uncertain climate, it incorporated "careful protection for large choirs and orchestra").[19] Even before this, Rutland Boughton had attempted to incorporate both landscape and audience into the drama in the first production of his *The Immortal Hour*, in 1912, staged in the grounds of his Surrey patron's home; not only would the opera's two acts be given on two successive days (timed for the full moon), but, he informed prospective audiences, "each scene will take place in a different part of the woods."[20]

For Boughton and Buckley, the musical and cultural regeneration they hoped to achieve at Glastonbury was tied up both with a rejection of industrialism and urban life and with the concomitant utopianism of a period when, as one fin-de-siècle progressive suggested, "social dreams are once more rife."[21] Utopian dreamers reacted against the uncertainties of their age (and the fundamental philosophical difficulty of conceptualizing "the murky smoked glass of the present condition," as William Morris put it) by seeking to understand it as Other, viewing it at a physical or temporal remove, whence "the silent movement of real history which is still going on around and underneath our raree show" might be made visible.[22] In novelistic terms, many (including Morris and H. G. Wells) offered this utopian overview from a putative future; others, such as Grant Allen in *The British Barbarians: A Hill-Top Novel* (1895), found their sense of perspective via what had been a trope of urban disquietude since the eighteenth century and "raise[d] a protest in favour of purity" through asserting geographic ("Hill-Top") distance.[23] Allen's opening conceit of the relationship between pastoral purity and clarity of insight is predicated on removal from the distantly visible "lurid glare . . . where the great wen of London heaves and festers," and from the "decadent sins and morbid pleasures" of this "urban age," of which theater and music hall are emblematic "sham idyls."[24]

Although appeals to purity through social and cultural retreat could have negative overtones at this time (particularly in light of the following that Oswald Spengler's *Der Untergang des Abendlandes* [1918 and 1922] attracted),[25] they could also be manifested in positive terms: a utopian retreat to the countryside and simultaneously to a more perfect past figured particularly in Morris's Arts and Crafts movement (though there the ideal was the cottage rather than the manor), of which Boughton and Buckley were both admirers.[26] Like Morris and John Ruskin (and Allen, quoted above), Boughton viewed industrialized urban life as dystopian, proclaiming in 1909 that "whatever is good in modern art comes into being in spite of modern civilization."[27] Like them too, and like pageant designers (and others, on the right), he sought to establish a communitarian musical drama expressing "the *oversoul* of a people," to call the nation back to (he felt) a truer sense of itself.[28] Such aspirations grew not only from the Arts and Crafts and pageant movements, but also from Wagnerian idealism; as Boughton was a composer, he focused on the (overtly) Wagnerian aim to create opera on a national mythical theme— the Arthurian legend. Also like Wagner, who rejected Munich as the site for his ideal opera house,[29] he sought a location removed from everyday city life, so that getting there would become something of a pilgrimage. Having attempted to establish a summer school in Surrey Woods in 1912 ("Opera for a Holiday. . . . Play by Moonlight," one paper called it),[30] Boughton and his collaborators turned their attention to Glastonbury, a place not only seen as dramatically fitting for such a project (as demonstrated in the "National Appeal" for funds for *The Birth of Arthur*), but where Boughton might also—in the tradition of the populist pageant— establish a "little army" of amateur (and professional) enthusiast collaborators.[31] Aside from Boughton's social (and socialist) aspirations, involving the local community in his operatic festivals was economically efficient—"a matter of importance in a new movement," the *Musical Standard* for 5 June 1915 noted (and as modern country-house opera companies, often reliant on [cheap] young singers, might agree).[32]

Community opera's practical difficulties and day-to-day compromises undoubtedly kept the utopian vision anchored—and avoided the potential for fascist and elitist associations engendered in this period by the connection of purity, rural life, and the sense of place and national idealism.[33] The Glastonbury Festival (as it came to be) was countryside rather than country-house opera: it lacked a glamorous venue (though Buckley wanted to commission a "Temple Theatre," they made do with the Assembly Rooms), and struggled with financial and practical limitations and the variable talents of the company.[34] But Boughton's productions—and his attempt through site-specific opera to resurrect a collective, national spirit—garnered increasing attention and praise, not only in the press, but also from the likes of Edward Elgar, who in July 1920 put his signature to an appeal for funds endorsing the movement's utopian, regenerative powers: "We believe that Glastonbury has in it the possible development of a movement of the greatest importance, both for British Music and for the regeneration of the life of the countryside. . . . We believe that at present Mr Boughton's work is unique, but we hope that it will also be an example."[35] Similarly, George Bernard Shaw praised Boughton's work in 1922: "Here in Glastonbury [you] have achieved a musical festival which is . . . the most important thing in England at the moment. All other festivals—like the Three Choirs Festival—are only marking time. There are no signs of any development in them."[36]

Boughton's (and, for a time, Buckley's) endeavors undoubtedly put countryside opera on the map, encouraging the enactment of the link between geography and history not only in the pageant, but also in the creation of powerful musical drama. In some senses they set the standard for subsequent ventures. So while the Glastonbury Festival came to an end in 1926, support withdrawn in horror at productions expressing Boughton's increasingly strident communism,[37] the germ of the idea did not wither—though its circumstances and ideals did undergo changes.

Glyndebourne

Glyndebourne, which opened its operatic doors in 1934, eight years after Glastonbury's demise, demonstrated a new phase of the countryside-opera idea, in which the venerable status of the country house, rather than the choice of operatic subject matter, provided a sense of connection to the past and the locale. Despite their very different circumstances, John Christie, the founder of Glyndebourne's operatic endeavor, must have been aware of Boughton's efforts, and of the gauntlet his success threw down.[38] Indeed, it might seem that Christie sometimes had Boughton's work in mind when setting out on his own project. In a letter of 1936 Christie intriguingly referred to his ambition in patriotic terms: "Our purpose is to give supremely good performances first of all of Mozart and then of most other operas. But this is merely a stepping stone towards the later and fuller purposes, which are to export British Music and British Musicians."[39] His continuation suggested a critique of the instability (and artistic quality) of earlier ventures: "I contend that this will not be expected or possible until England has succeeded in giving year after year a whole series of supremely good Festivals, which it has failed absolutely to do in the past." But however much Christie might have seen himself succeeding where Boughton and others had failed, Glastonbury and Glyndebourne still had much in common. The Glyndebourne "prospectus" of 1934, advertising the first season, shows similarities in aims to Boughton's for Glastonbury. Christie offered two "main intentions" for his enterprise: first, the unsurprising (though ambitious) aspiration to stage opera that in every element "can rank with the best that the European Opera Houses can offer"; and then, "to inspire in artists and audience alike by the charm and beauty of the theater's surroundings something of that spirit of Festival which characterizes the famous musical and dramatic Festivals abroad."[40] A comparison with those standards of festival excellence, Salzburg and Bayreuth, was explicitly drawn by Christie in the *Prospectus* and elsewhere, and so it also appeared in contemporary newspapers.[41] Boughton too had compared his festival with Bayreuth (not least on grounds of its social accessibility), in his national *Appeal* for funds for *The Birth of Arthur* (1913–14), and in the press (Boughton and Christie were, of course, not alone at this time in being ardent Wagnerians).[42]

Nature, too, was a common theme, as Christie's second "intention" demonstrates. The Glyndebourne *Prospectus* opens by evoking natural and architectural beauties for readers—"the ancient Tudor Manor House of Glyndebourne, situated in a beautiful wooded stretch of the Sussex Downland near Lewes"—with accompanying photographs; the 1934 program began similarly. The sense of Glyndebourne's utopian isolation, both physical and temporal, was also emphasized by the papers. A 1933 *Telegraph* article commences with poetic invocation of the English countryside, connecting it to a sense of removal from the concerns of modern life: "The poet longed 'to be in England now that April's here,' but he might as well have said November if he had known the balminess and beauty of recent windless, sunny days on the South Downs, where, once jerry-building is out of sight and main roads out of hearing, the England of the poets shows a scene as lovely as any place on earth."[43] The sense of escape (particularly from poorer, recent buildings and from roads—reminders of humdrum reality) was integral to Glyndebourne's appeal. Indeed, the rise of country-house opera seemed to belong to a strand of rural nostalgia distinct from that of Boughton, Morris, and Ruskin—part, instead, of an idealization of the old country order, which may have offered "a reassuring myth for upper and middle classes confronted with the rise of socialist and modernist values."[44]

This implicit aspiration to artistry and status was enhanced by the resonances and attractions of the area around Glyndebourne. The Sussex Downs and Kent were established Arcadian playgrounds for well-to-do Londoners, particularly those of an artistic disposition: Vanessa Bell and Duncan Grant, along with other members of the Bloomsbury set, had occupied the nearby Charleston House since 1916; Bell's sister, Virginia Woolf, and brother-in-law, Leonard, moved to Monk's House, also nearby, in 1919; Vita Sackville-West and her husband acquired Sissinghurst in Kent in 1930 (interestingly, because they feared development near their former home, Long Barn in Sevenoaks, Kent).[45]

On the other hand (and vital to Glyndebourne's success), this area of rural retreat was also eminently accessible: indeed, the south coast at this time was becoming ever easier to reach, with the newly amalgamated Southern Railway service's electrification and upgrading in the 1920s and '30s, and accompanying increase in population and rebuilding of stations—particularly London's Waterloo.[46] However, the modern practicality of such services as the 8:45 a.m. "stockbrokers' express" from Brighton to London Bridge did not dim the romanticized view of travel, which was nicely illustrated by the naming of railway locomotives on this network (or "grouping") after heroes of Arthurian legend.[47] Such gestures—odd as they may seem today—speak to the anxieties implicit in the *Telegraph* columnist's sniffiness about "jerry-building," demonstrating an attempt to smooth over the fissures between old and new. The longevity and careful construction of this romantic view of the Sussex countryside is demonstrated in Glyndebourne's first program book, of 1952, showing the road route from London on a mock-medieval map (see fig. 14.1); traveling to Glyndebourne, it implied, was returning to an earlier, perhaps better, time.

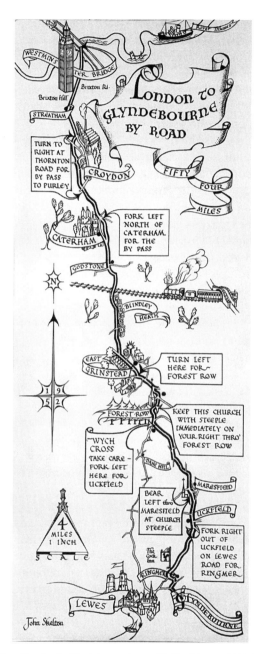

14.1 John Skelton, "London to Glyndebourne by road," *Glyndebourne 1952 Festival Programme Book* (London: Glyndebourne Festival Society, 1952), 76. Author's copy. Published by permission of John Skelton's family.

Of course, unlike village Glastonbury, the implicit exclusivity of country-house Glyndebourne was a corollary of its location. This translated into tangible considerations that would have reinforced the social signification of the operagoer's visit to Glyndebourne: the *Prospectus* informed readers that the train departing from Waterloo for the opera would have Pullman cars (Southern Railway's luxury carriages)[48] offering "light refreshments"; the return journey would be on a "Glyndebourne Opera Special" train.[49] Alternatively, the *Prospectus* suggested arriving by plane, as a landing strip was available. Those prepared to pay the eye-watering prices (Michael Kennedy said they were "unheard of in England at that time") were apparently mollified by "the gardens, the food and wines," and greeted with a gold-covered program and a half bottle of "Mrs Christie's Champagne" during the interval.[50]

In other words, Glyndebourne's separation from urban life, where there was increased public access to opera through the new BBC radio and cheap (unstaged) Proms performances, allowed—almost necessitated—a crystallization of the relationship between opera and power.[51] John Christie harnessed both the sense of removal from the quotidian and opera's implicit aristocratic associations in a 1933 manifesto, which seemed to consider something like Glastonbury's communitarian principles only to reject them:

The Glyndebourne Opera House has two possibilities:
1. To offer superb performances to people who will regard them as the chief thing in the day or week to be looked forward to, and who will not try to sandwich them between business interviews and a social party.
2. To give educational performances for the ordinary public, with the best possible stage setting and only English orchestras and lesser known singers.
I incline towards the superb performance.[52]

In practical terms, Christie's emphasis on "superb" opera was manifested in his focus on stage technology, an area in which his careful control and wealth were made visible—as well as his extraordinary enthusiasm and bent for engineering. Alongside his theater's demonstration of technical wizardry, the gregarious Christie also sought to encourage escape from the hurly-burly of city life, enjoining operagoers to move to the area, in an advertisement ("Where to live") that appeared regularly in the back of the early programs.[53] Whether he intended it or not, such preference for the "superb," and perhaps the appeal for removal to the idyllic countryside, would have attracted those fearful of the rise of "the

14.2 "Number Seven" advertisement, *Glyndebourne 1952 Festival Programme Book* (London: Glyndebourne Festival Society, 1953), 80. Author's copy.

masses," whose aspiration to artistic enlightenment would lead (it was thought) to the breakdown of social class.[54]

Newspapers picked up on Christie's manifesto, noting and thus reinforcing Glyndebourne's tendency to exclusivity; *The Times*, for instance, observed that Christie's "scheme is one not so much for the popularization of opera as for the attainment of ideal performances of selected works."[55] *The Times & Tide* offered an interesting theoretical division based, perhaps unwittingly, on nineteenth-century stereotypical distinctions between French grand opera and German high art, and the contemporaneous obsession with distinguishing high art from low:

The future destiny of opera [is] to diverge into two channels: one, "grand" opera, on a spectacular scale, appealing to the greater public, and needing an opera house with some thousands of seats at moderate prices to balance its budget; the other, opera in exquisite precision . . . [which] would need a small opera house, catering for the few at correspondingly high prices. The latter is, for the present, Mr. Christie's aim.[56]

A sure sign that Glyndebourne's exclusive status was well understood is found in the advertising appearing in the Festival program books, once they began to be produced in 1952, as figure 14.2 shows. Sounding very much like critics' nostalgic musings on Glyndebourne's attractions, the advertisement suggests that opera offers "the coloured gaiety of an Edwardian evening preserved within the greyer world of today."

The Immersive Aesthetic

It is clear that the relationship between genteel rural living and cultural exclusivity contributed enormously to Glyndebourne's initial appeal.[57] More particularly, I suggest, it did so because the location facilitated a sense of immersion in an opulent aristocratic aesthetic—which merged with the traditional view of opera in its heyday. At Glyndebourne this immersion depended on the Christies themselves: they created both the ambience and the welcome necessary to sustaining participation in Glyndebourne's elegant fantasy world.

Christie had long taken an interest in amateur dramatics, performing scenes from his beloved Wagner in the Organ Room before the theater was built and even, on one occasion, singing the role of Beckmesser.[58] The theatrical atmosphere extended to house parties: in 1965 Spike Hughes related that "the summer of 1933 was the climax of a great period at Glyndebourne when everybody was always dressing up in Austrian and Bavarian costume," with the men expected to wear lederhosen and the women dirndl dresses.[59] Christie's marriage to the soprano Audrey Mildmay emphasized the entwining of art and life still further: she influenced the choice of works for the fledgling house (Mozart rather than Wagner) and then sang in them.[60] After the opera house opened, Christie still roped in locals to perform as supernumeraries (bricklayers and painters as soldiers in the 1938 *Macbeth*, for instance).[61] In the first program book, of 1952, Christie fondly recalled these amateur dramatics, encouraging his society audiences (including the press) to come in on the joke:

[Childs, the butler,] took the part of the deaf mute in *Entführung* whilst the cook, Mrs Ashton, took the part of the cook in *Don Pasquale*, both with great success. Not to be outdone, the Glyndebourne pug dog, Tuppy, after "Tuppy Headlam," opened the second scene of *Don Pasquale* alone on the stage, where he awaited the arrival of Norina (Audrey Mildmay) and was also awarded a paragraph to himself in *The Times*, which said "he was certainly the best pug on the operatic stage of the present day."[62]

Perhaps Christie's appeal to fellow opera lovers to move to the vicinity also reflected that enthusiasm for amateur performances, allowing Christie and his collaborator-guests, he might have hoped, to continue living and breathing opera (just as Boughton had in Glastonbury, at a humbler level).

But Christie not only channeled his energies into making life more like art; he also worked to achieve the reverse, particularly in terms of the appearance of the buildings, which were "antiqued" in a manner later common within the heritage industry.[63] While the opera house had a traditional, notionally "classical" façade, the Green Room added in 1934–35 was in "Tudor" style (nominally, like the manor itself), an oak-paneled room with an oriel window at one end allowing views of the house and gardens and furnished with antiques and paintings from the house.[64] The Mildmay Dining Hall, built at the same time, was designed by Christie (according to the former archivist, Rosy Runciman) to have "the feeling of a converted barn that had been part of the estate for years rather than a new purpose-built hall."[65]

It seems that Christie sought to blur the line between the existing estate and the opera house, as early newspaper reports also noted. This slippage was visible in the festival program book—most colorfully, in its advertisements, capitalizing on the cachet of Glyndebourne to market their wares (figs. 14.3 and 14.4).[66] The press responded similarly, with spreads designed to promote high-society fantasy. A montage from 1937 by Bryan de Grineau (best known for drawing vintage cars) shows in one corner an over-generously proportioned opera-house foyer, resplendent with elegantly grouped couples, and suggests that "the foyer is an operatic setting in itself" (see fig. 14.5). On the other side, it shows "the tree growing through the dining hall [which] rather suggests the first act of Wagner's 'Valkyrie,'" while "Glyndebourne in the light of departing cars" offers a "final Mozartean effect."

As these various forms of marketing suggest, Glyndebourne's success reflected not just the audience's engagement with opera or even their personal participation in a theatrical world, but also their willing absorption in the aristocratic, country-house aesthetic. And in this regard too, the Christies' dedication and vision were essential.[67] Michael Kennedy, author of *Glyndebourne: A Short History* (2010), proposes that "what makes Glyndebourne unique is that it remains a family's private home in which one feels more like a guest than a patron."[68] He perhaps drew this idea from a retrospective appraisal by Audrey Christie herself, published in 1957, in which she explained that she and John Christie had "dreamed of an Opera Festival in these quiet, intimate and informal surroundings where the impression could be created of a great private party."[69] In the same 1957 issue of the festival program book, Siriol Hugh Jones proposed that "one of Glyndebourne's greatest charms is the way it offers sublimely professional all-star productions while still giving the impression that we are privileged guests at a select house-party and have just arrived in time for an impromptu game of Musical Clumps."[70] One suspects that many of Glyndebourne's first patrons would, indeed, have been guests at other times—like-minded friends, supporting the venture. Newspaper spreads show that it was early the resort of high society. (Michael Kennedy reports that Carl Ebert, theatrical director at Glyndebourne at the time, described the Glyndebourne audience as "seventy per cent '*die* Snobs' and thirty per cent music-lovers.")[71] Anecdotes about Christie's customary end-of-season speech, given before the last night's performance, suggest that he felt he was talking to an audience of intimates; in 1939, on the eve of war, he apparently began by referencing a cricket match between two famous and exclusive boarding schools: "I have some very serious news . . . for the first time since 1908, Harrow has beaten Eton at Lord's."[72] In various ways, then, Glyndebourne established a connection between country-house opera and high society.

But the illusion that Glyndebourne's visitors were aristocratic guests was most successfully achieved because they themselves were (as they are today) part of the spectacle. The operagoer's engagement with this

14.3   Moss Bros. advertisement, *Glyndebourne 1957 Festival Programme Book* (London: Glyndebourne Festival Society, 1957), 7. Author's copy. Published by permission of Moss Bros.

14.4   James Pascall Ltd. advertisement, *Glyndebourne 1957 Festival Programme Book* (London: Glyndebourne Festival Society, 1957), 8. Author's copy.

14.5 Bryan de Grineau, "Summer Opera in the South Downs: Mozart at Glyndebourne amid Country-House Charms," in *Illustrated London News*, 5 June 1937, pp. 1032–33. Mary Evans Picture Library.

reenactment of gracious living was facilitated by the structure of the event, which encouraged audience members to linger in the grounds before the performance and to dine or (later) picnic in the lengthy interval.[73] The often elaborate nature of the country-house opera picnics—with linen, candelabra, even butlers—testifies to the aestheticization and performativity of the dining experience, and thus to the audience's desire for immersion in the country-house aura.[74] Consumption's centrality within the Glyndebourne aesthetic has even been celebrated in a book of Glyndebourne picnics.[75] Indeed, the practice of the interval dinner at Glyndebourne is followed today by most country-house opera companies and recalls the degree to which the principle of conspicuous consumption has become necessary to the modern heritage industry, it being the simplest means by which visitors can incorporate themselves into the spectacle.

This desire for aesthetic integration, and the sense one was thereby reaching back to opera's more "natural" aristocratic environment, is neatly encapsulated in Vita Sackville-West's 1953 paean to Glyndebourne. For its acknowledgment of the connection between a sense of removal from urban trivia and facilitation of immersion in a fantasy of bygone operatic life, it is worth quoting at length:

> Lovers of music and the arts in general, should be sensitive also to the outside forum they are offered. Instead of battling through a crowd thronging a cocktail bar, they can wander at leisure on a summer evening between borders of blue delphiniums, blocked amongst cool grey-foliaged plants of artemesia, lavendar, santolina, stachys, and catmint. They can stroll down paved paths and imagine themselves back into the calm days of Dorothy Osborne and William Temple. They can gain the experience of walking back into a different century. It might be the Italian Renaissance, or it might be a little German court devoted to music and culture, a backwater in a bustling competitive world. You could imagine yourself in any of such pleasant places or epochs; but the truth is that you are in an English garden. . . . When the last chords of the music have died away, the garden and the Downs will still be there.[76]

Sackville-West's conclusion, reinserting the garden and the Sussex countryside into her brief dramatic scene, also reasserts this rural location's fitness as a home for opera, in the imaginative space it creates for audiences (as much as for performers) away from the press and bustle of city life. Such an assertion was escapist, but then that too is how opera was seen. The continued success of country-house opera in England—and its echoes abroad, in Glimmerglass, for instance—suggests that, for many, it is seen that way still.

FIFTEEN

The Opera House as Urban Exhibition Space

KLAUS VAN DEN BERG

In 1955 Theodor Adorno claimed that opera, with its massive artistic means, prefigured many practices of the twentieth-century culture industry.[1] In particular, its large orchestras and choruses, gigantic set designs, popular myths, and visualized scenarios prepared some of the "worst" of that industry, making culture ready for consumption by the masses. The cultural geography of contemporary opera appears to confirm Adorno's dictum, being shaped by a dialectical relationship between the loss of constituencies inside the auditoriums and the proliferation of glamorous opera houses in urban centers seeking to raise a city's profile. Visually stimulating buildings have been inserted by star architects into Copenhagen, Miami, Valencia, Dublin, and Dallas. These impressive opera houses respond to a need to create an urban scenography, or space for action, visualizing cultural inheritance today; but they also indicate how cities employ that legacy competitively.

With the nineteenth century's rise of capitalism and bourgeois culture, Adorno's "massive" means of opera emerged in full force off stage as performance venues assumed significance in the exponentially developing urban landscapes. The complexity of urban development in the twentieth century has transformed architects into scenographers who shape identity through the composition of dramatic spaces. As modern cities must fit opera houses into a spatial network defined by sophisticated capitalist practices and technological innovations, architects consider performance on urban sites in terms of visual plots, events, and stories.

This chapter explores three paradigmatic sites to pursue Adorno's suggestion that opera's massive means have been externalized. I trace the extension of opera's scenographic and narrative function in modern urban geography by way of New York's Metropolitan Opera, Paris's Bastille Opéra, and Dallas's Margot and Bill Winspear Opera House. These buildings continue the trend set by the Palais Garnier in 1875, the paradigm of the opera house as urban spectacle and agent of a cultural agenda: displayed as a public focal point, it elevated the arts to equal importance with politics and religion in urban design and turned architecture into cultural propaganda.[2] Focusing on the opera house's insertion into the diverse structures of these cities, I argue that all three buildings have become scenographic enactments of cultural identity.

Adorno's teacher and friend Walter Benjamin pioneered the critical analysis of nineteenth-century urban theatrical externalization, analyzing the pervasiveness of spectacle in Paris, from performances at the opera to the world fairs, panoramas, and, above all, the Arcades.[3] Precursors of department stores and shopping malls, the Arcades were an outgrowth of industrial luxury: in glass-covered, marble-floored hallways running through entire housing blocks, elegant luxury shops provided a stage for commodities. The Arcades' glass-and-iron roofs drew together residences and business into smaller urban areas. Theaters too were folded into this bourgeois context of commodified spectacle and representation: Jacques Offenbach's company, the Bouffes Parisiens, had its home in the Passage Choiseul, as the Théâtre des Variétés did in the Passage du Panorama.

The Arcades were the paradigm of a new space in which building and urban location formed an *Ausstellungsraum* (literally, exhibition site), whereby architecture became a performance, enacting a sweeping historical experience that turned the modern city into a "Schauplatz neuer unvorhergesehener Konstellationen" (theater of new, unforeseen constellations).[4] The Arcades thus offered a new public dramaturgy that reenvisioned theater, blurring "inside" and "outside" as their glass façades allowed for new viewing dynamics and transparencies that were also becoming central to modern self-understanding. Similarly, architects developed a language that reflected viewing and spectacle practices, with theatricalized scenography becoming ever more central to representation of cultural and political power.

Taking up Benjamin's view of urban geography as a space of cultural performance that expresses and conditions its historical moment, this chapter articulates the interrelationship between site and architecture (the site dramaturgy) for the Metropolitan Opera, the Bastille Opéra, and Dallas's Winspear.[5] As the opera houses' insertion into these cities has allowed them to become expressions of cultural identity representing different eras of urban development, so they all also demonstrate an inverting and unfolding of the arcades' principles of commodified theatrical display in the contemporary metropolitan landscape. They are all large urban-renewal projects grafted onto the potentiality of the modem city: the Met (completed in 1966) replaced an entire neighborhood to fit an opera house into a developing urban landscape; the Bastille Opéra (1989) emerged as one element of François Mitterrand's Grands Travaux at one of Paris's most emblematic locations, the Place de la Bastille; and Norman Foster's Winspear Opera House (2009), was placed in a large arts park, the Dallas Arts District, at the outskirts of downtown, one of the densest accumulations of major architectural projects in the world, surrounded by buildings designed by the likes of Renzo Piano, Rem Koolhaas, and I. M. Pei. Each of these houses moved opera into the public imagination through technological and planning means that allowed them to reflect the excitement and grandeur of the genre itself.

Palais Garnier: Paradigm of the Modern Exhibition Site

In the nineteenth century the opera house was fully realized as an "urban set piece"[6]—a structure generating complex narratives in modern metropolitan space. The thematization of the building as a "visual-spatial construct,"[7] delineating opera architecture as mass spectacle, began to take shape in the eighteenth and early nineteenth centuries.[8] However, it was the Paris Opéra, the Palais Garnier, that, as a key building block of modern Paris, initiated the Benjaminian exhibition space.

Begun in 1861 and completed in 1875, the Palais Garnier was conceived as a large-scale urban project, inaugurating the transformation of Paris into a modern city and becoming paradigmatic of the artistic institution as an exhibition site that not only structures but directs urban development. In his *Arcades Project*, Benjamin describes the Palais Garnier as a "stage" and "center of social life" on which Napoleon III's "imperial Paris could gaze at itself with satisfaction."[9] The theater's location at the intersection of several boulevards turned the Beaux-Arts building into the quarter's focal point. Indeed, the central Avenue de l'Opéra's connection with the Louvre area stages the building as a baroque set design, reflecting grand opera's aesthetic of overwhelming visual spectacle (see fig. 15.1).[10]

15.1   Façade of Paris Opéra, designed by Charles Garnier, as seen from the Avenue de l'Opéra. Photograph (c. 1880–1900).

15.2   Bon Marché department store interior as arcade, designed by Louis-Auguste Boileau (1869). Lithographie by Charles Fichot (1872).

15.3 Opulent grand staircase in the Garnier Opéra. Print (c. 1890–1905). Permission from Library of Congress Prints and Photographs Division, Washington, DC.

The Paris Opéra became the paradigm for twentieth-century opera houses, with its command of a massive space pioneering the city's transformation into a complex visual field, just as its emulation of arcades and department stores crystallized consumerist display strategies. Indeed, the Palais Garnier's grand foyer and staircase imitated the architecture of the newly inaugurated department store Bon Marché, which in turn had already reproduced the Arcades' strolling experience (see figs. 15.2 and 15.3). It became what Benjamin called a "dream house" of the rising bourgeoisie, its architecture indexing that dream by recalling at once the palazzo façades of early Italian opera houses, classical architecture (in its frieze), and vistas from Florence in its dome (then mimicked in subsequent glass versions by architects such as Norman Foster).[11] Although stage and city design had begun as early as the sixteenth century to reinforce each other in their use of perspectival vista to establish grandeur of scale,[12] the city as exhibition site, featuring dreams of the future, would be amplified at world's fairs such as the Columbian Exposition in Chicago in 1893. The bourgeoisie at the helm of the emerging nation-state staged its ideology by placing a high-tech building in the monumental style at a prominent site on the groundbreaking, confluent avenues that destroyed the historic fabric of *vieux Paris*.

New York's Met: Scenography and the Modern Grid

Unlike in Paris, where Napoleon III and Baron Haussmann ruthlessly cleared historical layers of the medieval city to stage the new social center, in New York the city's cultural and economic leaders created a similarly spectacular exhibition site out of the emptiness and monotony of a vast industrial grid of city blocks that had been laid out in anticipation of easy and rapid development by stores and apartment complexes. For example, Macy's department store originated in one building in the 1870s before expanding into others on the same block, with lit window displays capitalizing on the shopping experience's theatricality; then, in 1902, Macy's absorbed an entire block in the Harold Square location where it still stands. Bloomingdale's was the first 1880s department store to turn the arcade experience inside out, with large display windows, each staging a scene. Its façade also included large arched windows, later echoed in the Metropolitan Opera House façade. Thus, New York's tradition of expanding department stores inside city blocks shaped the thinking around inserting a large cultural institution into the city and influenced the visual language for opera buildings.

The inception of the Metropolitan Opera House at Lincoln Center is a story of wealthy businessmen pursuing a new opera house that ultimately produced two oversized urban complexes, Lincoln and Rockefeller Centers. The original building, which was located on Broadway between

Thirty-ninth and Fortieth Streets, near other Broadway theaters and the business district,[13] had been built and financed by the Astors, Vanderbilts, and Morgans in the late 1800s. While the expanse of the Met building was concealed at street level, aerial views attest to the grand dimensions of the architect J. Cleaveland Cady's opera house, which montaged two house structures, recalling the style of the Italian palazzo (the original operatic venue), on each side of the front, linked by the central entrance. Architecturally, however, the building had been a failure, an aesthetic hybrid that quickly became much too small both for ambitious artistic operations (due to inadequate backstage space) and for affluent patrons desiring a prestigious venue along European lines, like the Palais Garnier. The building's interior and exterior architecture drew public ire long before the board considered a replacement.[14] While aristocratic principles were clearly written into the interior's box system, the exterior and site dramaturgy failed to articulate a larger vision for the opera house's place within the emerging landscape of Manhattan's high-rise buildings.

As Rem Koolhaas has argued, New York set a precedent for American cities in creating a grid of equal blocks on which the city expanded over the decades. Noting the departure from traditional European city plans, he interprets the grid of blocks dividing Manhattan as 2028 "islands," contending that "the grid makes the history of architecture and all previous lessons of urbanism irrelevant," in that it acts like a dramatic plot that structures multiple episodes, forcing architects to make each block distinct from the others.[15] An entrepreneurial desire to reshape the city blocks dominates the spectacular game Koolhaas calls "Manhattanism." Thus, inside what is on the surface an egalitarian city structure, the grid seems to offer the possibility of a fantasy environment; however, this grid lacks significant space for public use, or even one that could be coopted for that purpose. In New York, semi-public squares only emerged in some areas where Broadway, the original north—south thoroughfare predating the grid, crossed one of the avenues—such as Times Square and Union Square. However, since those areas were also prime real estate, they left little space for cultural monuments.

In New York's grid, the old Met building (see fig. 15.4) was monumental in size but possessed neither the visibility nor the dramatic presence of a Palais Garnier. Beginning in the mid-1910s, downtown Manhattan underwent huge transformations that affected the opera company's cultural identity. Large mansions and villas that lined the more prestigious streets slowly gave way to skyscrapers,[16] which dwarfed even such

15.4 Façade of the old Metropolitan Opera House toward Broadway. Photograph, 1905.

mansions as that of John D. Rockefeller Sr. Given that the opera house lacked the "phantasmagorical" quality Koolhaas discerns in skyscrapers and modern hotels, it seemed destined for decline.

Under the leadership of the financier Otto Kahn, the Met's search for a new building beginning in the 1920s turned into what Koolhaas labeled an "architectural odyssey," "wander[ing] across the grid in a quest for an appropriate location."[17] The board's desire for a traditional cultural monument clashed with this grid of islands and set in relief the lack of a theatrical narrative and cultural identity for the Met congenial with New York City's development. Plans were drawn up on prospective locations, and models were built to see how those designs fitted into the overall conception of the city. In 1925 Kahn—with the help of such wealthy businessmen as George Eastman and Marshall Field—even managed to secure, for an astronomical sum, a property for the prospective site.[18]

The original 1928 plan, named Metropolitan Square, called for a European-American hybrid: a regular square as a European-style civic focal point with a new building at the center that the Met board deemed a proper cultural monument (see fig. 15.5). The architect, Benjamin Morris, tried to negotiate the patrons' cultural and business interests: four office towers flanked the central cultural monument to serve the business interests on this valuable piece of property. In Morris's model, the old Met's palazzo façades had multiplied and expanded into skyscrapers dwarfing the opera house. In its bulkiness the proposed Met building resembled an industrial power station or a fortress rather than a Beaux-Arts style cultural monument.[19] Rockefeller's site manager John Todd concluded that "the opera house would be a dead spot and greatly reduce shopping values in all properties facing it."[20] Todd discarded Morris's plan and turned the series of towers flanking the proposed Met into Rockefeller Center, at the time one of the most massive urban developments in the world.

The success of Rockefeller Center as a film, radio, television, and business center opened up the prospect for the Met to become part of a larger conceptual public space, part of a superblock with other cultural

15.5   Metropolitan Square: Benjamin Morris's rendering for the proposed Metropolitan Opera on the future site of Rockefeller Center (1929). Permission from Avery Architectural & Fine Arts Library, Columbia University.

institutions, forming a powerful exhibition site for New York's cultural offerings. Since the Met's board lacked Rockefeller's financial and political clout, they began to collaborate with powerful city officials such as Robert Moses to acquire sufficient blocks. Because the historic squares along Broadway were already occupied and the business district had steadily moved up from downtown to midtown Manhattan,[21] the planning commission looked further north to find a site at another significant intersection. The result, Lincoln Center for the Performing Arts, emerged as a result of aggressive urban renewal enabled by government legislation. City leaders settled on several blocks at one of the few intersections of Broadway with another north—south avenue, an area called Lincoln Square, which subsequently became "the site of the largest slum clearance and Urban Renewal projects in New York City."[22] The Met emerged, controversially, out of land acquisition at the expense of a mostly African American and Puerto Rican population. Robert Moses exploited Title 1 of the Federal Housing Act, which allowed the city of New York to purchase condemned land and sell it to private developers (so also replacing approximately 7000 low- to middle-income housing units with 4400 mostly luxury apartments, creating residential areas for future Met patrons). Since the New York Philharmonic and Fordham University also needed new space but could not afford the land prices in midtown in the mid-1950s, Moses's four-square-block plan for Lincoln Center for the Performing Arts included the Fordham campus, Philharmonic Hall, a new Metropolitan Opera House, the City Ballet, a repertory theater, a high school of the performing arts, a library and museum of the performing arts, and a new home for the Juilliard School of Music.[23]

Lincoln Center transformed what was a mid-to-low-end residential district into a modern performing-arts district.[24] The site for the Met and its adjacent theaters was the southwest corner of the intersection of Columbus Avenue and Broadway. Unlike European cities, in which streets often seem to radiate from squares and cultural monuments— urban focal points—the visual plot for the Met created a different form of cultural identity. Most notably, the Lincoln Center cultural complex and the street axis are set against each other in dynamic interaction. Unlike the Paris Garnier Opéra, which was conceived as a focal point at a site of urban confluence, the Met had not only to share the site with similar buildings but was also set off against an unusual street intersection in the grid, similar to the one at Times Square and Broadway (this at Broadway, Columbus Avenue, and West Sixty-fifth Street). This urban organization expressed a central axiom of American identity through

THE OPERA HOUSE AS URBAN EXHIBITION SPACE

15.6   Metropolitan Opera at Lincoln Center, New York. Master plan and design by Wallace Harrison (1966). Photograph, 2010.

the individualistic development of blocks and buildings without deference to the centrality of visual perspective that traditional squares could powerfully suggest.

In this spatial configuration, it became critical to raise the Met's and Lincoln Center's architectural profile against the surrounding blocks. The site dramaturgy created by Wallace Harrison and executed by a team of architects conceived the idea of a gigantic cultural arts center as an artificially produced square, emphasized by elevating the performance venues on a plinth, with many low-level arts buildings grouped around the square and in adjacent blocks.[25]

The slums gave way to a miniature elevated city, and the spectacle of the Met at Lincoln Square thus illustrated Adorno's claim regarding opera's externalization of massive means (see fig. 15.6). To counter the block system's lack of any kind of public space other than public parks or deliberately grouped religious or political institutions, the architect gave the buildings theatrical presence by symmetrically grouping the Metropolitan Opera, the City Ballet, and the New York Philharmonic (with the Met at the center) around a fountain plaza, echoing public space in European cities. There were European echoes, too, in the buildings' façades, scaled in the image of the palazzo but with colonnades in front of large glass windows instead of multiple layers of floors. Unlike

223

the palazzo style of the old Met, which concealed the opera and its patrons, or the Palais Garnier, which expresses splendor in its exterior but hides the interior celebration from view, the Met's windows display the operagoing public. Their logic of display recalls not just the department store, but Benjamin's dramaturgy of the exhibition space that aimed at transparency, permeability, and visibility, that advocated for a "glass culture" as part of modern experience.[26] Glass, he felt, forced architects to create a new type of room—transparent, auratically restrained, and offering a dynamic relationship between inside and outside—that would change society and the way it perceived itself.[27]

The new Lincoln Square followed the logic of Koolhaas's block dramaturgy, creating another "island of solitude," a cultural island set against the traffic flow of the avenues, demarcated from the rest of the city. This Koolhaasian island becomes, in the manner of bourgeois opera, a refuge for aesthetic reflection. In its creators' desire to fill the entire superblock and to compete with the New York cityscape, Lincoln Center became an early paradigm for an artistic theme park that embraces all central art institutions of the classical world in a baroque spectacle.

Paris's Bastille Opéra: Urban Integration and Historical Index

The Met's site dramaturgy hinged on applying a historically defined urban and theatrical concept to the New York grid, but the planners of the new Paris Bastille Opéra were confronted with an entirely different set of urban, cultural, and political challenges by their site: in Benjaminian mode, they sought to activate the site's historical layers and to create a building with transformative urban qualities. Unlike the grid of New York blocks, steadily expanding since the early 1800s, the new Paris Opéra at the Place de la Bastille had to weave itself into a historically charged fabric, standing on the site of the prison that became the iconic building of the French Revolution in 1789. While the Bastille Opéra, designed by the Canadian architect Carlos Ott and completed (pointedly) in 1989, has remained, at best, a controversial building, a scenographic reading of the site dramaturgy reveals the architect's complex task of responding to the site's historical layers, employing glass architecture, and solving the very practical needs of surrounding neighborhoods (see fig. 15.7).

The new opera house owes its existence to the French president François Mitterrand's goal of bestowing a grand architectural and cultural legacy on Paris via the Grands Travaux building program,[28] which was

THE OPERA HOUSE AS URBAN EXHIBITION SPACE

15.7 Opéra Bastille, Paris, designed by Carlos Ott (1989). Photograph, 2011.

designed to revitalize the city for the French Revolution's two hundredth anniversary. Rather than following Baron Haussmann's nineteenth-century approach of reshaping the network of Paris's streets, Mitterrand and his advisers decided to place cultural monuments at poignant spots across the city. While Haussmann's boulevards represented a demonstration of the state's power, reshaping the city structure via condemnation of properties, Mitterrand's projects aimed at rehabilitating distressed areas and reconnecting existing districts.[29] Thus, the Bastille Opéra was less a replacement for the Palais Garnier than a paradigmatic revitalization project that turned a very busy traffic intersection into a metaphoric nodal point, connecting the northwestern Marais district and the Seine to the south.

The site dramaturgy for the Bastille Opéra was conceptually and strategically complex. According to Alan Miller, the problem was that an opera house, a scenic but static urban prop, was built at a location that is a "genuine crossroads," a "point of transition," and an "accidental square."[30] First, the Bastille Opéra was a continuation of a grand Haussmann-like east-west axis, the Voie Triomphale, symbolically connecting the Grande Arche de La Défense via the Arc de Triomphe to the Grande Pyramide du Louvre.[31] Secondly, the site's historical narrative was highly charged, being the location of the former prison and the commemoration of the 1789, 1830, and 1848 revolutions. In 1803 the Place was laid out at the eastern edge of Paris, at the site where the fortress had been integrated into the city walls, running north-south along the Place's eastern side. The Place's Column de Juliet (commemorating the 1830 July Revolution) is a central focal point, just east of the original prison, around which the traffic flows. Thirdly, the Place de le Bastille was an icon of modern transportation, the site of the Gare de la Bastille, until 1969 the final destination for trains from the east and also a major eight-street traffic intersection. Thus the sense of speed and transition characteristic of modern urban areas is crucial to the site's narrative.

One key issue for the Bastille site scenography was how to interact sensitively with the historical footprint of the prison and existing developed neighborhoods, negotiating it as a space of cultural encounter. One of the major issues was whether to redesign the Place as a whole into a more traditional square or to insert the building somewhere into the existing space. The choice between the final two designs (selected from among 757 submissions for the original competition) was also between these two approaches: Christian de Portzamparc proposed radical remodeling, filling the entire triangle from the rue du Faubourg Saint Antoine to the rue du Lyon, building over the rue du Charenton; Carlos Ott (the winning designer) proposed inserting an opera house roughly into the preexisting footprint of the former train station, the Gare de la Bastille.

Turning against baroque spectacle, Carlos Ott created what one might call, following Benjamin, an indexical design, translating the temporal passage of the Place de la Bastille into the architecture of an urban project.[32] Above all, the Opéra's fortresslike structure stages the physical presence of two former massive buildings on the site: the prison and the train station.[33] The opera's footprint fills out the entire location of the former Gare de la Bastille, while the prerevolutionary prison is figured through the opera's sheer massiveness (its outlines remains visible in paving stones between the boulevard Henry IV and the column). The building towers over the historic neighborhoods, its presence particularly palpable during the day or after evening performances when the lights have been turned off: then, the opera appears as the foreboding container of the past rather than a palace of bourgeois entertainment. Within, Ott managed the transition between foyer and auditorium by creating tower-like structures, echoing the towers of the former Bastille. Ott also reworked the orientation of the train station's façade, replacing the sense of the train station as a cul-de-sac on the Place by borrowing its original glass frontage, facing the Place, for the Opéra, but curving that façade away from the Place, to echo the shape of the central traffic roundabout. As part of the urban revitalization project, shopping arcades were installed under what used to be the train's approach to the station, linking opera and shopping as major bourgeois activities.

The glass façade also moves the Bastille Opéra building, at least conceptually, beyond its indexical quality. Glass façades have become highly fashionable not only in skyscrapers and office towers but in opera buildings; they enact the supposed openness of Western societies and enable the voyeuristic spectacle of seeing the city from inside while being seen from the outside. Ott's façade, a mix of transparent and opaque fascia tiles that has been the object of much architectural criticism, stages historical interplay, with opacity evoking the closed prison walls and transparency both gesturing to democratic permeability and creating a Benjaminian extended space of action. It is a strategy to activate a contemporary, non-elitist image of opera—certainly a concern of the socialist President Mitterrand—and to create a reflective relationship to the past prison.

The indexical design of the Bastille accomplished one of the principal goals of Mitterrand's Grands Travaux, assisting integration in a Parisian neighborhood that had been rather neglected and helping to connect the fourth and eleventh arrondissements, historically divided by the city wall. The plan for the Opéra left the neighborhood of Faubourg St. Antoine intact and (like the Centre Pompidou, at the other end) helped the Marais district become fashionable, as well as connecting with neighborhoods to the south. Instead of highlighting the Bastille's original function as a fortress at the boundary between Paris and its supporting farmland, Ott planned the opera as a force of integration, transforming a potential cultural monument into a viaduct of energies and an exhibition site of historical forces.

Winspear Opera House, Dallas: International Scenography and Cultural Destination

Unlike Paris and New York, Dallas is a new type of city, lacking rich urban history but made wealthy by oil. It presents a new set of challenges for urban site dramaturgy but also further scenographic opportunities for opera houses as civic exhibition sites. The city, sprawling across the plains of the southwestern United States, has grown exponentially since it began developing as an oil and transportation hub in the 1960s. Until the 1970s central Dallas was basically a business district without significant residential neighborhoods or stable artistic destinations. In a city with a relatively weak core, a smaller grid than New York's, but extensive suburbs, exurbs, and edge cities, the creation of an opera house as an exhibition site became a new challenge, its reference point a more diverse, fluid economy and a globalized culture. Dallas's Winspear Opera House became what Benjamin called a dialectical image: because the city's urban core lacked coherence and magnitude, the architect had to create a high-tech building as a means of staging Dallas itself.

In the mid-1970s Dallas began planning an extensive arts district— including the new Winspear Opera House, designed by Norman Foster—to shift a range of arts institutions from perceived wastelands located outside the transportation network.[34] The Dallas Museum of Art had already begun to search for a more respectable space so that it could eventually house the valuable paintings owned by affluent Dallas citizens.[35] To stimulate interest in urban planning, the city requested a study from Kevin Lynch, the author of an influential book on city design, and his partner, the urban designer Stephen Carr.[36] Their report suggested that Dallas's other major arts organizations should move to the same area as the museum, as this clustering would "benefit the organizations involved while also helping to revitalize the central business district."[37] For Dallas, already a city with a perceived "edifice complex," the call for more buildings presented an opportunity the city leaders readily accepted, adopting a master plan assigning each institution a block of its own.

Like other urban renewal plans in the United States, the Dallas Arts District represents a reenactment of 1950s policies, in which cultural monuments were deployed as neighborhood-clearing programs that accentuated racial divides. The planners for the Dallas Arts District, like those clear-cutting for modern suburban subdivisions, carved out a seventeen-block section on the edge of downtown Dallas and assigned locations to arts venues.[38] Then, following the utopian paradigm of Frank Lloyd Wright's *Broadacres* (1932), which had guided much suburban development,[39] this area was to be entirely landscaped and artificially developed as an entity separate from the existing urban fabric. The planners envisioned organizing the arts district into three areas: Flora Street, a pedestrianized area, was considered the "front door to major cultural institutions and to private and semi-private developments" (Sasaki). The street was then subdivided along an east-west orientation: "Museum Crossing is a collection of boutiques, galleries, and art shops, Concert Lights centers on the Dallas Concert Hall with theater-oriented restaurants and clubs, and Fountain Plaza creates an artists' quarter ambience with gourmet shops and open air markets" (Sasaki). Over a period of twenty years, ambitious business leaders acted on Lynch and Carr's recommendation and the city's aspirations and took advantage of generous land deals. In a race with commercial developers, who also wanted the cheap land available near downtown, wealthy businessmen such as Bob Kilcullen (the art museum), Raymond Nasher (the sculpture garden), Morton Meyerson (Symphony Hall), and Bill Winspear (the opera house) bought plots cheaply or provided significant start-up funds (in Winspear's case, $30 million). In each case, donors and art institutions approached prestigious architects, which created an area of international architectural interest. Thus, Foster's opera house, in the "Concert Lights" section of the plan, is surrounded by venues designed by I. M. Pei (Symphony Hall), Rem Koolhaas (Dallas Theater Center), Renzo Piano (Nasher Sculpture Garden), and Edward Barnes (Dallas Museum of Art), each occupying a block of its own.

The massive means of nineteenth-century opera that Adorno envisioned as a harbinger of the modern culture industry erupted in Dallas in full force. The Dallas Art District reenacts the nineteenth-century world's fair concept by featuring exemplary design solutions as exotic commodities inside a new type of exhibition space, becoming a high-culture architectural theme park. Thus, while Rem Koolhaas described New York's grid of independent blocks as a system of "solitudes," in the Dallas Arts District in general, arts organizations float free of the urban fabric in spaces that become at once megablock and (particularly apposite for the opera house) megatheater. In this conglomeration of edifices, it is particularly challenging to achieve urban reference, architectural identity, and cultural narrative. The opera house, removed from a traditionally mixed urban context, is exhibited amid several other cultural institutions and required to assert itself. While Koolhaas (the architect of the neighboring theater) built a skyscraper to solve this problem, Norman Foster decided to exhibit his opera inside a massive glass casing that, much as Benjamin theorized in his reflection on glass architecture, created a voyeuristic viewing dynamic. This was a challenge Foster must have relished: throughout his career he has been adept in creating powerful architecture in congested environments and in particular has found imaginative scenographic solutions for juxtaposing modernism and historical styles.[40]

The Dallas Arts District allowed Foster to work in a highly planned urban environment and to draw on a visual language developed in other contexts to create a new narrative. Thus Foster stated that he wanted to create a "building that is a model for the future," and that the project was about an architecture that "offers a truly democratic experience of opera for the 21st century."[41] Fundamental to Foster's approach to Dallas's opera house is his sense of architecture as a matter of performance

15.8 Winspear Opera House, Dallas, designed by Norman Foster (2009). Photograph, 2014.

and narrative. Accordingly, Foster laid bare the traditional container of operatic performances, the horseshoe auditorium, exhibiting it as a large red drum that glows at night inside a glass case, a façade covered by a canopy radiating away from the building (see fig. 15.8). Since the Dallas opera community is small by comparison with those of other metropolises, the auditorium was designed for multipurpose events, including opera, ballet, musicals, and large-scale concerts. Inside the foyer (entirely visible through the glass façade) Foster wrapped the auditorium in a system of stairs, enabling the audience to promenade before the show and during intermissions. The auditorium itself is as spectacular as it is conventional: the horseshoe, draped in red and gold, contains four open balconies with the traditional sight-line issues; it sports a gigantic chandelier that recedes into the ceiling and a flexible orchestra pit that adjusts to take the shape of either a Bayreuth-type sunken pit or an eighteenth-century-style pit.

The auditorium's glass casing is the building's core concept, making the auditorium visible within the arts district but also graphing the building onto a larger international context, arising from more than a century of glass architecture. While Benjamin admired glass architecture for its sobriety and transparency, it has assumed its own aura to a point where the style has itself become a historical index. It is also an index of Foster and Partners' trademark style: this large-scale international firm reconfigures successful design elements from Berlin's Reichstag and London's British Museum. In his reworkings of the Reichstag, for which he created a new glass dome, and in the courtyard for the British Museum in particular, Foster showed his talent for reinterpreting existing buildings and turning them into stylish architectural statements while at the same time applying ecological ideas. One of the main architectural elements featured at both the Reichstag and the British Museum is a central structure from which a glass canopy radiates out to allow both visibility and opportunities for walking inside the structures. This strategy reappears in his design for the Dallas opera house.[42] In the Berlin Reichstag this idea was an ingenious solution to provide transparency and mutual visibility between the building's interior and its Berlin environment. This modern intervention into the building's Beaux-Arts architecture of the late nineteenth century has been appreciated by public and critics alike; in the case of the British Museum, it also served to link various spaces that display its ancient cultural treasures, collected from around the globe. In both cases Foster augmented sites stemming from the imperialistic past with an image of democracy and openness. For his design of the Dallas opera house, Foster recalled these images, but in a substantially different context. He established a similarly strong link between inside and outside—the shell of the auditorium and transparent lobby— allowing the red drum, Foster's architectural image of opera, to be the focal point. But the red glow of the auditorium and opera patrons strolling alongside the building's core also assured the building a wider visibility within the Arts District, while recalling the strolling visitors in the glass dome of the Berlin Reichstag, the courtyard of the British Museum, or operagoers at Lincoln Center.

The site dramaturgy created for the Winspear Opera House is more fragmented than the one for the Met and less site-specific than the one for the Bastille Opéra. The building contains very little reference to its Dallas location. Because the Arts District is entirely unconnected to either residential life or the business district, the building's immediate reference system is to the other surrounding high-class, high-tech buildings. Foster's opera house and its immediate neighbor, Koolhaas's Theater Center, appear as the Janus faces of modernity, expressions of a contemporary split consciousness about theater architecture. While Koolhaas's building is conceived as a skyscraper, experimenting with spaces stacked on top of each other, Foster's opera spills out into the district, its transparent canopy extending to the front and rear; while Koolhaas experiments with a curtain-like shell veiling the inside space (with the exception of the ground floor, which contains the performance space), Foster deliberately enacts openness and transparency, using a glass façade (which, with its opportunities for gazing into and out of the building, corresponds to the glass employed by Renzo Piano for the nearby Nasher Sculpture Garden); while Koolhaas works with rectangular forms, Foster uses round ones (in the adjacent block Pei manages to find a combination of both shapes in his Symphony Hall); while Koolhaas employs a very detailed gray surface texture, Foster uses a deep red.

Like the Bastille Opéra, the historical index of Foster's architecture facilitates the building's dramatic storytelling, lending it agency in the urban landscape. Aside from its interplay with the architectural vocabulary employed by Koolhaas, Pei, and Piano, the building's core, the red drum, enacts a new, visceral appeal of opera. The red horseshoe of traditional opera is not only made visible but seductive, even erotic; by turning the conventional colors of an opera auditorium inside out, attention is drawn to the genre through visual experience. Furthermore, Foster links the deeply expressive form of the drum to the well-known symbol of new democracy, the glass dome of the Berlin Reichstag, which allows visitors to experience the layers of history while walking and, like the courtyard in the British Museum, refreshes the museum experience; in the same manner, Dallas's opera house enables a contemporary audience to re-experience the genre of opera. The dramaturgy of walking, advocated by modern theoreticians from Walter Benjamin to Michel de Certeau as a form of active experience and visual perception, links the visual experience from Berlin and London to Dallas, an ironic gesture in a city that relies on the car—indeed, one whose sheer size makes the city inhospitable to pedestrians.

Foster's site dramaturgy treads a fine line between excavating the interior of what many might consider an outdated genre and invoking the emotional atmosphere of opera through visual display. His opera architecture has become, from Benjamin's perspective, a display cabinet, exhibiting a venerable genre yet (seemingly) making it available for contemporary consumption, since the structure is, like the other art venues, visible from outside the Arts District. However, he also reverses the perspective by making the downtown area visible from the opera's lobby, allowing the audience members a critical look at their own city: the transparency of the glass surrounding the auditorium allows for a dialectic of seeing and being on mannequin-like display.

Exhibition Site and Cultural Identity

Just as Paris, New York, and Dallas represent different historical stages of and physical constraints on modern urban development, the exploration of selected site dramaturgies offers one way to describe historically a cultural geography of twentieth-century opera. In modern societies, the political and cultural coalitions supporting opera have become more complex, so that the demands for a scenographic realization of a cultural narrative have steadily increased. In a complex modern urban topography, contemporary opera buildings have become an exhibition site for those forces and a means of describing and debating cultural identity, while at the same time supplementing and potentially enhancing the distinctive attractions of the urban environment as a whole, as a cultural destination.

One might argue, with Adorno, that contemporary cities have enlisted architects to turn opera's grand effects into large-scale narratives for and in the contemporary city. I have examined three different types of urban development and their respective efforts to co-opt the urban sphere as a large-scale exhibition site. In these three examples, opera architecture extends beyond the individual building and turns into an urban concept: the Met constructed an archetypal environment by transforming a segment of New York's modern block system into a small, neo-Baroque city. The architecture and site dramaturgy of the Bastille Opéra may be read as a historical index of past and present forces of urban development. The Winspear Opera House in Dallas, in a variation of segregating urban districts, is detached from its actual urban fabric but integrated into a new kind of cultural district; it draws its narrative from global contexts, which, in turn, are on display and ready for consumption in Dallas. In each case, architecture has moved opera into the public imagination not only in the hope of bringing audiences back to the actual artwork; it has also employed modern technologies and urban planning to create a scenography that turns the theatricality of grand opera inside out, displaying its cultural inheritance to and for the city.

SIXTEEN

Underground in Buenos Aires: A Chamber Opera at the Teatro Colón

ROBERTO IGNACIO DÍAZ

On 25 May 1908 the Teatro Colón in Buenos Aires opened its doors to the public for the first time, and the men and women of the Argentine oligarchy, including some ardent believers in the art of opera, must have listened with varying degrees of rapture to Verdi's *Aida*. Sitting in dark-red velvet seats, the audience could have imagined itself transported to the ancient Egypt of that tragic story, but also to the heart of modern Europe—to Paris or Vienna. They may have been geographically removed from the traditional centers of Western culture, but the extravagant art of opera carried them aurally and visually across the Atlantic.[1] Indeed, growing prosperity in the early twentieth century meant that, as Alan Gilbert states, "Buenos Aires began to transform itself into the first European city in Latin America." Argentina's program of urban modernization was strikingly European:

Electricity and water systems were improved and major port works began. City beautification included the building of wide avenues, most notably the *Avenida de Mayo*, intended to rival the *Champs Elysées*. An electric tramway was opened in 1897 and, in 1913, the region's first underground railway—the *subte*. The city's elite aspired to turn Buenos Aires into one of the world's main cultural centres and after its opening in 1908 the Colón theatre became a regular stop for top opera singers.[2]

234

The Colón would soon be regarded not only as one of the great twentieth-century opera houses, but also as an urban icon that, to this day, conjures the European origins of opera as well as its expansion across the globe. Citizens of Buenos Aires cherish the opulent edifice in the city center, and many visitors from the rest of Argentina and abroad flock to it for tours that proudly detail the building's transatlantic lineage. Indeed, the design of the Colón mimics that of grand opera houses in the capitals of the so-called Old World—an act of architectural citation confirming the city's reputedly European airs. There is all that marble from Italy and Portugal used in its construction; there are the Salón Dorado's stained-glass windows reminiscent of Versailles and Schönbrunn, according to various websites, visually retelling the myths of Apollo and Sappho. And there are also, in the Salón de los Bustos, the effigies of eight prominent European composers—Beethoven, Bellini, Bizet, Gounod, Mozart, Rossini, Verdi, and Wagner—whose classic works continue to operate their magic on a continent far from the birthplace of opera.[3]

Through much of its history, the Teatro Colón has functioned brilliantly as a space for the display of operatic trophies from Europe. If Argentine composers have been well represented in its programs,[4] nonetheless the bulk of each performing season is devoted to works from the other side of the Atlantic. The Colón, along with other similar theaters, contributed to Buenos Aires's self-fashioning as a cosmopolitan outpost of European cultural forms from the mid-nineteenth century into the twentieth. If distance from other operatic centers might have acted as an impediment for the transport of opera, Argentina's antipodean location meant that its operatic "season" neatly dovetailed with that of the northern hemisphere—affording European singers the opportunity to schedule performances at the Colón during what was their fallow period. Perhaps this is why the Colón was regarded as part of an exclusive sextet of great operatic venues—along with Milan's Teatro alla Scala, London's Royal Opera House, the Vienna Staatsoper, the Paris Opéra, and New York's Metropolitan Opera House—that routinely presented the most brilliant programs.[5] The management, ever up-to-date in its knowledge of what was happening abroad, crafted seasons in which works were often performed shortly after their European premieres. Thus, the Argentine public enjoyed—at least in the realm of opera—what could hardly be considered a peripheral experience; for the few hours it took a performance to unfold at the Colón, the citizens of Buenos Aires could regard their hometown as a world-class city. By means of opera, the stories and histories of Europe came alive, and the city partook of the cultural authority vested in the Old World, especially when

ROBERTO IGNACIO DÍAZ

compared to other Latin American capitals, where the practice of opera was far less common.

Despite some major troubles through its history—political strife, reduced programs, delayed renovation around its 2008 centennial—the Teatro Colón remains a steadfast purveyor of the grand traditions of European opera to a loyal (if not obsessive) band of enthusiasts.[6] But in recent times, a different kind of music drama, including locally created chamber works, has been staged under its roof at the Centro de Experimentación del Teatro Colón (CETC), a venue with a curious double status. Founded in 1990, the CETC is (as its name indicates) a nucleus for theatrical innovation, but it is also a peripheral center—an oxymoron of sorts. Its activities do not include the Colón's best-known repertoire, and even its physical location in the building, far removed from the plush main auditorium and having only two hundred seats, is modest and marginal—an underground space. The CETC has produced a wide range of European musical works from various periods and genres—from Monteverdi and Bach to Berg, Britten, and Henze—but its commitment to Argentine composers and new operas is unwavering.[7]

Among the works created for the CETC is *V.O.*, a chamber opera by Beatriz Sarlo and Martín Bauer for an ensemble of six strings, piano, and percussion. Premiered in 2013, the one-act opera unfolds in fourteen scenes that chronicle the life and times of the Argentine author Victoria Ocampo, best known as the publisher of *Sur*, one of the most prominent Spanish-language literary journals of the twentieth century. Sarlo, whose previous work as a cultural historian of Buenos Aires and critic of Ocampo's life has focused on ideas of modernity, takes up that analysis again in her libretto. Beyond its biographical plotline, *V.O.* may be read as a subtle meditation on the transport of European forms to Argentina—not just music and opera, but literature and architecture as well—and their place in the geography of Buenos Aires. Sarlo's choice of a chamber opera staged underground at the Colón as her medium is undoubtedly significant. The traditional repertoire continues in the main auditorium. Yet the building's smaller, seemingly clandestine operatic stage tells a different story of transatlantic liaisons—one in which modernity, as understood by Ocampo and Sarlo, disrupts the certainties of a well-guarded canon and the idea of Buenos Aires as a simulacrum of European tradition and grandeur.

A little background will explain why *V.O.* is ideally placed to perform this unsettling work. Born in 1890 to a wealthy patrician family in Buenos Aires, Victoria Ocampo (1890–1979) had a lifelong, versatile relationship with music and the Teatro Colón, but, more importantly, with literature. A cosmopolitan reader, she envisioned *Sur*, which she founded in 1931 and edited until its cessation in 1970, as a space for presenting Argentine and Latin American literature to the world, and for translating mostly European and North American works into Spanish.[8] An author educated in French and English and never completely at home writing in Spanish, she penned numerous essays and a six-volume autobiography (in French), becoming the first woman to be admitted as a member of the Academia Argentina de Letras.[9] Ocampo was a passionate devotee and promoter of music and opera. Like other members of her social milieu, she regularly attended performances at the Colón and often wrote eloquently about what she saw and heard there, but that was hardly the extent of her musical fervor or her presence at the theater. A gracious hostess to foreign authors such as Rabindranath Tagore and Graham Greene at her family's villa in San Isidro, just north of the city, Ocampo was also involved in bringing distinguished musical figures to Buenos Aires. In 1925 she helped attract the Swiss conductor Ernest Ansermet to the country and even performed under his direction as Speaker in Arthur Honegger's *Le roi David* at the Teatro Politeama.[10] For a few months in 1933 she served as a member of the Colón's board of directors (before resigning), and in 1936 she even performed there, taking on the role of Speaker in the Latin American premiere of *Perséphone*, by Igor Stravinsky, a personal friend whose visit she also sponsored.[11] In these multiple pursuits Ocampo created her own cultural geography—a space in which the Americas and Europe acted jointly through the pleasures afforded by literature, music, and the other arts.

That Ocampo should become the subject of an opera staged at the Colón may be seen as a homecoming of sorts, but also, arguably, as a reevaluation of her work in the cultural history of Argentina. In this, Sarlo's analysis of a "mixed" modernity in 1920s and '30s Buenos Aires, in her influential *Una modernidad periférica* (1988), is a useful key. The culture of intellectuals—writers and artists—was in her view a blend of opposites: "European modernity and Rio Plata difference, speeding up and anguish, traditionalism and a spirit of renewal; regionalism and the avant-garde" (*modernidad europea y diferencia rioplatense, aceleración y angustia, tradicionalismo y espíritu renovador; criollismo y vanguardia*).[12] Sarlo calls Buenos Aires "the great Latin American stage for a culture of mixings" (*el gran escenario latinoamericano de una "culturas de mezclas"*),[13] a phrase that invokes not only the notion of hybridity, so central for thinking about culture in the region, but, perhaps less obviously, a vision of the city as a site for the performance of a qualified modernity. This is the background against which *V.O.* is set and in which Bauer and Sarlo's Victoria performs her story.[14]

In its dramatic portrayal of the author's life and times, Sarlo's libretto (spoken and sung) often cites Ocampo's own words. By means of impressionistic tableaux, the plot chronicles Victoria's childhood, filled with rigorous piano lessons; her wish to become an actress, despite her father's interdiction; her ardent liaisons with the forms of modern art, especially the music of Stravinsky (discovered in Paris); her complicated affair with Pierre Drieu La Rochelle; her passionate engagement with the arts; and her gradual self-realization. As Sarlo described it in *Una modernidad periférica*, Ocampo, despite her privileged social status, was deprived of real and symbolic freedom because of her gender, lacking "intellectual freedom, sexual freedom, affective freedom" (*libertad intelectual, libertad sexual, libertad afectiva*, 87). In Sarlo's account, the struggle ended only when Ocampo founded *Sur* and began also to reign over her own body with "the freedom of men" (*la libertad de los hombres*, 87). By staging Victoria's progress in an opera produced under the aegis of the Colón and within its premises, Sarlo's libretto vividly performs the symbolic elevation of a cultural figure at times dismissed in Argentina and abroad as just a wealthy woman with interesting contacts. Although—or perhaps because—it is only a chamber work, *V.O.* implicitly advocates for a kind of opera that can tell stories about Argentina that have been unseen and unheard in the main auditorium's opulent production. Likewise, the modest space in which *V.O.* is performed emerges implicitly as an underground commentator on the goings-on upstairs and on Ocampo's role as a harbinger of aesthetic modernity in Argentina.

Indeed, creative bonds tie *V.O.* to a network of past and present writers and artists that include not only Ocampo and Stravinsky, but also Bauer and Sarlo themselves.[15] Although my focus here is on the libretto, one should note Bauer's role in founding (in 1997) the Ciclos de Conciertos de Música Contemporánea at the Teatro San Martín, which shaped and supported an appreciation for modern works among music lovers in Buenos Aires—something that Ocampo herself had championed when on the Colón's board of directors, announcing plans for the opera's renewal to *El mundo*, but being thwarted by concerted opposition.[16]

If Sarlo's critical writings on Ocampo have played a major role in raising the author's perceived value, the libretto goes even deeper as a critical tool. The words—like Bauer's music—act out the scholar's vision of her subject by replicating what she identifies as the defining trait of Ocampo's style: the practice of citation. Indeed, Sarlo's longest and most substantial study of Ocampo's work, included in *La máquina cultural* (1998), is titled "Victoria Ocampo o el amor de la cita" (Victoria Ocampo or the Love of Citation).[17] Sarlo's portrait of the writer in *V.O.*

thus cites Ocampo's own words as well as her citations of other authors, while Bauer's score engages in musical citation—particularly of composers Ocampo admired—including several chords of Stravinsky's *Rite of Spring* and Debussy's prelude "Hommage à S. Pickwick Esq. P.P.M.P.C.," which in its turn famously quotes the opening refrain of the British national anthem. That Sarlo follows Ocampo's voice so closely might tempt one to read in *V.O.* a (female) librettist's anxious dependence on her subject's authority—a state of affairs, perhaps, not unlike the Colón itself as an architectural space and theatrical enterprise reliant on previous models. Yet Sarlo's mimicry allows us to view Ocampo's literary and musical flights not as mere Europhilia, of which she was often accused, but as the crafting of a cultural geography of her own. Although the opera's plot moves vertiginously across time and space, traversing decades and following Victoria (the first name matters) across the Atlantic several times, its narrative thrust is firmly situated in Buenos Aires. Through the focus on Victoria's aesthetic and amatory bonds, the supposedly peripheral city emerges, if only belatedly, as a thoroughly modern place in which the Colón—its programs, its opulent edifice—is no longer the principal center of things, and where other kinds of opera, such as *V.O.* itself, also have a place.

Sarlo engages in many instances of citation throughout her libretto. These often invoke Ocampo's love for European letters, especially English and French literature, but they also refer to her passionate embrace of modern music, which she discovered in Europe. As the opera begins, for instance, the audience listens to string music, and words are projected onstage and read by an actor: a brief text by Ocampo that describes the legendary premiere of Stravinsky's and Vaclav Nijinsky's *Rite of Spring* in Paris in May 1913. The projected citation ends by underscoring that these are Ocampo's own words: "Victoria said: it has the brutal rhythm of a cataclysm" (*Victoria dijo: tiene el ritmo brutal de un cataclismo*)[18]—a phrase taken from *La rama de Salzburgo* (1981), the passionate third volume of her posthumously published autobiography.[19] This is a significant starting point: as is well-known, the first performances of the ballet triggered riots and accusations of barbarism; this watershed moment in European music was (as Sarlo underscores in her critical works) also a revelation for Ocampo, who was in the audience. Later, Ocampo returned to the Ballets Russes—to *Le spectre de la rose*—in the company of both her husband and her husband's cousin, Julián Martínez, whom she had recently met in Rome, and who would become her lover back in Buenos Aires shortly thereafter. Public and private "cataclysm" are therefore interwoven at the opera's opening, with Ocampo treated, through the use of deliberate citation ("Victoria dijo"), as an authoritative historical source, and her words as a valued record of her feelings and a key to interpreting the opera about to unfold.

Beyond the story of *The Rite of Spring*, the act of citation reappears powerfully in the libretto's account of Ocampo's thwarted acting career. As Ocampo tells it in her letters to her friend, Delfina Bunge, cited in full in *El imperio insular* (1980), the second volume of her autobiography, she was an eager and talented student of French diction and acting with the renowned actress Marguerite Moreno (who appears as a character in *V.O.*). Ocampo's dream was to perform: "Ah, to be able to perform Musset's Muses and Racine's Princesses—and Doña Sol!" (*¡Ah! Poder representar a las Musas de Musset y a las Princesas de Racine. ¡Y Doña Sol!*)[20] The process of acting, of course, entails citation on various levels, but in Sarlo's play this becomes particularly pointed—a multilayered act built on translations visible and invisible whereby cultural identity is performed. Later in the same scene, Victoria says to her teacher, "I'd like to be like you" (*Quisiera parecerme a usted*), and then recites the first three lines of a monologue in Racine's *Andromaque*: "Where am I and what have I done? What should I do now? / What emotion possesses me? What pain devours me? / Errant and without a destination, lost in the palace" (*¿Dónde estoy y qué hice? ¿Qué debo hacer ahora? / ¿Qué emoción me domina? ¿Qué pena me devora? / Errante y sin destino, perdida en el palacio*). In a libretto rife with citation, the choice of such lines seems emblematic of Victoria's broader search for a place of her own, a search Sarlo had already touched on. In *La máquina cultural* she analyzes Ocampo's writings on acting, relating them to the author's penchant for citing the words of others, often in foreign tongues:

El actor ocupa el texto con su voz y su cuerpo. Se sustituye. Este movimiento por el cual las palabras ajenas se convierten en palabras propias es el que Victoria Ocampo elige, más tarde, como movimiento típico de su escritura donde las citas y los *glissandos* al francés, el inglés y el italiano son tan importantes como el texto propio y la propia lengua.[21]

[An actor occupies the text with her voice and body. She substitutes herself. This movement through which someone else's words become one's own is what Victoria Ocampo would later choose as the habitual movement of her writing, in which citations and glissandos into French, English, and Italian matter as much as one's own text and tongue.]

It is unsurprising that Sarlo engages in a similar act, whereby someone else's words—Ocampo's, Moreno's, or even Racine's as recited by Victoria—

UNDERGROUND IN BUENOS AIRES

become the fabric of her own libretto. Sarlo's own translation of Racine's text into Spanish is also in itself an extension of Ocampo's literary and editorial practice—a "máquina de traducir" (translating machine), as Sarlo puts it, given Ocampo's frequent self-translations and *Sur*'s mission of publishing foreign texts in Spanish. In multiple ways, then, Sarlo offers a kind of mimicry and ventriloquism that serves to both represent and validate Ocampo's practice of writing.

Likewise, when set against the history of Buenos Aires as an importer of opera, Sarlo's libretto may be read as a tale of cultural naturalization. Transported to new shores even as it, in turn, transports the city's audiences elsewhere (Europe, Egypt), opera is first and foremost about the performance of strongly affecting European forms. But the importation of those works also requires the establishment of literal and figurative spaces—opera houses, cultural discourses—in which they can be deployed and housed in their new geographical context. The singing of Italian or German opera may at first be heard as an invocation of alien things, an act of citation, but the repetition and normalization of operatic performances in Buenos Aires insinuates these foreign words and stories into the local systems of culture. These transatlantic passages, in which Europe and Argentina mix, resonate with Ocampo's interlingual writing and Sarlo's operatic recreation thereof.

In her critical analysis of Ocampo's corpus, Sarlo employs the Italian word *glissando* a second time (after the instance cited above) to refer to the author's effortless movement from one language to another:

Jamás se le ocurrió que la idea de que esa mezcla de lenguas, convocadas cada vez que las siente necesarias, no sea sólo un emblema de riqueza sino también la imposibilidad de decir las cosas de otro modo: una servidumbre al francés o al inglés, en lugar de una libertad de transitar *glissando* de uno al otro.[22]

[Never did it occur to her that the idea of this mixture of languages, brought together whenever she felt necessary, could be not just an emblem of wealth, but also of the impossibility of saying things in any other way: a servitude to French or English, instead of the freedom to transit, glissando, from one to another].

Sarlo's use of a musical term to describe what some may view as Ocampo's linguistic inelegance, or even colonial dependence, retrieves *Sprachmischung* as a fruitful authorial practice. Indeed, the notion of glissando also appears to describe unwittingly the nature of *V.O.* as its own kind of opera. In Bauer and Sarlo's work, song and speech, or music and words, slide against and away from each other in unconventional manner. A new staging, arguably, of the old debate about the primacy of one code over the other in the operatic text. *V.O.* demonstrates that operas come in many different forms. Here, singers and actors sometimes share the stage, the musical and the verbal work together in three arias, and a piano solo and a player's monologue glide harmoniously together. These fluctuations may remind some operagoers of the ardent interplay between literature and music in Ocampo's autobiography.[23]

Like many of the European operas seen on the main stage of the Colón, *V.O.* touches on the subject of love—specifically, Ocampo's short affair with La Rochelle. That he is identified both as "Drieu" and "Hombre" (Man) suggests that his character might be read, generically, as any of the men with whom Ocampo was romantically involved, including, most prominently, her husband's cousin, Julián Martínez. Conventions of romantic—and particularly operatic—love were powerful signifiers in Ocampo's own analysis of her story. She resorted to words from Stendhal's *De l'amour* (1822) to recount and scrutinize her feelings for Martínez. Not only is the third volume of her autobiography, which focuses on the affair, titled *La rama de Salzburgo*, but the book also opens with an epigraph from *De l'amour* on the process of love as a form of crystallization. Ocampo's use of the French author's words and concepts alternates with her relentless invocation of operas by Wagner and Debussy—works that deal with what she calls, invoking Stendhal's erotic lexicon, "amor pasión," even as they chronicle stories of forbidden love with tragic outcomes.

As it happens, Ocampo and Martínez, upon returning to Buenos Aires from Paris, first encountered each other by chance at a performance of *Parsifal* at the Colón, which (according to Ocampo's autobiography) kindled their passion, repressed until then.[24] Throughout her narrative, allusions to *Parsifal* as well as to *Tristan und Isolde* and *Pelléas et Mélisande* serve to explain various aspects of that relationship. All three operas chronicle stories of sexual transgression, and in the latter two the protagonist is a married woman who, like Ocampo herself, loves a man who is not her husband. In *La rama de Salzburgo*, Ocampo resorted to the *Tristan* Prelude and the opera's motifs to capture the sense of dramatic expectation as her affair with Martínez unfolds. Aware of her parents' and her husband's almost certainly negative reaction to news of the affair, she invoked the story of Pelléas and Mélisande, citing Arkel's words and transposing the opera's plot to her own circumstances. She also mentioned a series of literary lovers, including Paolo and Francesca and the Princesse de Clèves, but kept returning to opera, focusing, for instance, on Amfortas's wound in *Parsifal*, mentioning *Tristan und Isolde* several times, and even alluding to the real-life affair of Wagner and Mathilde Wesendonck. Furtively, Ocampo and Martínez sought out discreet locations in Buenos Aires where they could see each other; in a strange botanical displacement, the autumnal *tipas* (a local type of rosewood) at the Paseo de Julio reminded her of the Cornish pine tree featured in some versions of the Tristan and Isolde legend.[25]

The passions of European opera, then, served Ocampo as a meditation on the affects of her own South American love story. Opera fans reading about her travails in the autobiography, aware of the stories she alludes to as well as other well-known operatic plot lines, might expect a tragic denouement—one in which Ocampo is defeated, like so many other operatic heroines before her. Yet Ocampo, as her autobiography eventually reveals, is saved. Ultimately, all passion is spent, reason prevails, and the text ends up chronicling the rise of a strong woman who becomes the publisher of a storied journal—and the author of a compelling literary oeuvre.

In her libretto, too, Sarlo empowers the character of Victoria with the ability to verbalize and act out her own story. In their famous operas, Isolde exits the stage after her "Liebestod" and Lucia di Lamermoor also dies after her mad scene. Unlike the heroines of many of such operatic warhorses, Sarlo's Victoria brings about an altogether different denouement. The central moment in *V.O.* is arguably Victoria's passionate aria (one of three) in the ninth scene—a perfect hendecasyllabic sonnet in which she invokes through music and words aspects of her erotic life and love for a man—Drieu, but also, arguably (because of Sarlo's choice of citations from Ocampo), Martínez—even as she asserts a sense of victory over the social impediments of gender:

Otros cuerpos nunca conocieron
mayor entendimiento que los nuestros,
mayor placer, tampoco más ternura,
cuando el deseo, cumplido, se dormía.
No dejes que me vaya, no me pierdas.
No sé nada de vos. ¿De dónde viene
esta locura? Restos del naufragio
de un verde paraíso en la llanura.
¿Quién llora cuando yo estoy llorando?
¿El viento y los diamantes de la noche?
¿De quién es esta mano que me toca?

ROBERTO IGNACIO DÍAZ

Amor, voy a olvidarte y recordarte.
Allá voy. Te doy la espalda y te miro.
Fui Isolda y Melisande y fui Victoria.

[Other bodies never knew
a greater understanding than our bodies,
or greater pleasure, or tenderness,
when desire, fulfilled, would fall asleep.
Don't let me go, don't lose me.
I haven't heard from you. Where does it come from,
this madness? Residues of the shipwreck
of a green paradise on the plain.
Who cries when I cry?
The wind and the night's diamonds?
Whose hand is this that touches me?
Love, I will forget and remember you.
I'm going there. I turn away and look at you.
I was Isolde and Mélisande and I was Victoria.]

Like much else in *V.O.*, the sonnet rests largely on the practice of citation. Its first two lines are Sarlo's version of a sentence from *La rama de Salzburgo* in which Ocampo openly extols her physical relationship with Martínez: "Dudo que otros cuerpos hayan tenido, jamás, mayor entendimiento, mayor placer en tutearse y más ternura que prodigarse cuando el deseado saciado se alejaba" (I doubt that any other bodies could ever have had a better understanding, a bigger pleasure in treating each other with familiarity, or more tenderness to share after the fulfillment of desire).[26] Although neither Ocampo's husband nor her lovers are named directly in the opera, the words in the aria are nevertheless Victoria's meditation on her amatory liaisons. Placed at its center, two rather cryptic lines powerfully evoke this narrative of love and creation: "Residues of the shipwreck of a green paradise on the plain" (*Restos del naufragio de un verde paraíso en la llanura*). This lost green paradise contains a complicated story of real and literary geographies. In the context of Ocampo's chronicle of her erotic life, the phrase "verde paraíso" alludes to "Le vert paradis," the last section in the first volume of the autobiography, which in turn cites three non-sequential lines from Baudelaire's "Moesta et errabunda" in its epigraph: "But the green paradise of childhood loves, / . . . That sinless paradise, full of furtive pleasures, / Is it farther off now than India and China?" (*Mais le vert paradis des amours enfantines, / . . . L'innocent paradis, plein de plaisirs furtifs, / Est-il déjà*

244

plus loin que l'Inde et que la Chine?). *Le vert paradis* is also the title of a French-language book by Ocampo, published during World War II, in which she collected four essays on her passionate entanglements with England and France, idealized European locations whose languages and literatures had shaped her since childhood. By contrast, "Le vert paradis" of the autobiography focuses on her early life in Argentina and girlhood infatuation with a boy identified simply as L.G.F., whom she likes first and foremost for his physical beauty. Likewise, the sonnet-aria's invocation of *la llanura* as both green paradise and site of metaphorical shipwrecks refers not to Europe, but to the actual landscape of her native land (where she met L.G.F.) and to Victoria's tumultuous life as a reader, a writer, and a lover. Ocampo's appropriation of Baudelaire highlights her accomplishments in the permutation of experience into literature, and vice versa. That the same phrase reappears at the very center of *V.O.* is yet another instance of multilayered citation in crafting an operatic Victoria who can sing and speak for herself.

In the latter part of the sonnet, Sarlo's words for Victoria resonate with the grand operatic narratives seen season after season at the Colón. But, unlike the "Liebestod" and Lucia's mad scene, Victoria's sonnet aria is performed not as the character is about to exit the stage, but rather at the work's midpoint. Yet, like those two famous moments in opera, it defines her story most clearly. On a superficial level, the sonnet-aria might be read as a response to the character of Drieu, played in the opera by an actor (not a singer). At the end of the previous scene, Drieu, identified as Hombre, addresses to his Argentine lover a series of insulting phrases, ending: "Tan rica, tan pobre, tan colmada y vacía. Victoria, con esa tricota pareces un muchachito" (So rich, so poor, so satiated, so empty. Victoria, you look like a little boy in that sweater). Against those words, the sonnet-aria is also Sarlo's oblique reflection on the potential of the chamber opera to tell a story of empowerment. Unlike Isolde and Mélisande, Victoria celebrates her physical passion but then sings exultantly about her decision to leave her lover in order to preserve her sanity and life.

Preceding the sonnet-aria, Drieu's cruelty to Victoria resonates with earlier scenes in which she, or her name, is belittled. In the first scene, for instance, she says, "Victoria me llamo, por algo será" (My name is Victoria, there must be a reason), but the character of María Inés replies, "Tu nombre es nada, estás atrapada" (Your name is nothing, you're trapped). In the third scene, still a young girl, she returns to this theme, wondering about the appropriateness of her imperial appellation when she wields little power. Her parents seek to control her body, manage her readings, and impede her physical movements:

Me llamo Victoria,
por algo será.
Me encierran,
me mandan,
me visten,
me peinan,
me traen y llevan
de aquí
para allá.

[My name is Victoria,
there must be a reason.
They don't let me go out,
they order me around,
they dress me,
they comb my hair,
they take me
from here
to there.]

Yet, in the sonnet-aria, as she rejoices in the pleasures of love and laments its sorrows, she finally proclaims the name "Victoria" both as her ultimate identity and as the word that defines her most clearly. The preterite tense—"fui Victoria"—suggests that her struggle has now come to an end; when all is said and done, she has lived up to her name's signification. She may have started on the path of opera's tragic heroines, but in the end she plays her own triumphant role.

Victoria's tale, as it diverges from the grand tradition of operas performed at the Colón, concerns the cultural landscapes of Buenos Aires— musical and literary, for sure, but also the potential role of architecture, as advanced by Ocampo, in forging the city's modernity. Beyond words and music, opera is also about what audiences see onstage, and here too *V.O.* showcases Ocampo's passion for modernity by exerting a striking relocation of the urban landscape of Buenos Aires. Throughout the opera as designed for the Centro de Experimentación, Matías Sendón and Minou Maguna's set depicts only one place: the very modern house Ocampo had built in Buenos Aires's Barrio Parque in the late 1920s and where she lived for just a few years. In the opera, as we saw, Victoria acts out Racine's Hermione walking around an imaginary palace, and she frequently recalls a performance at a theater in Paris. But the audience at the CETC sees no palace or European theater—nor, for that matter, anything approaching the grandeur of Egyptian pyramids seen at the Colón's first night. What they see is a structure located in their own city: the living room of a house within which Ocampo assertively contested the traditional styles of architecture prevailing in 1920s Buenos Aires. A lover of Le Corbusier's vision, she persuaded Alejandro Bustillo, an architect whose work until then had been classically conventional, to design a house that expressed her love of clean lines and minimal decor. Importantly, as Pablo Gianera observes, the house's contents were devoid of historical cachet—an act in which, for once, Ocampo appears to eschew references.[27] Ocampo herself, writing in 1935 about her house in Mar del Plata, yet another modernist building that Le Corbusier praised but that others sarcastically dismissed as "una usina o un establo" (a factory or a stable),[28] addresses her own city with a warning about the perils of replicating the architecture of other places: "Líbrate para siempre de la tentación de imitar las viejas y magníficas ciudades de Europa, o parecerás eternamente una grotesca caricatura. Otro destino te aguarda" (Deliver yourself forever from the temptation of imitating the old and magnificent cities of Europe, or you will eternally appear to be a grotesque caricature. A different destiny awaits you).[29] The old architecture she describes is invoked in operatic language, the adjective "magnificent" easily applied to the Teatro Colón's grand spectacles, while "destiny" too resounds with opera's extravagant plotlines. At the same time, Ocampo's call to avoid imitation strikes a rather odd note when considered alongside her own enthusiasm for quotation; moreover, to ask an Argentine architect to imitate the work of a (modern) European may appear to contradict her appeal to abandon mimicry. Yet, the fact that that her new house boasts a modern style—akin to the musical world of Debussy and Stravinsky—bespeaks a new sensibility in which imitation, paradoxically, conveys originality.

Ocampo's well-heeled neighbors on the calle Rufino de Elizalde were troubled when plain white cubes appeared amidst their Beaux-Arts mansions in Barrio Parque, and she eventually sold the house in order to preserve her parents' villa in San Isidro, a far more traditional Dutch-style structure. In her writings on Ocampo, Sarlo views the audacious architectural experiment as an ambiguous accomplishment: a lesser act of imitation of Le Corbusier's vision, but also Ocampo's "first great translation" (*primera gran traducción*),[30] the initial link in what would become in the pages of *Sur*, founded shortly thereafter, a long chain of transits and transports between languages and cultures. Displayed—or visually cited—on the stage of the CETC, the house may be seen, then, as the

ROBERTO IGNACIO DÍAZ

theatrical embodiment of Ocampo's aesthetic affinities and projects—
even if its starkness somehow belies her own love of citation. If only
temporarily and obliquely, the modernist structure could also be interpreted as a visual commentary on the nature of opera in a city like Buenos Aires and a theater like the Colón. Ocampo's minimalist house, rising again in a peripheral site inside the grand old edifice, like the music
and words of *V.O.*, could only suggest the design of another kind of opera for Argentina.

248

Acknowledgments

The translation of this project from a 2014 conference to its present form has occasioned many debts. I should like to thank the University of Oxford's John Fell Fund, Jesus College, Oxford, the Oxford Music Faculty, and the Royal Musical Association for financial and other support for the conference, and the American Musicological Society for a publication subvention. I would like to thank the Oxford Music Faculty for allowing me to allocate my entire annual research allowance as a further subvention for the publication of this volume. Those conference participants unable to contribute to the volume generously acted as readers, as have other scholars in the field: Peter Borsay, Jérôme Brillaud, Daniel M. Grimley, Christopher Morris, Roger Parker, Arman Schwartz, Emanuele Senici, Gavin Williams, and Flora Willson deserve thanks in this regard. I'm especially grateful to Marta Tonegutti of University of Chicago Press for guiding this project to publication with such care and wisdom, to the Press's anonymous readers, and to Marta's assistants, Evan White and Susannah Engstrom, for their facilitation. Finally, I must thank my long-suffering fellow contributors, many of whom acted as readers for other chapters, and all of whom put up with a good deal of editorial tinkering in the process of pulling the volume together.

Contributors

Rebekah Ahrendt is an associate professor of Musicology in the Department of Media and Culture Studies, Utrecht University. Prior to joining Utrecht's faculty, she was an assistant professor in the Yale University Department of Music and a Mellon postdoctoral scholar in the humanities at Tufts University. Much of Ahrendt's recent scholarship has focused on music and international relations. She is the coeditor of *Music and Diplomacy from the Early Modern Era to the Present* (Palgrave Macmillan, 2014). Her current monograph project illuminates the musical networks maintained by the refugees, exiles, and migrants who traversed the landscape of the Dutch Republic.

Amanda Eubanks Winkler is an associate professor of music history and cultures at Syracuse University. Her research and performance activities focus on English theater music, with articles on topics ranging from seventeenth-century didactic masques to Lloyd Webber's *The Phantom of the Opera*; a book, *O Let Us Howle Some Heavy Note: Music for Witches, the Melancholic, and the Mad on the Seventeenth-Century English Stage* (Indiana University Press, 2006); two editions of Restoration-era theater music; and a co-edited collection, *Beyond Boundaries: Rethinking Music Circulation in Early Modern England* (Indiana University Press, 2017). She has been awarded fellowships and grants from the Folger Shakespeare Library and the AHRC (UK).

Michael Burden is professor of opera studies at the University of Oxford and a fellow in music at New College, where he is also dean. His research is on the music of Purcell, and on dance and music in the eighteenth-century London theaters. His five-volume collection of documents on the staging of opera in London, *London Opera Observed*, was published in 2013 by Pickering & Chatto;

CONTRIBUTORS

The Works of Monsieur Noverre, edited with Jennifer Thorp and published by Pendragon Press, appeared in 2014; and a jointly edited volume, *Staging History*, published by the University of Chicago Press, appeared in 2016. He is the director of productions of New Chamber Opera, www.newchamberopera.co.uk

Margaret R. Butler is an associate professor of musicology at the University of Florida. Her research focuses on eighteenth-century opera in Europe. Her books include *Operatic Reform in Turin: Aspects of Production and Stylistic Change in the 1760s* (LIM, 2001) and *Musical Theater in Eighteenth-Century Parma* (University of Rochester Press, in press). Her articles have appeared in the *Journal of the American Musicological Society*, *Cambridge Opera Journal*, *Eighteenth-Century Music*, *Early Music*, and *Music in Art*, and she authored the chapter on Italian opera in *The Cambridge History of Eighteenth-Century Music*. Her work has been supported by the AMS, the Fulbright and Delmas Foundations, and the Newberry Library.

Jonathan Hicks is a research fellow at Newcastle University's Humanities Research Institute. Prior to joining Newcastle he held fellowships at King's College London and Lincoln College, Oxford. He is completing a monograph about music and mobility in nineteenth-century London, and has co-edited (with Katherine Hambridge) *The Melodramatic Moment: Music and Theatrical Culture, 1790–1820* (University of Chicago Press, 2018).

Susan Rutherford is a professor of music at the University of Manchester. Her publications include *The New Woman and Her Sisters: Feminism and Theatre, 1850–1914* (co-editor; University of Michigan Press, 1992), *The Prima Donna and Opera, 1815–1930* (Cambridge University Press, 2006), and *Verdi, Opera, Women* (Cambridge University Press, 2013), as well as numerous essays on nineteenth-century opera. She is the recipient of the Pauline Alderman Prize (IAWM), and the Premio Internazionale: Giuseppe Verdi (Istituto Nazionale di Studi Verdiani). Her current project (funded by a three-year Leverhulme Trust Major Research Fellowship, 2016–19) is entitled "A History of Voices: Singing in Britain 1690 to the Present."

Charlotte Bentley is a junior research fellow at Emmanuel College, Cambridge. She completed her doctorate ("Resituating Transatlantic Opera: The Case of the Théâtre d'Orléans, New Orleans, 1819–1859") at the University of Cambridge in 2017 under the supervision of Benjamin Walton. Her research interests include cultural transnationalism in the nineteenth century, the role of human agency in globalization, and opera in Paris at the fin de siècle. Her work on the significance of French grand opera for the Francophone and Anglophone populations of New Orleans in the 1830s has appeared in *Cambridge Opera Journal*.

Benjamin Walton teaches at the University of Cambridge, where he is a fellow of Jesus College. His book *Rossini in Restoration Paris: The Sound of Modern*

Life was published by Cambridge University Press in 2007, and he has co-edited two collections of essays: *The Invention of Beethoven and Rossini* (with Nicholas Mathew; Cambridge University Press, 2013), and *Nineteenth-Century Opera and the Scientific Imagination* (with David Trippett; Cambridge University Press, 2017). Since 2014 he has served as editor, with Stefanie Tcharos, of *Cambridge Opera Journal*. He is currently writing a book about the first opera troupe to go around the world, during the 1820s and 1830s.

Kerry Murphy is the head of musicology at the Melbourne Conservatorium of Music, the University of Melbourne. Her research interests focus chiefly on nineteenth-century French music and music criticism and colonial Australian music history; she has published widely in these areas. She is currently researching the impact of traveling virtuosi and opera troupes to Australia and the Australian music publisher and patron Louise Hanson-Dyer.

Yvonne Liao is a Leverhulme early career fellow at the University of Oxford and an early career fellow at TORCH, The Oxford Research Centre in the Humanities. Her current research is situated at the intersection of music, political geography, and global history, with a particular focus on multi-colonialism and postcolonialism in Chinese port cities. For her article in *The Musical Quarterly* on European-Jewish refugees and a "Little Vienna" in Japanese-occupied Shanghai, she was awarded the Royal Musical Association 2017 Jerome Roche Prize. New essays are forthcoming in the edited volumes *Music History and Cosmopolitanism* and *Rethinking Bach*.

Peter Franklin was a professor of music at the University of Oxford until his retirement in 2014; he is an emeritus fellow of St. Catherine's College, Oxford. His publications include *Mahler Symphony No. 3*, *The Life of Mahler* (both Cambridge University Press, 1991 and 1997 respectively) and *Seeing Through Music: Gender and Modernism in Classic Hollywood Film Scores* (Oxford University Press, 2011). His most recent book, based on his 2010 Bloch Lectures at the University of California, Berkeley, is *Reclaiming Late-Romantic Music: Singing Devils and Distant Sounds* (University of California Press, 2014).

Katharine Ellis is the 1684 Professor of Music at Cambridge and a specialist on France in the long nineteenth century. Her books cover canon formation and the press (*Music Criticism in Nineteenth-Century France*, Cambridge University Press, 1995), the early music revival (*Interpreting the Musical Past*, Oxford University Press, 2005), and Benedictine musical politics ca. 1900 (*The Politics of Plainchant in* fin-de-siècle *France*, Routledge, 2013). Her current project reappraises the history of French musical life from provincial viewpoints. Elected to the Academia Europaea in 2010, she became a Fellow of the British Academy in 2013 and was elected a Member of the American Philosophical Society in 2017.

CONTRIBUTORS

Suzanne Aspden is an associate professor in the faculty of music, University of Oxford, and a fellow of Jesus College, Oxford. Her research focuses on expressions of identity in eighteenth-century music, from the personal to the national. Singer identity construction is the subject of her monograph *The Rival Sirens: Performance and Identity on Handel's Operatic Stage* (Cambridge University Press, 2013), and a guest-edited issue of *Cambridge Opera Journal*. Several articles have examined music and national identity in eighteenth-century Britain, which is the subject of her next monograph. She is also preparing a monograph on the culture of country-house opera in twentieth-century Britain, and an edited volume on representations of consciousness in eighteenth-century opera. She is a former editor of the *Cambridge Opera Journal*.

Klaus van den Berg is a theater historian, professional dramaturg, and urban scenographer based in Atlanta. After studies in Berlin and earning his doctorate from Indiana University, he taught theater for over two decades, most recently at the University of Tennessee. His research centers on scenography in performance, spectacle in contemporary society, and performance venues in urban landscapes. He is working on a book, "Image Space: Walter Benjamin and Cultural Performance," presenting Benjamin's foundational contributions to performance theory. Dramaturgical credits include the US premieres of three Georg Tabori plays and a professional production of his own translation and adaptation of Schiller's *The Robbers* at Asolo Conservatory.

Roberto Ignacio Díaz is an associate professor of Spanish and comparative literature at the University of Southern California. His teaching and scholarship engage Latin American and literary and cultural history, with a focus on transatlantic and north-south relations. He is the author of *Unhomely Rooms: Foreign Tongues and Spanish American Literature* (Bucknell University Press, 2002) and is completing a book manuscript on historical and textual convergences of opera and Latin America from the eighteenth century to the present.

Notes

CHAPTER ONE

1. On the importance of opera houses as nationalist civic institutions, see Jim Samson, "Nations and Nationalism," in *The Cambridge History of Nineteenth-Century Music*, ed. Jim Samson (Cambridge: Cambridge University Press, 2002), 568–600, at 580.
2. On *lieux de mémoire* as an essential aspect of French collective memory and collective identity, see Pierre Nora, ed., *Les lieux de mémoire* (Paris: Gallimard, 1984–c. 1992).
3. Marvin Carlson, *Places of Performance: The Semiotics of Theatre Architecture* (Ithaca, NY: Cornell University Press, 1989), 2.
4. Ibid., 5–6.
5. Ibid., 74–75.
6. Anselm Gerhard, *Die Verstädterung der Oper: Paris und der Musiktheater des 19. Jahrhunderts* (1992), trans. Mary Whittall as *The Urbanization of Opera* (Chicago: Chicago University Press, 1998).
7. Hans Erich Bodeker, Patrice Veit, and Michael Werner, eds., *Éspaces et lieux de concert en Europe, 1700–1920: Architecture, musique, société* (Berlin: BWV Berliner Wissenschafts-Verlag, 2008).
8. Carlson, *Places of Performance*, 7.
9. For an example of this problematization, see Sarah Whatmore, *Hybrid Geographies: Natures, Cultures, Spaces* (London: Sage, 2002).
10. Barney Warf and Santa Arias, "Introduction: The Reinsertion of Space in the Humanities and Social Sciences," in *The Spatial Turn: Interdisciplinary Perspectives*, ed. Barney Warf and Santa Arias (New York: Routledge, 2009), 1–10, at 1.
11. Kay Anderson, Mona Domosh, Steve Pile, and Nigel Thrift, *Handbook of Cultural Geography* (London: Sage, 2003), 2.

NOTES TO PAGES 3–4

12. On the decontextualizing effects of the work concept in relation to ideas of place, see Andrew Leyshon, David Matless, and George Revill, *The Place of Music* (New York: Guilford Press, 1998). By way of example of geographically contextualized studies, see Craig Wright, *Music and Ceremony at Notre Dame of Paris, 500–1550* (Cambridge: Cambridge University Press, 1989); Daniel Grimley, *Grieg: Music, Landscape and Norwegian Identity* (Woodbridge: Boydell, 2006); Sheila Whiteley, Andy Bennett, and Stan Hawkins, eds., *Music, Space and Place: Popular Music and Cultural Identity* (Aldershot: Ashgate, 2004); and Martin Stokes, ed., *Ethnicity, Identity and Music: The Musical Construction of Place* (Oxford: Berg, 1994).

13. For example, see Reinhard Strohm, *Music in Late Medieval Bruges* (Oxford: Clarendon Press, 1985), 1–5; Craig Wright, "The Palm Sunday Procession in Medieval Chartres," in *The Divine Office in the Latin Middle Ages*, ed. Margot Fassler and Rebecca A. Baltzer (Oxford: Oxford University Press, 2000); Kathleen Ashley and Wim Hüsken, eds., *Moving Subjects: Processional Performance in the Middle Ages and the Renaissance* (Amsterdam: Rodopi, 2001); Carol Symes, *A Common Stage: Theater and Public Life in Medieval Arras* (Ithaca, NY: Cornell University Press, 2007); and Emma Dillon, *The Sense of Sound: Musical Meaning in France, 1260–1330* (Oxford: Oxford University Press, 2012).

14. See Hervé Lacombe, *Géographie de l'opéra au XXe siècle* (Paris: Fayard, 2007), relevant here for its consideration of European opera's global reach. See also Hugh McDonald, "De l'Opéra au Conservatoire: La géographie musicale de Paris sous la Monarchie de Juillet et le Second Empire," in *La maison de l'artiste: Construction d'un espace de représentations entre réalité et imaginaire (XVIIe–XXe siècles)* (Rennes: Presses Universitaires de Rennes, 2007), 164–69; Benjamin Pintiaux, "L'abbé Pellegrin et les lieux du chant: Esquisse d'une géographie des spectacles lyriques parisiens au XVIIIe siècle," in *L'Opéra de Paris, la Comédie-Française et l'Opéra-Comique: Approches comparées (1669–2010)*, ed. Sabine Chaouche, Denis Herlin, and Solveig Serre (Paris: École des Chartes, 2012), 154–66; Michel Foucher, ed., *Les ouvertures d'opéra: Un nouvelle géographie culturelle?* (Lyon: Presses Universitaires de Lyon, 1997), in which the essays by Foucher ("Géographie des maisons d'opéra: Variations sur la culture lyrique et la géopolitique en Europe"), Nadine Gelas ("L'opéra et ses effets de mode") and (to a degree) Christian Montès ("Les lieux mythiques de l'opéra") touch on geography.

15. Nuala C. Johnson, Richard H. Schein, and Jamie Winders, "Introduction," in *The Wiley-Blackwell Companion to Cultural Geography* (Chichester: Wiley-Blackwell, 2013), 26–39. Anderson, Domosh, Pile, and Thrift note the overarching importance of understanding culture as an expression of power in their delineation of the different ways cultural geographers approach the concept of "culture"; *Handbook of Cultural Geography*, 2.

16. Among numerous studies one might cite, two distinctive approaches are offered by Jane Fulcher, *The Nation's Image: French Grand Opera as Politics and Politicized Art* (Cambridge: Cambridge University Press, 1987); and Curtis

256

Price, Judith Milhous, and Robert D. Hume, *Italian Opera in Late Eighteenth-Century London*, Vol. 1, *The King's Theatre, Haymarket, 1778–1791* (Oxford: Clarendon Press, 1995).

17. Jacques Attali, *Noise: The Political Economy of Music*, trans. Brian Massumi (Manchester: Manchester University Press, 1985), 6.

18. For examples, see, on medieval soundscapes, n. 13 above; Bruce R. Smith, *The Acoustic World of Early Modern England* (Chicago: University of Chicago Press, 1999); and John Connell and Chris Gibson, eds., *Sound Tracks: Popular Music, Identity and Place* (London: Routledge, 2003).

19. The sixteenth-century aspiration to recapture the ancient unity of music and words, and with it regain control over auditors, which led to opera, also inspired Charles IX of France to form the Académie de Poésie et de Musique, which hoped through *vers mesurés* to exploit music's power by performing it in civic spaces to foster order and harmony.

20. Foucher, *Les ouvertures d'opéra*.

21. Nigel Thrift, *Non-Representational Theory: Space, Politics, Affect* (Routledge: London, 2007); and H. Lorimer, "Cultural Geography: The Busyness of Being 'More-than-Representational," *Progress in Human Geography* 29 (2005): 83–94.

22. Emanuele Senici, *Landscape and Gender in Italian Opera: The Alpine Virgin from Bellini to Puccini* (Cambridge: Cambridge University Press, 2005).

23. Daniel M. Grimley, "Carl Nielsen's Carnival: Time, Space and the Politics of Identity in *Maskarade*," in *Art and Ideology in European Opera: Essays in Honour of Julian Rushton*, ed. Rachel Cowgill, David Cooper, and Clive Brown (Woodbridge: Boydell, 2010), 241–61.

24. Christopher Morris, *Modernism and the Cult of Mountains: Music, Opera, Cinema* (Farnham: Ashgate, 2012).

25. Arman Schwartz, *Puccini's Soundscapes: Realism and Modernity in Italian Opera* (Florence: Olschki, 2016).

26. David Charlton, "Hearing through the Eye in Eighteenth-Century French Opera," in *Art, Theatre, and Opera in Paris, 1750–1850: Exchanges and Tensions*, ed. Sarah Hibberd and Richard Wrigley (Farnham: Routledge, 2014).

27. On popular music and "myths of place," see Connell and Gibson, *Sound Tracks*, 6.

28. Samson, "Nations and Nationalism," 580, notes a succession of such openings: Warsaw (1765), Vienna (1776), Berlin (1786), Pest (1837), Bucharest (1852), Belgrade (1869), Zagreb (1870), Prague (1881), Brno (1884), and Pozsony (1886).

29. Nicholas Abercrombie and Brian Longhurst, *Audiences: A Sociological Theory of Performance and Imagination* (London: Sage, 1998), 81–82; and T. J. Clark, *The Painting of Modern Life* (London: Thames & Hudson, 1984), 9.

30. Chris Philo, "More Words, More Worlds: Reflections on the 'Cultural Turn' and Human Geography," in *Cultural Turns/Geographical Turns: Perspectives on Cultural Geography*, ed. Ian Cook, David Crouch, Simon Naylor, and

NOTES TO PAGES 9–12

James Ryan (Harlow: Prentice Hall, 2000), 33. Philo is critical of the dematerialization of cultural geography brought about by this focus, but it is nonetheless useful for those working in the humanities.

31. Alison Blunt, "Cultural Geographies of Migration: Mobility, Transnationality and Diaspora," *Progress in Human Geography* 31, no. 5 (2007): 684–94; and Mike Featherstone, ed., *Global Culture: Nationalism, Globalisation and Modernity* (London: Sage, 1990).

32. Mimi Sheller and John Urry, "The New Mobilities Paradigm," *Environment and Planning A* 38 (2006): 207–26; see also Tim Cresswell, *On the Move: Mobility in the Modern Western World* (London: Routledge, 2006).

33. Nigel Thrift, "Space: The Fundamental Stuff of Geography," in *Key Concepts in Geography*, ed. Nicholas J. Clifford, Sarah Holloway, Stephen P. Rice, and Gill Valentine, 2nd ed. (London: Sage, 2009), 95–107, at 96, and "Space," *Theory, Culture & Society* 23 (2006): 139–55. See also Georgina Born, "Introduction—Music, Sound and Space: Transformations of Public and Private Experience," in *Music, Sound and Space: Transformations of Public and Private Experience*, ed. Georgina Born (Cambridge: Cambridge University Press, 2013), 1–70, at 20–21.

34. The challenge to colonial hegemony has particularly followed Nicolas Thomas's anthropological work in *Colonialism's Culture: Anthropology, Travel and Government* (Oxford: Polity Press, 1994); see Catherine Nash, "Cultural Geography: Postcolonial Cultural Geographies," *Progress in Human Geography* 26, no. 2 (2002): 219–30, at 221.

35. Dipesh Chakrabarty, *Provinicialising Europe: Postcolonial Thought and Historical Difference* (Princeton, NJ: Princeton University Press, 2000).

36. Richard Leppert, *The Sight of Sound: Music, Representation, and the History of the Body* (Berkeley and Los Angeles: University of California Press, 1993), 26.

37. Erkki Huhtamo and Jussi Parikka, "Introduction: An Archaeology of Media Archaeology," in *Media Archaeology: Approaches, Applications, and Implications*, ed. Erkki Huhtamo and Jussi Parikka (Berkeley and Los Angeles: University of California Press, 2011), 1–21, at 3; and Marshall McLuhan and Quentin Fiore, *The Medium Is the Massage* (London and New York: Penguin, 1967), 8–10.

38. Huhtamo and Parikka draw Benjamin and the "picture-historians" Warburg, Panofsky, and Gombrich together in their "Introduction," 6–7.

39. H. Aram Veeser, "Introduction," *The New Historicism*, ed. H. Aram Veeser (New York: Routledge, 1989), xi; cited in Huhtamo and Parikka, "Introduction," 9.

CHAPTER TWO

1. The Paris Opéra has long been a favorite case study. See, for example, Raymond de Pezzer, *L'opéra devant la loi et la jurisprudence* (Paris: Alb. Manier,

NOTES TO PAGES 12–16

1911); and Mathieu Touzeil-Divina and Geneviève Koubi, eds., *Droit et opéra* (Poitiers: LGDJ, 2008).

2. For example, Judith Milhous and Robert D. Hume, "Eighteenth-Century Equity Lawsuits in the Court of Exchequer as a Source for Historial Research," *Historical Research* 70, no. 172 (1997): 231–46; and Cheryll Duncan, "Castrati and Impresarios in London: Two Mid-Eighteenth-Century Lawsuits," *Cambridge Opera Journal* 24 (2012): 43–65.

3. See, for example, Gerald Frug, "A Legal History of Cities," in *The Legal Geographies Reader: Law, Power, and Space*, ed. Nicholas Blomley, David Delaney, and Richard T. Ford (Oxford and Malden, MA: Blackwell, 2001), 154–76; David Delaney, "Seeing Seeing Seeing the Legal Landscape," in *The Wiley-Blackwell Companion to Cultural Geography*, ed. Nuala C. Johnson, Richard H. Schein, and Jamie Winders (Chichester and Malden, MA: John Wiley & Sons, 2013), 238–49.

4. Naomi Mezey, "Law as Culture," *Yale Journal of Law and the Humanities* 13 (2001): 35–67.

5. Lesley Murray and Sara Upstone, "Conclusion," in *Researching and Representing Mobilities: Transdisciplinary Encounters*, ed. Lesley Murray and Sara Upstone (Basingstoke: Palgrave Macmillan, 2014), 191–93.

6. Henri Lefebvre, *The Production of Space*, trans. Donald Nicholson-Smith (Oxford: Blackwell, 1991).

7. See esp. Manlio Bellomo, *The Common Legal Past of Europe, 1000–1800*, trans. Lydia G. Cochrane (Washington, DC: Catholic University of America Press, 1995), chap. 1.

8. Britain was an exception that presented special challenges to Continental performers. A comparative study will be pursued more fully elsewhere.

9. Transcribed in C. N. Wybrands, *Het Amsterdamsche Toneel van 1617–1772* (Utrecht: Beijers, 1873), 232. See also Jan Fransen, *Les comédiens français en Hollande au XVIIe et au XVIIIe siècles* (Paris: H. Champion, 1925), 164–65.

10. Rudolf Rasch, "De moeizame introductie van de opera in de Republiek," in *Een muziekgeschiedenis der Nederlanden*, ed. Louis Peter Grijp (Amsterdam: Amsterdam University Press, 2001), 311–16.

11. On The Hague's preference for French opera, see Rebekah Ahrendt, "A Second Refuge: French Opera and the Huguenot Migration, 1685–1713" (PhD diss., University of California, Berkeley, 2011).

12. Georges Mongrédien, *La vie quotidienne des comédiens au temps de Molière* (Monaco: Hachette, 1966), 215–17; and John S. Powell, *Music and Theatre in France, 1600–1680* (Oxford: Oxford University Press, 2000), chap. 1.

13. The phrase comes from Ludovico Guicciardini's *Description de tout les Pais Bas autrement dit la Germanie Inferieure ou Basse Allemaigne* (Antwerp, 1567) and was repeated in travelogues for the next two centuries.

14. For a map illustrating these jurisdictional divisions in 1683, see F. Pieter Wagenaar, "Haagse bestuurders en ambtenaren," in *Den Haag: Geschiedenis*

259

NOTES TO PAGES 16–18

van de Stad, ed. Thera Wijsenbeek and Elisabeth van Blankenstein (Zwolle: Wanders, 2005), 90–120, at 91.

15. F. Pieter Wagenaar, *"Dat de regeringe niet en bestaet by het corpus van de magistraet van Den Hage alleen": De Sociëteit van 's-Gravenhage (1587–1802): Een onderzoek naar bureaucratisering* (Hilversum: Verloren, 1999), chap. 2.

16. Kornee van der Haven, *Achter de schermen van het stadstoneel: Theaterbedrijf en toneelpolemiek in Amsterdam en Hamburg, 1675–1750* (Zutphen: Walburg Pers, 2008), 161.

17. For the theater's seventeenth- and eighteenth-century histories, see respectively Fransen, *Les comédiens français en Hollande*; and Aldo Lieffering, "De franse comedie in Den Haag 1749–1793" (PhD diss., Utrecht University, 1999), 17ff.

18. Fransen, *Les comédiens français en Hollande*, 43–44.

19. The Hague, Gemeentearchief (hereafter NL-DH-ga), Resolutiën van Baljuw (Schout), Burgemeesters en Schepenen, inv. 350–01 no. 57.

20. On Kremberg's and Schott's legal disputes, see Walter Schulze, *Die Quellen der Hamburger Oper (1678–1738): Eine bibliographisch-statistische Studie zur Geschichte der ersten stehenden deutschen Oper* (Hamburg: Stalling, 1938), 139–58.

21. "Memorie in't frans, uyt den naam en uyt ordre van monsr. Schot, soo geseyt wierd 'senateur et premier juge de la ville de Hambourg,' aan haar Ed[ele] Achtb[aarheyt] overgeleverd." NL-DHga, Resolutiën van Baljuw, Burgemeesters en Schepenen, 26 October 1700, inv. 350–01 no. 58.

22. Not much is known of Colombel; all other sources on this opera company omit his name. According to documents conserved in the Amsterdam city archives, it appears that Colombel was "in the family" of French-born musicians working in the Dutch Republic.

23. Rudolf Rasch, "Om den armen dienst te doen: De Amsterdamse Schouwburg en de godshuizen gedurende het laatste kwart van de 17e eeuw," *Holland* 23 (1991): 243–67; and Charles H. Parker, "Poor Relief and Community in the Early Dutch Republic," in *With Us Always: A History of Private Charity and Public Welfare*, ed. Donald T. Critchlow and Charles H. Parker (Lanham, MD: Rowman & Littlefield, 1998), 13–33.

24. Lorenzo Bianconi and Thomas Walker, "Production, Consumption and Political Function of Seventeenth-Century Opera," *Early Music History* 4 (1984): 209–96, at 265–66.

25. Jérôme de La Gorce, *L'Opéra à Paris au temps de Louis XIV: Histoire d'un théâtre* (Paris: Éditions Desjonquères, 1992), 111.

26. NL-DHga, Diaconie van de Hervormde gemeente 's-Gravenhage, inv. 0133–01 no. 575, fol. 9v; no. 576, fol. 16v. Actual audience numbers may have varied, as faulty accounting practices were constantly alleged.

27. See Reinhard Zimmermann, *The Law of Obligations: Roman Foundations of the Civilian Tradition* (Oxford: Oxford University Press, 1996), chap. 15, "Societas."

NOTES TO PAGES 18–21

28. On the rise of the entrepreneurial model, see Beth L. Glixon and Jonathan E. Glixon, *Inventing the Business of Opera: The Impresario and His World in Seventeenth-Century Venice* (Oxford and New York: Oxford University Press, 2006), 8–10.

29. On the Comédie Française, exemplifying the acting troupes' model with the principal performers holding equal shares and responsibilities, see H. Carrington Lancaster, *The Comédie Française, 1680–1701: Plays, Actors, Spectators, Finances* (Baltimore, MD: Johns Hopkins Press, 1941); and Lancaster, "The Comédie Française 1701–1774: Plays, Actors, Spectators, Finances," *Transactions of the American Philosophical Society* n.s. 41, pt. 4 (1951): 593–849.

30. Bellomo, *Common Legal Past*, chapter 7.

31. I use the masculine pronoun because women were excluded from the notarial profession in this period.

32. Ernst Ferdinand Kossmann, *De boekverkoopers, notarissen en cramers op het Binnenhof* ('s-Gravenhage: Martinus Nijhoff, 1932), 200.

33. NL-DHga, notarial archive of Samuel Favon, inv. no. 749, contract of 31 May 1701; partially transcribed in Jean-Jacques Quesnot de la Chenée, *L'Opéra de La Haye: Histoire instructive et galante* (Cologne: Les Heritiers de Pierre le Sincere, 1706), 183–92.

34. Rebekah Ahrendt, "*Armide*, the Huguenots, and The Hague," *Opera Quarterly* 28 (2012): 131–58.

35. NL-DHga, Favon 752, attestation of 2 February 1704.

36. For a 1698 contract between Beaujean and Denis Nanteuil's theater company, see Émile Biais, "Le théâtre à Angoulême (quinzième siècle—1904)," *Réunion des Sociétés des Beaux-Arts de Départements* 28 (1904): 279–33, at 318–19. If Beaujean traveled to Warsaw with Nanteuil in 1699, he could have met Deseschaliers there and then joined the company in The Hague.

37. On the brothel, see Paul R. Sellin and Augustus J. Veenendaal Jr., "A 'Pub Crawl' through Old The Hague: Shady Light on Life and Art among English Friends of John Donne in the Netherlands, 1627–1635," *John Donne Journal* 6 (1987): 236–60, at 237, 251.

38. Fransen, *Les comédiens français en Hollande*, 165–66. See also Ahrendt, "A Second Refuge," chap. 1.

39. Fransen, *Les comédiens français en Hollande*, 174, 203.

40. NL-DHga, Favon 749, *Bail de Maison* of 6 June 1701. See also Fransen, *Les comédiens français en Hollande*, 203.

41. Deseschaliers and Dudard signed a rental contract on 3 August 1708, with occupancy set to begin on 1 May 1709. See Fransen, *Les comédiens français en Hollande*, 226–27.

42. The pair had journeyed together to Warsaw in 1699 to perform at the court of August the Strong; see Louis Delpech, "*Frantzösische Musicanten*: Musique et musiciens français en Basse-Saxe et en Saxe" (PhD diss., Université de Poitiers, 2015), 265–72.

NOTES TO PAGES 21–23

43. Two thousand florins in 1701 would have the "purchasing power" of over twenty thousand euros in 2015. See http://www.iisg.nl/hpw/calculate.php.

44. For a summary of these rights, see Antoine Despeisses, *Les Œuvres de M. Antoine Despeisses . . . Derniere Edition, Nouvellement Reveüe & Corrigée* (Lyon: Huguetan, 1685), 368.

45. "Renonçeant pour cette effet à tous exceptions et bénéfices, spécialement celles du Senatus Consultum Velljanum et les ordonnances de cette ville qui dictent que nulles femmes ou filles ne peuvent cautionner ou s'obliger pour autrui, dont elle reconnoit avoir esté instruit par moy le susd[i]t Notaire, obligeant à cela sa personne et tous et chacuns ses biens, presents et avenirs." Stadsarchief, Amsterdam (hereafter NL-Aa), notarial archive of H. de Wilde no. 6474, contract of 30 September 1701.

46. Simon van Leeuwen, *Het Rooms-Hollands-Regt . . .* , 6th ed. (Amsterdam: Boom, 1686), 316–17; and Zimmermann, *Law of Obligations*, 145–52.

47. Ariadne Schmidt, "Gelijk hebben, gelijk krijgen? Vrouwen en vertrouwen in het recht in Holland in de zeventiende en achttiende eeuw," in *Het Gelijk van de Gouden Eeuw: Recht, onrecht en reputatie in de vroegmoderne Nederlanden*, ed. Michiel van Groesen, Judith Pollmann, and Hans Cools (Hilversum: Verloren, 2014), 109–25, at 110–12. On Venetian female singers, see Glixon and Glixon, *Inventing the Business of Opera*, 514–18.

48. Laurens Winkel, "Forms of Imposed Protection in Legal History, Especially in Roman Law," *Erasmus Law Review* 3, no. 2 (2010): 161. Dutch women did not become legally competent until 1957 and achieved full legal equality in family law only in 1984.

49. Zimmermann, *Law of Obligations*, 53–54, 418–19.

50. "A comparu en sa personne Demoiselle Catherine du Dar femme et procuratrice de Louis Deschaillieres directeur de l'Academie Royalle de musique a Lille en Flandres fondée de sa procuration generalle pour touttes leurs affaires, passé par le S.r Pierre Roland notaire Royal de la residence de la ville de Lille en date du 10 du mois de febvrier 1699." NL-Aa, De Wilde 6474, 30 September 1701.

51. On Venice, see Glixon and Glixon, *Inventing the Business of Opera*, 199–201.

52. Jérôme de La Gorce, "La musique et la danse dans les spectacles donnés par la troupe de Rosidor à Stockholm autour de 1700," in *"L'esprit français" und die Musik Europas: Entstehung, Einfluss und Grenzen einer ästhetischen Doktrin /"L'esprit français" et la musique en Europe: Émergence, influence et limites d'une doctrine esthétique; Festschrift für Herbert Schneider*, ed. Michelle Biget and Rainer Schmusch (Hildesheim and New York: Olms, 2007), 219–27, at 220–21.

53. Glixon and Glixon, *Inventing the Business of Opera*, 200–201.

54. Marie-Françoise Limon, *Les notaires au Châtelet sous le règne de Louis XIV, étude institutionelle et sociale* (Toulouse: Presses Universitaires du Mirail, 1992). On Rochois's contract, see Jérôme de La Gorce, "Contribution des Opéras de Paris et de Hambourg à l'interprétation des ouvrages lyriques

donnés à La Haye au début du XVIIIe siècle," in *Aufklärungen: Studien zur deutsch-französischen Musikgeschichte im 18. Jahrhundert; Einflusse und Wirkungen*, ed. Wolfgang Birtel and Christoph-Helmut Mahling (Heidelberg: Winter, 1986), 90–103.

55. Ibid., 92–93.

56. "Pour l'accomplissement et execution de tout ce qui est susd[i]t ont Lesd[it]s Comparants par ensemble et chacun en particulier obliger et obligent par ces presentes leurs personnes et biens rien exceptez, les soumettants à la Rigueur de toutes Cours Juges et Justices. Fait et passé à la Haye en presence de Philippe Beaujean et Samuel Favon Lejeune Clercq tesmoins à ce requis." NL-DHga, Favon 752, employment contract of 3 November 1704.

57. NL-DHga, Rechterlijk Archief, inv. 351–01, nos. 164–65.

CHAPTER THREE

1. Accessed 15 August 15, 2016, http://www.localhistories.org/population .html; see also "The Demography of Early Modern London circa 1550 to 1750," accessed 15 August 2016, http://www.geog.cam.ac.uk/research /projects/earlymodernlondon/.

2. Bryan White, "Letter from Aleppo: Dating the Chelsea School Performance of *Dido and Aeneas*," *Early Music* 37, no. 3 (2009): 417–28.

3. Christopher Marsh, *Music and Society in Early Modern England* (Cambridge: Cambridge University Press, 2010), chaps. 2–4.

4. English opera's fuzzy generic issues in this period are elucidated by, for example, Robert Hume, "The Politics of Opera in Late Seventeenth-Century London," *Cambridge Opera Journal* 10 (1998): 15–43, at 16–22; Michael Burden, "Aspects of Purcell's Operas," in *Henry Purcell's Operas: The Complete Texts*, ed. Michael Burden (Oxford: Oxford University Press, 2000), 1–27; and Richard Luckett, "Exotick but Rational Entertainments: The English Dramatick Operas," in *English Drama: Forms and Development: Essays in Honour of Muriel Clara Bradbrook* (Cambridge: Cambridge University Press, 1977), 123–141. Andrew R. Walkling has also tackled the issue in *Masque and Opera in England, 1656–1688* (London: Routledge, 2017).

5. In 1649 John Evelyn visited the *"Schooles, or Colledges* of the Young Gentlewomen" in Putney (*The Diary of John Evelyn*, ed. E. S. De Beer, vol. 2 [Oxford: Clarendon Press, 1955], 555), and during the Commonwealth the Perwich family ran a school at Hackney, as documented in John Batchiler's *The Virgin's Pattern* (London, 1661).

6. Julie Sanders, *The Cultural Geography of Early Modern Drama, 1620–1650* (Cambridge: Cambridge University Press, 2011), 163; and Mark Jenner, "Circulation and Disorder: London Streets and Hackney Coaches, c. 1640– c. 1740," in *The Streets of London: From the Great Fire to the Great Stink*, ed. Tim Hitchcock and Heather Shore (London: Rivers Oram Press, 2003), 40–53.

NOTES TO PAGES 28–30

7. Susan E. Whyman, *Sociability and Power in Late-Stuart England: The Cultural Worlds of the Verneys, 1660–1720* (Oxford: Oxford University Press, 1999), chap. 4.

8. Sanders, *Cultural Geography*, 163; and Hitchcock and Shore, "Introduction," in *The Streets of London*, 1–9.

9. Pelling, "Skirting the City," in *Londinopolis: Essays in the Social and Cultural History of Early Modern London,* ed. Paul Griffiths and Mark S. R. Jenner (Manchester: Manchester University Press, 2000), 154–75, at 154, 156.

10. Steven Mullaney, *The Place of the Stage: Licence, Play and Power in Renaissance England* (Chicago: University of Chicago Press, 1988); Jeremy Boulton, *Neighbourhood and Society: A London Suburb in the Seventeenth Century* (Cambridge: Cambridge University Press, 1989).

11. For a map, see Mullaney, *The Place of the Stage*, 28–29. See also Boulton, *Neighbourhood and Society*, 62.

12. "Settlement and Building: Chelsea up to 1680," in *A History of the County of Middlesex*, vol. 12, *Chelsea*, ed. Patricia E. C. Croot (London, 2004), 14–26, accessed 7 January 2016, http://www.british-history.ac.uk/vch/middx/vol12/pp14–26.

13. On similar issues for West End academies of the 1630s see Jean E. Howard, *Theater of a City: The Places of London Comedy, 1598–1642* (Philadelphia: University of Philadelphia Press, 2009), 162–208. On "Town" decadence see J. F. Merritt, "Introduction," in *Imagining Early Modern London: Perceptions and Portrayals of the City from Stow to Strype, 1598–1720*, ed. J. F. Merritt (Cambridge: Cambridge University Press, 2001), 1–24, at 15.

14. Eleanor Chance, Christina Colvin, Janet Cooper, C. J. Day, T G Hassall, Mary Jessup, and Nesta Selwyn, "Early Modern Oxford," in *A History of the County of Oxford*, vol. 4, *The City of Oxford*, ed. Alan Crossley and C. R. Elrington (London, 1979), 74–180. *British History Online*, accessed 6 May 2016, http://www.british-history.ac.uk/vch/oxon/vol4/pp74–180.

15. The dancing master John Weaver and a "Mr. Banister" both had Oxford establishments during the late seventeenth century. Richard Ralph, *The Life and Works of John Weaver* (London: Dance Books, 1985), 3.

16. Robert Thompson, "Playford, John (1622/3–1686/7)," in *Oxford Dictionary of National Biography* (hereafter *ODNB*), accessed 23 February 2016, http://www.oxforddnb.com.libezproxy2.syr.edu/view/article/22374.

17. Jennifer Thorp, "Dance in Late 17th-Century London: Priestly Muddles," *Early Music* 26, no. 2 (1998): 198–210, at 202–4.

18. Dorothy Gardiner, *English Girlhood at School: A Study of Women's Education through Twelve Centuries* (Oxford: Oxford University Press, 1929), 216.

19. White, "Letter from Aleppo," 422.

20. T[homas] Duffett, *Beauties Triumph; A Masque* (London, 1676), title page.

21. Walter H. Godfrey, "Milman's Street," in *Survey of London*, vol. 4, *Chelsea, Pt II* (London, 1913), 45, plates 1 and 22. *British History Online*, accessed 2 August 2016, http://www.british-history.ac.uk/survey-london/vol4/pt2/p45.

NOTES TO PAGES 30–31

22. "Settlement and Building."
23. Randall Davies, *Chelsea Old Church* (London: Duckworth, 1904), 131. On Priest's rental and then purchase of the property see Thorp, "Dance in Late 17th-Century London," 204–5.
24. Peter Holman, "Banister, Jafery [Jeffrey]," in *Grove Music Online* (hereafter *GMO*), accessed 9 November 2015, http://www.oxfordmusiconline.com .libezproxy2.syr.edu/subscriber/article/grove/music/42774pg1; and Andrew Ashbee, David Lasocki, Peter Holman, and Fiona Kisby, *A Biographical Dictionary of English Court Musicians, 1485–1714* (hereafter *BDECM*), vol. 1 (Aldershot: Ashgate, 1998), 63–65.
25. Ian Spink, "Hart, James," in *GMO*, accessed 23 February 2017, http://www .oxfordmusiconline.com.libezproxy2.syr.edu/subscriber/article/grove /music/12464 and *BDECM*, 1:550–51.
26. *London Gazette*, no. 1567 (22–25 November 1680), 2.
27. Thorp, "Dance in Late 17th-Century London," 203.
28. There were, however, multiple dancing Priests; see ibid.
29. This epilogue first appeared in *New POEMS, Consisting of SATYRS, ELEGIES, AND ODES: Together with a COLLECTION Of the Newest Court Songs, Set to MUSICK by the best Masters of the Age* (London, 1690): "*Epilogue to the Opera of DIDO and AENEAS, performed at Mr.* Preist's *Boarding School at* Chelsey, *spoken by the Lady* Dorothy Burk," 82. The table of contents lists the location as "*Mrs.* Priest's *Boarding School*" (with gender and spelling changed).
30. See Thomas Hearne's comment and n. 60 below.
31. *The Life and Times of Anthony Wood, Antiquary, of Oxford, 1632–1695, described by Himself*, ed. Andrew Clark, vol. 3, *1682–1695* (Oxford: Clarendon Press, 1894), 471.
32. *The Itinerary of John Leland in or about the years 1535–1543: Newly Edited from the MSS. By Lucy Toulmin Smith*, 5 vols (London: G. Bell and Sons, 1910), 5: 72.
33. Stephen K. Roberts, "Lenthall, William, appointed Lord Lenthall under the protectorate (1591–1662)," *ODNB*, accessed 8 January 2016, http://www .oxforddnb.com.libezproxy2.syr.edu/view/article/16467.
34. Mrs. Clarke, GB-Ob Ms. Top. Berks e. 5.
35. *Musica Oxoniensis: A Collection of Songs* (Oxford, 1698), 134–38; and Robert Thompson, "Goodson, Richard (i)," *GMO*, accessed 30 March 2017, http:// www.oxfordmusiconline.com.libezproxy2.syr.edu/subscriber/article/grove /music/43316.
36. See Neal Zaslaw, "An English 'Orpheus and Euridice' of 1697," *Musical Times* (1977): 805–8.
37. Olive Baldwin and Thelma Wilson, "Lindsey, Mary (fl. 1697–1713)," *ODNB*, accessed 21 February 2016, http://www.oxforddnb.com.libezproxy2.syr .edu/view/article/70111.
38. Zaslaw, "An English 'Orpheus and Euridice,'" 807.
39. Baldwin and Wilson, "Lindsey, Mary."

NOTES TO PAGES 32–34

40. On the pedagogical use of Och MS Mus. 389 see Rebecca Herissone, *Musical Creativity in Restoration England* (Cambridge: Cambridge University Press, 2013), 113; and John Milsom, *Christ Church Music Catalogue*, accessed 26 January 2017, http://library.chch.ox.ac.uk/music/page.php?set=Mus.+389. "Stop, O ye Waves" is not in Goodson's hand.

41. On these earlier entertainments, see Amanda Eubanks Winkler, "Dangerous Performance: Cupid in Early Modern Pedagogical Masques," in *Gender and Song in Early Modern England*, ed. Katie Larson and Leslie Dunn (Farnham: Ashgate, 2014), 77–91.

42. Richard Luckett, "A New Source for 'Venus and Adonis,'" *Musical Times* 130, no. 1752 (1989): 76–79, at 76; and GB-Cu Sel.2.123 (6).

43. Amanda Eubanks Winkler, "Madness 'Free from Vice': Musical Eroticism in the Pastoral World of the Fickle Shepherdess," in *The Lively Arts of the London Stage, 1675–1725*, ed. Kathryn Lowerre (Farnham: Ashgate, 2014), 149–69.

44. Judith Peraino analyzes the boarding-school potential for slippage between homosociality and same-sex desire in "I Am an Opera: Identifying with Henry Purcell's *Dido and Aeneas*," in *En Travesti: Women, Gender, Subversion, Opera* (New York: Columbia University Press, 1995), 99–132.

45. Bruce Wood suggests that Goodson made this arrangement for performance by Oxford choristers; John Blow, *Venus and Adonis*, ed. Bruce Wood, Purcell Society Edition Companion Series 2 (London: Stainer & Bell, 2008), xvii, xxiv.

46. Ellen Harris, *Henry Purcell's 'Dido and Aeneas'* (Oxford: Clarendon Press, 1987), 17: "The moral, of course, is that young girls should not accept the advances of young men no matter how ardent their wooing or how persistent their promises." See also the second edition (2018), 52.

47. As Margaret Laurie and others have noted, the sole surviving libretto for *Dido* does not correspond with the Tenbury manuscript, the earliest extant source. Margaret Laurie, "Preface," in *The Works of Henry Purcell*, vol. 3, *Dido and Aeneas*, ed. Margaret Laurie (Borough Green: Novello, 1979), ix. See also Andrew Walkling's examination of the libretto (GB-Lcm D144), "The Masque of Actaeon and the Antimasque of Mercury: Dance, Dramatic Structure, and Tragic Exposition in Dido and Aeneas," *Journal of the American Musicological Society* 63, no. 2 (2010): 191–242.

48. Lcm D144, 5.

49. See Eubanks Winkler, "Dangerous Performance."

50. As Zaslaw notes, it is difficult to see how satyrs, creatures known for sexual excess, would have been appropriate roles for schoolgirls; "An English 'Orpheus and Euridice,'" 806.

51. Lcm D144, 7.

52. John Dennis makes this argument in *The Usefulness of the Stage*, answering Jeremy Collier's 1698 excoriation of the stage's "immorality"; on the Col-

NOTES TO PAGES 34–37

lier controversy, see Robert D. Hume, "Jeremy Collier and the Future of the London Theater in 1698," *Studies in Philology* 96, no. 4 (1999): 480–511.

53. Lcm D144, 1.

54. See Eubanks Winkler, "'O Ravishing Delight': The Politics of Pleasure in *The Judgment of Paris,*" *Cambridge Opera Journal* 15, no. 1 (2003): 15–31, at 17–19.

55. Cf. the portrayal of Venus in Dryden's *Albion and Albanius* and *King Arthur*: James A. Winn, *"When Beauty Fires the Blood": Love and the Arts in the Age of Dryden* (Ann Arbor: University of Michigan Press, 1992), 254; and Amanda Eubanks Winkler, "From Whore to Stuart Ally: Musical Venuses on the Early Modern English Stage," in *Musical Voices of Early Modern Women: Many-Headed Melodies,* ed. Thomasin LaMay (Aldershot: Ashgate, 2005), 171–85, at 179–83.

56. Linda Phyllis Austern, "'Alluring the Auditorie to Effeminacie': Music and the Idea of the Feminine in Early Modern England," *Music and Letters* 74, no. 3 (1993): 343–54; and Amanda Eubanks Winkler, *Singing at School: Performance and Pedagogy in Early Modern England* (forthcoming).

57. Somerset Record Office (Taunton), DD/SF 3106. See also Mark Goldie, "The Earliest Notice of Purcell's *Dido and Aeneas,*" *Early Music* 20, no. 3 (1992): 393–400.

58. Thomas D'Urfey, *Love for Money: Or, The Boarding School* (London, 1691), [A3v].

59. D'Urfey and his play were satirized in the anonymous *Wit for Money: Or Poet Stutter* (London, 1691).

60. Thomas Hearne, *Remarks and Collections,* vol. 6 (Oxford: Clarendon Press, 1902), 92, Monday, 23 September 1717.

61. Duffett, *Beauties Triumph,* unpaginated prologue.

62. Ibid.

63. Ibid, unpaginated epilogue. On actresses' sexual liaisons, see Elizabeth Howe, *The First English Actresses: Women and Drama, 1660–1700* (Cambridge: Cambridge University Press, 1992), 32–37.

64. Thomas D'Urfey, "Epilogue to the Opera," in *New Poems, Consisting of Satyrs, Elegies, and Odes Together with a Choice Collection of the Newest Court Songs Set to Musick by the Best Masters of the Age* (London, 1690), 82.

65. Whyman, *Sociability and Power*; and White, "Letter from Aleppo."

66. GB-Lbl M636/38. Microfilm of letters dated July 1683–84.

67. Lbl M636/40.

68. Whyman, *Sociability and Power,* xiv, 21.

69. On the fashionability of Kynaston's location, see Howard, *Theater of a City,* 186–87.

70. Francis Kynaston, *The Constitvtions of the Mvsaeum Minerva* (London, 1636), 12.

71. On the area and Gerbier's lease, see "Bethnal Green: The Green, Bethnal Green Village," in *A History of the County of Middlesex: Volume 11, Stepney,*

NOTES TO PAGES 37–44

Bethnal Green, ed. T. F. T. Baker (London, 1998), 95–101. *British History Online*, accessed 6 May 2016, http://www.british-history.ac.uk/vch/middx /vol11/pp95–101.

72. *Several Proceedings IN PARLIAMENT,* 10 (30 November–7 December 1649), 122; 12 (14–21 December 1649), 147.

73. *Perfect Diurnall* 10 (11–18 February 1650), sig. K3; quoted in Jason Peacey, "Print, Publicity, and Popularity: The Projecting of Sir Balthazar Gerbier," *Journal of British Studies* 51, no. 2 (2012): 284–307, at 302.

74. Peacey, "Print, Publicity, and Popularity," 302.

75. *London Gazette* no. 1154 (7–11 December 1676).

76. *MUSICK: OR A PARLEY OF Instruments. The First Part* (London, 1676), [15].

77. Peter Holman, *Four and Twenty Fiddlers: The Violin at the English Court, 1540–1690* (Oxford: Clarendon Press, 1993), 352.

CHAPTER FOUR

1. See Jay P. Pederson, *International Directory of Company Histories* 71 (London: St. James Press, 2005).

2. *Carlton House: The Past Glories of George IV's Palace* (London: Queen's Gallery, Buckingham Palace, 1991).

3. Dana Arnold, *Rural Urbanism: London Landscapes in the Early Nineteenth Century* (Manchester: Manchester University Press, 2005), 77–96; Hermione Hobhouse, *A History of Regent Street* (London: Queen Anne Press, 1975); and Hobhouse, *Regent Street: A Mile of Style* (London: History Press, 2008).

4. *Maps of Old London* (London: Adam & Charles Black, 1908), 16–17.

5. Colley Cibber, *An Apology for the Life of Colley Cibber* (London: By John Watts for the Author, 1740), 183–84.

6. Judith Milhous, "New Light on Vanbrugh's Haymarket Theatre Project," *Theatre Survey* 17 (1976): 143–61, at 155.

7. Ibid., 156–59.

8. John Stow, *A Survey of the Cities of London and Westminster: Containing the Original, Antiquity, Increase, Modern Estate and Government of those Cities,* 2 vols. (London: A. Churchill et al., 1720), 2:83.

9. John Entick, *A New and Accurate Survey of London, Westminster, Southwark, and Places Adjacent,* 4 vols. (London: Edward & Charles Dilly, 1766), 4:424.

10. M.R.C.O. Haymarket Miscellaneous Accounts, in *Survey of London,* ed. F. H. W. Shepherd, 52 vols. (London: P. S. King and Son, 1900–): *The Parish of St. James Westminster, I: South of Piccadilly,* 29 & 30 (London: Athlone Press, 1960), 29:211–12.

11. *Public Advertiser,* 17 March 1785.

12. See, for documentary evidence: Michael Burden, ed., *London Opera Observed 1711–1844,* 5 vols. (London: Pickering and Chatto, 2013); and for a modern narrative: Curtis A. Price, Judith Milhous, and Robert D. Hume, *Italian*

268

NOTES TO PAGES 45–52

Opera in Late Eighteenth-Century London, Vol. 1, *The King's Theatre, Haymarket, 1778–1791* (Oxford: Clarendon Press, 1995).

13. John Vanbrugh, *The Complete Works of John Vanbrugh*, ed. G. Webb, 4 vols. (London: Nonesuch Press, 1928), 4:8–9. On the early financial state of the concern, see: Milhous, "New Light," 143–61. The "spare ground" appears to refer to land not covered by the theater building itself, out of which Vanbrugh hoped to make money, but which the lessee might argue was excluded from the lease.

14. Graham Barlow, "Vanbrugh's Queen's Theatre in the Haymarket, 1703–9," *The Baroque Stage I, Early Music* 17, no. 4 (1989): 515–21.

15. Engraved for Gabriel Dumont, *Parallele de plans des plus belles salles de spectacles d'Italie et de France* and published in 1764, it apparently shows the building after alterations of 1709. Michael Burden, "Where Did Purcell Put His Theatre Band?," *Early Music* 28, no. 3 (2009): 429–43.

16. Cibber, *Apology*, 183.

17. Ibid., 183; and Barlow, "Vanbrugh's Queen's Theatre," 520.

18. James Ralph, *A Critical Review of the Public Buildings, Statues, and Ornaments, in and about London and Westminster* (London: For John Wallis, 1783), 177.

19. Daniel Defoe, *Selected Writings of Daniel Defoe*, ed. James T. Bolton (Cambridge: Cambridge University Press, 1975), 117–18.

20. *London Chronicle*, 16–18 June 1789; and Edward Wedlake Brayley, *London and Middlesex*, 2 vols. (London: Longman, Hurst, Rees, Orme, & Brown, 1810), 1:545.

21. *Oracle*, 19 June 1789.

22. Henry Angelo, *Reminiscences of Henry Angelo*, 2 vols. (London: Henry Colburn, 1830), 2:100–101.

23. James Pellar Malcolm, *Londinium Redivivum Or an Antient History and Modern Description of London*, 4 vols. (London: Nichols & Son, 1802), 4:315.

24. Curtis Price, Judith Milhous, and Robert D. Hume, "A Royal Opera House in Leicester Square (1790)," *Cambridge Opera Journal* 2, no. 1 (1990): 1–28.

25. *World*, 20 March 1790; also announced as *Ideas on the Opera, Offered to the Subscribers, Creditors, and Amateurs of that Theatre. By Mr. Le Texier. Translated from the French* (London: Printed for J. Bell, 1790).

26. *World*, 20 November 1789.

27. *English Chronicle* or *Universal Evening Post*, 27–30 March 1790.

28. Michael Burden, "A Return to London's Opera House in 1782, with an English Translation of Jean-Georges Noverre's *Observations sur la construction d'une nouvelle salle de l'Opéra*," *Music in Art: International Journal for Music Iconography* 42, nos. 1–2 (2017): 11–26.

29. Marvin Carlson, "Theatre as Civic Monument," *Theatre Journal* 40 (1988): 12–32; and Carlson, *Places of Performance: The Semiotics of Theatre Architecture* (Ithaca, NY: Cornell University Press, 1989).

30. Voltaire, *Oeuvres complètes*, 92 vols. ([Kehl]: Société Littéraire-typographique 1785[–89]), 47:500.

NOTES TO PAGES 52-55

31. These included Charles-Nicolas Cochin, Pierre Patte, and Chevalier de Chaumont, *Projet d'une salle de spectacle pour un theatre de comedie* (1765); Chevalier de Chaumont, *Véritable construction d'un théâtre d'opéra à l'usage de France* (1766); Cochin, *Lettres sur l'Opéra* (1781); and Patte, *Essai sur l'architecture théâtrale, ou, De l'ordonnance la plus avantageuse à une salle de spectacles, relativement aux principes de l'optique & de l'acoustique* (1782). Chaumont also produced two other treatises on theater buildings.

32. Jean-Georges Noverre, *Observations sur la construction d'une nouvelle salle de l'Opéra* (Amsterdam: D. J. Changuion and Paris: P. de Lormel, 1781), 7–8, trans. Tom Hamilton in Burden, "A Return to London's Opera House," 13.

33. Noverre, *Observations*, 8–9; and Burden, "A Return to London's Opera House," 13.

34. Jean-Baptiste Colbert served Louis XIV, managing France's finances from 1665 to 1683 and taking responsibility for a number of economic reforms including the growth of commerce. He played a part in establishing the Académie d'Opéra (later the Académie Royale de Musique) when he granted Pierre Perrin the privilège for the Opéra.

35. Noverre, *Observations*, 36–37; and Burden, "A Return to London's Opera House," 22.

36. The Théâtre de l'Odéon was designed by Marie-Joseph Peyre and Charles De Wailly on an island site linked to the buildings to the right and left by masonry bridges over the side streets. Allan Braham, *The Architecture of the French Enlightenment* (Berkeley and Los Angeles: University of California Press, 1980), 98–100. Noverre refers to the Théâtre de l'Odéon in his 1804 edition of *Observations*.

37. Briant Hamor Lee, "Pierre Patte, Late 18th-Century Lighting Innovator," *Theatre Survey* 15 (1974): 177–83.

38. Michael Burden, "Visions of Dance at the King's Theatre: Reconsidering London's 'Opera House,'" *Music in Art: International Journal for Music Iconography* 36 (2011): 92–116.

39. Michèle Sajous D'Oria, "Les observations sur la construction d'une nouvelle salle de l'Opéra de Noverre," in Laurine Quétin, *Jean-Georges Noverre (1727–1810), danseur, choréographe, théoricien de la danse et du ballet: Un artiste européen au siècle des Lumières*, Musicorum 10 ([Tours]: Université de François Rabelais, 2011), 97–113. Bernard Poyet won second prize in the 1768 Prix de Rome with a theater design. Translation by Tom Hamilton.

40. Christopher Mead, "Urban Contingency and the Problem of Representation in Second Empire Paris," *Journal for Society of Architectural History* 54 (1995): 138–74, at 152.

41. Ibid., 167.

42. Ibid., 171.

43. John Summerson, *The Life and Works of John Nash, Architect* (Cambridge, MA: MIT Press, 1980), 78.

NOTES TO PAGES 57–58

CHAPTER FIVE

1. The metaphor of opera as a "theatrical conversation" is Mary Hunter's in *The Culture of Opera Buffa in Mozart's Vienna: A Poetics of Entertainment* (Princeton, NJ: Princeton University Press, 1999), especially 3–24; see also Mary Hunter and James Webster, eds., *Opera Buffa in Mozart's Vienna* (Cambridge: Cambridge University Press, 1997).

2. For example, Ellen Rosand, *Opera in Seventeenth-Century Venice: The Creation of a Genre* (Berkeley and Los Angeles: University of California Press, 1991); and Martha Feldman, *Opera and Sovereignty: Transforming Myths in Eighteenth-Century Italy* (Chicago: University of Chicago Press, 2007).

3. Thomas F. Gieryn, "A Space for Place in Sociology," *Annual Review of Sociology* 26 (2000): 463–96.

4. Geoffrey Symcox, *Victor Amadeus II: Absolutism in the Savoyard State, 1675–1730* (London: Thames & Hudson, 1983), 18; Martha D. Pollak, *Turin, 1564–1680: Urban Design, Military Culture, and the Creation of the Absolutist Capital* (Chicago: University of Chicago Press, 1991).

5. A well-known and widely reproduced painting of the theater's interior is "Interno del Teatro Regio di Torino," attributed to Giovanni Michele Graneri (mid-1750s, Turin, Palazzo Madama, Museo Civico d'Arte Antica, inv. 534/D). Margaret R. Butler, "'Olivero's' Painting of Turin's Teatro Regio: Reevaluating an Operatic Emblem," *Music in Art: International Journal of Musical Iconography* 34, nos. 1–2 (2009): 137–51. Documents are held at Turin, Archivio Storico della Città di Torino (hereafter I-Tac). Marie-Thérèse Bouquet, *Il teatro di corte dalle origini al 1788* (Turin: Cassa di Risparmio di Torino, 1976); and Margaret R. Butler, *Operatic Reform at Turin's Teatro Regio: Aspects of Production and Stylistic Change in the 1760s* (Lucca: LIM, 2001).

6. On Turin's theater as representative of a type (including Naples and Milan), see Margaret R. Butler, "Italian Opera in the Eighteenth Century," in *The Cambridge History of Eighteenth-Century Music*, ed. Simon P. Keefe (Cambridge: Cambridge University Press, 2009), 203–71.

7. Franco Piperno, "Opera Production up to 1780," in *Opera Production and Its Resources*, ed. Lorenzo Bianconi and Giorgio Pestelli, trans. Lydia G. Cochrane, *The History of Italian Opera*, Part 2, *Systems*, Vol. 4 (Chicago: University of Chicago Press, 1998), 1–79; Piperno, "Impresariato collettivo e strategie teatrali: Sul sistema produttivo dello spettacolo operistico settecentesco," in *Civiltà teatrale e Settecento emiliano*, ed. Susi Davoli (Bologna: Il Mulino, 1986), 345–56 (esp. 345–48).

8. Marie-Thérèse Bouquet, "Public et répertoire aux Théâtres Regio et Carignano de Turin," *Dix-Huitième Siècle* 17 (1985): 229–40; and Bouquet, "Role du Théâtre Carignan dans l'histoire des spectacles à Turin au XVI–IIe siècle," in *Culture et pouvoir dans les états de Savoie du XVIIe siècle à la Révolution: Actes du colloque d'Annecy-Chambéry-Turin, 1982* (Geneva: Slatkine,

NOTES TO PAGES 58–62

1985), 161–75. On the Cavalieri's engagement of French troupes and their European cultural impact, see Rahul Markovits, *Civiliser l'Europe: Politiques du théâtre français au XVIIIe siècle* (Paris: Fayard, 2014), esp. 49–50, 375. In addition to marionette plays, displays of animals occurred in 1778 and 1781, of tightrope walkers in 1765 and 1780, of fireworks in 1765, of life-like wax figures in 1776, and of acrobats and dancers in 1780. I-Tac, Carte sciolte 5499.

9. "Memoriale a capi della Nobile Società dei Cavalieri," 1738 (and renewed every six years through the 1770s); and Bouquet, *Il teatro di corte*, 113–15.

10. Bouquet, *Il teatro di corte*; and Butler, *Operatic Reform*.

11. "Non potranno essi Sig.ri Contraenti rappresentare d.o Spettacolo inferior-mente all'Isola della Torre verso Piazza Castello, ma superiormente verso Porta Susina, cioè in retta Linea dalla Torre di qu.sta Città tendente verso porta Palazzo." Contract for Carnival 1763: Pietro Maria Deferari del fu Eusebio, Michele Mazzolino del fu Franco, Pietro Dematuri del fu Gio Domenico, Vinardo [Vinardi], 20 December 1762. I-Tac, Carte sciolte 5499.

12. *Nuova Pianta della Reale Città di Torino* (Milan: Carlo Giuseppe Ghislandi, 1751), I-Tac, Coll. Simeom, Serie D, inv. 56.

13. Margaret R. Butler, "Time Management at Turin's Teatro Regio: Galuppi's *La clemenza di Tito* and Its Alterations, 1759," *Early Music* 40, no. 2 (2012): 279–89, at 285.

14. Alberto Basso, ed., *Storia del Teatro Regio di Torino*, 5 vols. (Turin: Cassa di Risparmio di Torino, 1976–88); Basso, ed., *Il nuovo Teatro Regio di Torino* (Turin: Cassa di Risparmio di Torino, 1991); and *L'arcano incanto: Il Teatro Regio di Torino, 1740–1990*, ed. Alberto Basso (Milan: Electa, 1991).

15. *Novelle letterarie di Firenze* (30 vols., Florence, 1740–69; n.s., 23 vols., Florence, 1770–92; *Alcina* reviewed in 1775); Ottaviano Diodati, *Biblioteca teatrale italiana*, 12 vols. (Lucca: Giovanni della Valle, 1761–65: *Ifigenia in Aulide*, vol. 6, 1762; *Enea nel Lazio*, vol. 8, 1762; *Motezuma*, vol. 11, 1765).

16. *Nuovo almanacco de' teatri di Torino* (Turin: Onorato Derossi, 1780).

17. Roberto Verti, ed., *Un almanacco drammatico: L'indice dei teatrali spettacoli, 1764–1823*, 2 vols. (Pesaro: Fondazione Rossini, 1996).

18. Luciano Tamburini, *I teatri di Torino: Storia e cronache* (Turin: Edizioni dell'Albero, 1966). The Gallo Ughetti gave *drammi giocosi* in 1775.

19. Riccardo Giusti, "Il Teatro d'Angennes di Torino: Profilo storico, cronologia e catalogo dei libretti (1765–1848)," *Fonti musicali italiane* 15 (2010): 229–64.

20. Renamed the Teatro Gianduia in the nineteenth century, the theater moved location and today houses an important marionette museum.

21. I-Tac, Ordinati vol. 5, p. 277, 5 January 1765; Ordinati vol. 6, p. 242, 9 December 1768.

22. I-Tac, Libri conti vol. 53 (1768–69), 20. The comparison table presents dates, numbers of performances, and nightly earnings. Gugliemone's earn-

NOTES TO PAGES 63–65

ings exceed Vinardi's, but details needed to assess the data's significance (such as location and venue size) are lacking.

23. I-Tac, Coll. XI vol. 106 (1769), receipt for purchase of materials.

24. Non-performing nights were Fridays and occasionally Tuesdays.

25. There were six to eleven balls (most often eight) per season. I-Tac, Libri conti, vol. 35 (1757–58) through vol. 55 (1770–71).

26. At least some instrumentalists seem to have played in both operas and balls, including the Celoniats, a Turinese family of violinists who led ball ensembles from the 1750s to 1770s. Butler, "Time Management," 285.

27. Ibid., 281–85.

28. A table of profit and losses for 1753–57 shows the scale of the problem: I-Tac, Libri conti, vol. 32 (1756–57), 114.

29. Verti, *L'indice*, Turin entry for 1776: "Si danno le solite Feste di Ballo due volte la settimana" (The usual balls are given twice a week) (p. 162); Turin entry for 1783 (p. 453).

30. Contracts for one "Decarli" for display of a crocodile, 11 October 1775, and for "Guerra" for the display of a creature with two heads, 24 November 1774, I-Tac, Carte sciolte 5502; see Silvia Cariani, "Torino a teatro: Il Carignano e il Regio nel Settecento" (thesis, Università degli studi di Torino, 2001–2).

31. Contract for La Gagneur and Trevisiani for the display of an elephant for one month, 14 July 1774. I-Tac, Carte sciolte 5502.

32. Ignazio Sclopis del Borgostura, *Passeggio della Cittadella con l'elefante venuto in Torino l'anno 1774*.

33. On shops and vendors see Bouquet, *Il teatro di corte*, 151–54; and Butler, "Time Management," 285. The shops selling refreshments were adjacent to the casino, the entrance to the theater's main stairway lying beyond them both; the audience thus encountered these venues before reaching the auditorium. (The space for the refreshment shops might have been in fact a single two-room "shop," similar to a café.) Diagram of the theatrical complex in *Nuovo almanacco de' teatri di Torino* (Turin: Onorato Derossi, 1780), n.p., I-Tac, Coll. Simeom F, no. 271. See Butler, *Operatic Reform*, 9. The marchese d'Angennes, who sponsored one of Turin's minor theaters, brokered arrangements with the jewelry vendors at the Teatro Regio and Carignano: "Il Sigr marchese d'Angennes ha riferto d'avere convenuto il permesso concesso agli fratelli Sereni di vendere Galanterie, o sia Bijoux sopra il ripiano della Scala del Regio Teatro in tempo delle Recite delle Opere nel prossimo Carnevale, e de' Balli in Maschere del Teatro Carignano." I-Tac, Ordinati vol. 5, pp. 275–76, 1 December 1764.

34. Butler, *Operatic Reform*, 16.

35. "Si farà gloria e pregio la Società di dare le più belle e le più sontuose opera, che le sarà possibile." Statuti della Nobile Società dei Cavalieri (approved 15 June 1770), Article 11, I-Tac, Carte sciolte 6188; and Bouquet, *Il teatro di corte*, 453–56.

NOTES TO PAGES 65–67

36. Bouquet, *Il teatro di corte*, 164; and anonymous commentary on *L'isola di Alcina ossia Alcina e Ruggero* (Turin, 1775), *Novelle letterarie di Firenze*, n.s. 6 (1775): 173–76.

37. Francesco Algarotti, *Il saggio sopra l'opera in musica: Le edizioni di Venezia (1755) e di Livorno (1763)*, ed. Annalisa Bini (Lucca: LIM, 1989).

38. Margaret R. Butler, "L'esotismo nell'opera torinese del Settecento: *Motezuma* e il suo contesto," in *Le arti della scena e l'esotismo in età moderna/The Performing Arts and Exoticism in the Modern Age*, ed. Francesco Cotticelli and Paologiovanni Maione (Naples: Turchini, 2006), 329–42; Pierpaolo Polzonetti, "Opera as Process," in *The Cambridge Companion to Eighteenth-Century Opera*, ed. Anthony R. DelDonna and Pierpaolo Polzonetti (Cambridge: Cambridge University Press, 2009), 3–23; and Pierpaolo Polzonetti, "Montezuma and the Exotic Europeans," in *Italian Opera in the Age of the American Revolution* (Cambridge: Cambridge University Press, 2011), 107–32. Further interest in Montezuma settings is attested in papers given by John A. Rice in 2009–11 (https://sites.google.com/site/johnaricecv/lectures) and Marita P. McClymonds in 2002 (cited in Polzonetti, *Italian Opera*, 110).

39. Butler, *Operatic Reform*, 105.

40. Butler, "L'esotismo nell'opera torinese," 340–41.

41. Bouquet, *Il teatro di corte*, 322.

42. Butler, "L'esotismo nell'opera torinese," 340–41.

43. The dragon appeared in *Tigrane* of 1761 and the animals in *Didone abbandonata* of 1750; Bouquet, *Il teatro di corte*, 304, 284.

44. Ibid., 294, 295, 318.

45. I-Tac, list of supernumeraries (with total numbers) for *Ifigenia in Aulide* (Ordinati vol. 5, p. 116, 25 May 1761) and *Tigrane* (Ordinati, vol. 5, p. 2; cited in Bouquet, *Il teatro di corte*, 304, without date).

46. I-Tac, Ordinati vol. 5, p. 54, 20 November 1760: "Si è stabilito di formare il bilancio sul piede di un'opera delle più spettacolose sia a riguardo del scenario, che del vestiario" (It has been decided that we will develop a budget based on an opera exhibiting the greatest spectacle possible in regard to both scene and costume design).

47. I-Tac, Ordinati vol. 5, p. 254, 4 April 1764.

48. Charles Burney, *The Present State of Music in France and Italy* (London: T. Becket, 1771, 1773), 72–73; Joseph-Jerôme Le Français de La Lande, *Voyage d'un François en Italie*, 8 vols. (Venice and Paris: Desaint, 1769), 1:110.

49. For example, an illustration of the *Assedio della città di Filippi* presents a monumental scene similar to those featuring large numbers of supernumeraries on the Teatro Regio's stage.

50. "In tale circostanza si eseguí una fastosa illuminazione 'con tremila circa Torcie, e Candele di Cera, distribuite lungo i cinque ordini di Palchetti, ed disposte su braccia a foglie dorate con sí bella simmetria, che venivano a rialzare insieme la maestà, e la richezza del Teatro'; mentre gli scenari, rischiarati la 'lumi di cristallo' rifrangevano il fulgore degli addobbi di gala."

Tamburini, *I teatri di Torino*, 35, quoting Luigi Mussi, *Torino nel 1761* (Turin: Arduini, 1961), n.p.

51. Tamburini, *I teatri di Torino*, 35: "Ma anche fuori di tali occasioni la città offriva, di notte, uno spettacolo gradevole: 630 lanterne ad olio appese a sostegni metallici rompevan l'oscurità e davano, fatto nuovo, sicurezza ai passanti" (undated).

52. Symcox, *Victor Amadeus*, 77–78. See also Paolo Cornaglia, "Torino nel Settecento e la sua immagine perfezionata: Risplasmazioni urbanistiche, vedute incise, matrimoni dinastici tra corte e città," in *Città nel Settecento: Saperi e forme di rappresentazione* (Rome: Edizioni di storia e letteratura, 2014), 219–45.

53. "La più bella città d'Italia, forse d'Europa, per le strade diritte, la regolarità degli edifici e la bellezza delle piazze." Charles de Brosses, *Lettres historiques et critiques sur l'Italie*, An VII, vol. 3 (Paris: Chez Ponthieu, 1799), 384; cited in Cornaglia, "Torino nel Settecento," 219.

54. "La città è di per sé molto bella; le strade sono tutte perfettamente allineate e le case edificate con grande regolarità." Élisabeth Vigée Le Brun (1789) in *Ricordi dall'Italia* (Palermo: Sellerio, 1990) 43; cited in Cornaglia, "Torino nel Settecento," 219.

55. Giovanni Gaspare Craveri, *Guida de' Forestieri per la Real Città di Torino* (Turin: n.p., 1753; facs. rpt., Le livre précieux, 1969).

56. Cornaglia, "Torino nel Settecento," 231.

57. Craveri, *Guida de' Forestieri*, 40: "Questo teatro è giudicato da tutti il più grandioso, e compito d'Europa, ed è meritevolmente l'oggetto della maraviglia de' Forestieri, per la vastità, e ampiezza sua, e per l'architettura, e comodità dell'edifizio, e per l'interna bellezza degli ornamenti, per lo più dorati. È rimarchevole la pittura della Volta. Ivi si recitano ogni Carnevale i Drammi musicali con tali magnificenza di apparato, quale si conviene alla grandezza della Real Corte, che v'interviene sulla Loggia spaziosa a lei destinata, che poi si suole illuminare. Vi si chiamano sempre i migliori Musici d'Europa."

58. Ibid., 38. "Che oggidì serve d'abitazione ai Principi Reali. Fu nel 1416. edificato da Amedeo VIII., Primo Duca di Savoia, con quattro torri, cioè una per ogni angolo. E qui terminava allora il recinto della Città. Fu poscia nel 1720. dalla munificenza di Madama Reale, Madre del Re Vittorio Amedeo abbellito di una superbissima Facciata di pietra, la quale ne copre tutto l'esterno verso ponente, ed è ornata di Vasi, e di Statue, con gran fenestroni, e gallerie. Nell'ingresso ha tre grandi Porte uniformi, chiuse da ferrate, che non impediscono la vista del superbo Atrio, e delle due grandi, e spaziosi Scaloni di marmo uniformi: il tutto costrutto con estrema magnificenza. Ne diede il disegno il celebre Cavaliere D. Filippo Juvara [sic], e si può con sicurezza dire, che questo sia il miglior pezzo d'Architettura, che si veda nella Città, e che può gareggiare con qualunque de' più belli edifizj d'Italia."

NOTES TO PAGES 69–75

59. The *Guida de' Forestieri*'s original publication by Rameletti (1753; not extant) was as a guide for religious pilgrims (hence its detailed descriptions of Turin's churches); it was immediately reissued as the court-sponsored publication known today (perhaps with expansions on secular venues such as the Teatro Regio). Cornaglia, "Torino nel Settecento," 232–33.

60. *Nuova Pianta della Città*: "Questa città è assai nota per la sua bellezza, e regolarità de' suoi Edifizj, e Contrade a livello. I suoi Cittadini si dilettano di varie Lingue particolarmente della Francese, sono manierosi, di bel tratto, civili, amanti delle Virtù, cortesi verso li Forastieri, e fedelissimi Sudditi de' suoi Sovrani. Ella è posta nel cuore del Piemonte qual è abondantissimo di tutte le cose necessarie al vitto umano onde vien chiamata il Giardino dell'Italia."

CHAPTER SIX

1. Mozart's opera was first fully staged in London in 1817, though excerpts had been heard earlier. See Rachel Cowgill, "'Wise Men from the East': Mozart's Operas and Their Advocates in Early Nineteenth-Century London," in *Music and British Culture, 1785–1914: Essays in Honour of Cyril Ehrlich*, ed. Christina Bashford and Leanne Langley (Oxford: Oxford University Press, 2000), 39–64, at 47–48; Alfred Loewenberg, "'Don Giovanni' in London," *Music & Letters* 24 (1943): 164–68; and Alexander Hyatt King, "'Don Giovanni' in London before 1817," *Musical Times* 27 (1986): 487–93.

2. Mobility was a key theme as long ago as 1969 in Stephan Thernstrom and Richard Sennett, eds., *Nineteenth-Century Cities: Essays in the New Urban History* (New Haven, CT: Yale University Press, 1969). For an overview of mobility studies in contemporary geography, see Tim Cresswell, *On the Move: Mobility in the Modern Western World* (New York: Routledge, 2006).

3. An example of the former might be Benjamin Walton, *Rossini in Restoration Paris: The Sound of Modern Life* (Cambridge: Cambridge University Press, 2007). On the city as soundscape see Arman Schwartz, "Rough Music: *Tosca* and *Verismo* Reconsidered," *19th-Century Music* 31, no. 3 (2008): 228–44. For a discussion of *Don Giovanni*'s premiere in eighteenth-century Prague, see Thomas Forrest Kelly, *First Nights at the Opera* (New Haven, CT: Yale University Press, 2004), 63–131.

4. Anonymous, "A scene from Don Giovanni as perform'd at the Kings Theatre" (London: H. Fores, 23 July 1820).

5. Gregory Dart, *Metropolitan Art and Literature, 1810–1840: Cockney Adventures* (Cambridge: Cambridge University Press, 2012), esp. chap. 4.

6. See Jane Rendell, *The Pursuit of Pleasure: Gender, Space and Architecture in Regency London* (London: Athlone Press, 2002), esp. chaps. 2 and 5 on *Life in London* and the Italian opera house.

7. See Jane Moody, *Illegitimate Theatre in London, 1770–1840* (Cambridge: Cambridge University Press, 2000).

NOTES TO PAGES 75–80

8. Rachel Cowgill, "Re-Gendering the Libertine; or, The Taming of the Rake: Lucy Vestris as Don Giovanni on the Early Nineteenth-Century London Stage," *Cambridge Opera Journal* 10, no. 1 (1998): 45–66. Gillen D'Arcy Wood, *Romanticism and Music Culture in Britain, 1770–1840: Virtue and Virtuosity* (Cambridge: Cambridge University Press, 2010), esp. chap. 4.

9. Jennifer Hall-Witt, *Fashionable Acts: Opera and Elite Culture in London, 1780–1880* (Durham: University of New Hampshire Press, 2007).

10. For details of *Life in London*'s publication history see Dart, *Metropolitan Art and Literature*, 115–16. The version I cite in this chapter is the modern reprint of an 1822 edition: Pierce Egan, *Life in London; or, The Day and Night Scenes of Jerry Hawthorne, Esq., and His Elegant Friend Corinthian Tom* (London: Sherwood, Neely, and Jones, 1822; repr., Cambridge: Cambridge University Press, 2011).

11. On this binary, see Michel de Certeau, *The Practice of Everyday Life*, trans. Steven Rendall (Berkeley and Los Angeles: University of California Press, 1984), esp. chap. 7.

12. The Corinthian column frontispiece is discussed in Dart, *Metropolitan Art and Literature*, 109–13.

13. Egan, *Life in London*, 13.

14. Ibid., 35.

15. Ibid., 32.

16. Egan, *Life in London*, 334; discussed in Deborah Epstein Nord, "Night and Day: Illusion and Carnivalesque at Vauxhall," in Jonathan Conlin, ed., *The Pleasure Garden, from Vauxhall to Coney Island* (Philadelphia: University of Pennsylvania Press, 2013), 177–94, at 179.

17. Egan, *Life in London*, 333.

18. When Jerry is not on hand to demand explanations, Egan adds notes for his linguistically curious readers.

19. Such an attitude toward fighting is evident throughout the many "Tom and Jerry" iterations, all the way up to the Hanna-Barbera cartoons in the twentieth century.

20. Egan, *Life in London*, 118.

21. Ibid., 125.

22. It was purportedly "Written and set to Music by CORINTHIAN TOM."

23. See the modern Cambridge edition, facing p. 118.

24. A note to Voight's song acknowledges that "the first and last Strains of these three verses are not original." The tune was associated with Thomas D'Urfey's comic song of 1719, beginning "Oh London is a fine town," which described a trip to Greenwich. It was also used to set "Our Polly is a sad slut" in John Gay's *The Beggar's Opera* (1728), a work mentioned in Egan's text and used by nineteenth-century critics as a reference point for *Life in London*.

25. Egan, *Life in London*, 30.

26. Ibid., 90.

NOTES TO PAGES 81–89

27. Ibid.
28. Ibid., 91.
29. Ibid., 192.
30. Ibid., 323.
31. John Camden Hotten, "Introduction," in Pierce Egan, *Life in London; or, the Day and Night Scenes of Jerry Hawthorn, Esq. and His Elegant Friend Corinthian Tom in Their Rambles and Sprees Through the Metropolis* (London: John Camden Hotten, 1881), 12–13.
32. Ibid., 12. "The author" here refers to Egan, who struggled to maintain the intellectual property rights to his work.
33. William Thomas Moncrieff, preface to *Tom and Jerry; or, Life in London, an Operatic Extravaganza, in Three Acts, first performed at the Adelphi Theatre, Monday, November 26, 1821*, quoted in Hotten's edition, 13. Hotten does not give the date of the preface but states that Moncrieff's words were "written long, long before he was a blind old man."
34. Hotten, "Introduction," 15.
35. In one of Egan's later attempts in his own genre—*Life in Paris* (1824)—he added a Captain O'Shuffleton, who, we might imagine, was rather less dashing than Tom.
36. The parallels are not only with Don Giovanni, pursued by Donna Elvira, but also with *The Rake's Progress*, in which Hogarth's Tom Rakewell is followed to the city by his jilted lover.
37. At least six of the cast supporting Lucy Vestris at Drury Lane (Messrs. Keeley, Emery, Atkins, and Henry, Mrs. Daly, and Miss Hammerley) appeared elsewhere in Moncrieff's adaptation of *Tom and Jerry*, either at the Adelphi or at Covent Garden or both. One Mrs. Weston, a member of the original cast of *Giovanni in London* at the Olympic, later played alongside Keeley et al. in the Covent Garden version of *Life in London*.
38. George Daniel, introduction to William T. Moncrieff, *Tom and Jerry; or, Life in London in 1821, a Drama in Three Acts* (London: Thomas Hailes Lacy, n.d.), 6.
39. Anonymous, "Don Giovanni the XVIII: A Musico-Burlesque-Comico-Nonsensical Opera," *London Magazine* 5, no. 29 (May 1822): 436–37, at 436.

CHAPTER SEVEN

1. Mercedes Viale Ferrero, "Stage and Set," in *Opera on Stage*, ed. Lorenzo Bianconi and Giorgio Pestelli (Chicago: University of Chicago Press, 2002), 1–124, at 95–96.
2. See also Margaret Butler's chapter in this volume.
3. Ellen Rosand, *Opera in Seventeenth-Century Venice: The Creation of a Genre* (Berkeley and Los Angeles: University of California Press, 2007), 139.
4. Francesco Algarotti, *Saggio sopra l'opera in musica* (1755; repr., Livorno: Marco Coltellini, 1763), 19–20.

NOTES TO PAGES 89–92

5. Stanislao Gatti, "Letteratura," *Il saggiatore giornale romano* II/3, nos. 9–10 (1845): 310–18, at 312–13.

6. Rosand, *Opera in Seventeenth-Century Venice*, 135.

7. Ellen Rosand, "Venice: Cradle of (Operatic) Convention," in *Operatic Migrations: Transforming Works and Crossing Boundaries*, ed. Roberta Montemorra Marvin and Downing A. Thomas (Aldershot: Ashgate, 2006), 7–20; at 10.

8. Elena Povoledo, "Regular Comedy and the Perspective Set," in *Music and Theatre from Poliziano to Monteverdi*, ed. Nino Pirrotta and Elena Povoledo (Cambridge: Cambridge University Press, 1982), 311–34, at 316–18.

9. Julia Cartwright, *Baldassarre Castiglione: The Perfect Courtier. His Life and Letters 1478–1529*, 2 vols. (London: John Murray, 1908), Vol. I, 336–37.

10. Jack D'Amico, "Drama and the Court in *La Calandria*," *Theatre Journal*, 43 (1991), 93–106; at 94.

11. Rosand, *Opera in Seventeenth-Century Venice*, 100.

12. Catherine Keen, "Boundaries and Belonging: Imagining Urban Identity in Medieval Italy," in *Imagining the City*, vol. 2, *The Politics of Urban Space*, ed. Christian Emden, Catherine Keen, and David R. Midgley (Bern: Peter Lang, 2006), 65–86, at 68.

13. Paul Zumthor, "L'espace de la cité dans l'imaginaire médiéval." Ibid., 68.

14. Rosand, *Opera in Seventeenth-Century Venice*, 142.

15. Ibid.

16. Jonathan Keates, *The Siege of Venice* (London: Chatto & Windus, 2005), 22–23.

17. Particularly influential was Pierre-Antoine-Noël Daru's *Histoire de la République de Venise*, 28 vols. (Paris: Didot, 1819), which drew heavily on the *Capitolare*, published again in the *Gazzetta privilegiata di Venezia* on 3 August 1819. See Paolo Preto, *I servizi segreti di Venezia: Spionaggio e controspionaggio ai tempi della Serenissima* (Milan: Il Saggiatore, 2010), 597–61, at 598–99; and Claudio Povolo, "The Creation of Venetian Historiography," in *Venice Reconsidered: The History and Civilisation of an Italian City-State, 1297–1797*, ed. John Martin and Dennis Romano (Baltimore, MD: Johns Hopkins University Press, 2000), 491–520.

18. James H. Johnson, "The Myth of Venice in Nineteenth-Century Opera," *Journal of Interdisciplinary History* 36, no. 3 (Winter 2006): 533–54. On the "black legend" formulation, see Preto, *I servizi segreti di Venezia*.

19. Gaetano Donizetti, *Lucrezia Borgia: Melodramma da rappresentarsi nell'Imp. Regio Teatro alla Scala il Carnevale, 1833–34* (Milan: Pirola, 1833), 9. The stage designers were listed as Baldassarre Cavallotti, Carlo Ferrari, and Domenico Menozzi.

20. Letter from Byron to John Murray, 25 November 1816, in Thomas Moore, ed., *The Life, Letters and Journals of Lord Byron* (London: J. Murray, 1866), 332.

21. Ferrero, "Stage and Set," 90.

NOTES TO PAGES 92–96

22. Giuseppe Mazzini, *Filosofia della Musica* (1836), in *Scritti letterari di un italiano viventi* (Lugano: Tipografia della Svizzera Italiana, 1847), 2: 268–318, at 299.

23. *L'eco* 6, no. 155 (27 December 1833): 620. In *Il censore universale dei teatri* (1 January 1834, p. 3), Luigi Prividali offered more comment, criticizing the change from the actual Palazzo Barbarigo in Hugo's original drama to the fictitious Palazzo Grimani.

24. *Teatri, arti e letteratura* 31, anno 17 (27 June 1839): 800.

25. Martinelli is listed as a scenographer at La Fenice from December 1839 to March 1840.

26. *Gazzetta di Venezia*, 14 January 1840; repr., Tommaso Locatelli, *L'appendice della Gazzetta di Venezia*, 16 vols. (Venice: Tipografia del Commercio, 1871), 7:223–24.

27. Ibid., 227.

28. *Al fausto arrivo di Ferdinando Primo Imperatore e Re, Venezia nel gran teatro della Fenice, plaudiva con questa lirica azione di Giovanni Peruzzini da Gio. Battista Ferrari di musicali note* (n.p., 1838).

29. Fabio Mutinelli, *Dell'avvenimento di S.M.I.R.A Ferdinando I d'Austria in Venezia, e delle civiche solennità d'allora* (Venice: Tipi del Gondoliere, 1838), 14.

30. Translation adjusted from Helen Greenwald, "*Son et lumière*: Verdi, *Attila*, and the Sunrise over the Lagoon," *Cambridge Opera Journal* 21, no. 3 (2010): 267–77, at 268.

31. Ibid., 269.

32. Marcello Conati, *La bottega della musica: Verdi e La Fenice* (Milan: Il Saggiatore, 1983), 141.

33. Conati observes that this letter was initially dated erroneously 1844. Ibid., 142–45.

34. Ibid., 143.

35. Francesco Izzo, "Verdi, Solera, Piave and the Libretto for *Attila*," *Cambridge Opera Journal* 21, no. 3 (2010): 257–65.

36. Conati, *La bottega della musica*, 159.

37. Verdi to Piave, 22 July 1848; cited in Julian Budden, *The Operas of Verdi*, 3 vols. (London: Cassell, 1973–81), 1:179.

38. G. Brenna to Verdi, 26 July 1843; cited in Conati, *La bottega*, 62–63.

39. *Teatri, arti e letteratura* 43, anno 23 (3 April 1845): 49 n. 1104. In 1846 the Teatro Apollo also staged the opera.

40. Cited in Bruno Cagli, "'. . . Questo povero poeta esordiente': Piave a Roma, un carteggio con Ferretti, la genesis di Ernani," in *Ernani ieri e oggi: Atti del convegno internazionale di studi, Modena, Teatro San Carlo, 9–10 dicembre 1984*, Bollettino dell'Istituto di studi verdiani 10, ed. Pierluigi Petrobelli (1987), 1–20, at 11.

41. Anonymous, *Ferdinando I, nel Tirolo, nella Lombardia e nel Veneto o sia descrizione di tutte le feste publicche e private datesi nel principato e nel regno in occasione della venuta in Italia delle LL. MM. austriache e dell'incoronazione*

NOTES TO PAGES 97–102

di Ferdinando I a Re del Regno L.V. l'anno 1838 (Milan: Placido Maria Visaj, 1838), 58. Solera's poem was published in *Cosmorama pittorico 6, no. 43* (1838): 342–43.

42. See Anselm Gerhard, "'Cortigiani, vil razza bramata!' Reti aristocratiche e fervori risorgimentali nella biografia del giovane Verdi," *Acta musicologica* 84, no. 2 (2012): 199–223, at 200–201.

43. Keates, *The Siege of Venice*, 33–39.

44. Letter from Giovanni Berti to Francesco Bagnara, 29 September 1845; Archivio storico del Teatro La Fenice. On Bertoja's designs for Verdi, see Evan Baker, "Verdi's Operas and Giuseppe Bertoja's Designs at the Gran Teatro la Fenice, Venice," in *Opera in Context: Essays on Historical Staging from the Late Renaissance to the Time of Puccini*, ed. Mark A. Radice (Portland, OR: Amadeus Press, 1998), 209–40. On Verdi's opera scenography, see *"Sorgete! Ombre serene!" L'aspetto visivo dello spettacolo verdiano*, ed. Pierluigi Petrobelli, Marisa Di Gregorio Casati, and Olga Jesurum (Parma: Istituto Nazionale di Studi Verdiani, 1994).

45. Flaminio Corner, *Notizie storiche delle chiese e monasteri di Venezia, e di Torcello* (Padua: Giovanni Manfrè, 1758), 214–15.

46. Tommaso Temanza, *Antica pianta dell'inclita città di Venezia delineata circa la metà del XII secolo, ed ora per la prima volta pubblicata, ed illustrata* (Venice: Carlo Palese, 1781).

47. Edward Muir, *Civic Ritual in Renaissance Venice* (Princeton, NJ: Princeton University Press, 1986), 68–70.

48. Ibid., 70. Muir comments: "The Annunciation Day procession and high mass in San Marco permanently bound the destiny of Venice to the veiled will of God, the harmony of nature, and the imperial authority of Rome."

49. Domenico Crivelli, *Storia dei veneziani* (Venice: Tipi del Gondoliere, 1839), 53. For earlier works, see Alessandro Graziani, *Raggioni parochiali della chiesa di San Giacomo di Rialto* (Venice: Francesco Zane, 1725), v–vi; and Corner, *Notizie storiche delle chiese e monasteri di Venezia*, 369.

50. Margaret Plant, *Venice: Fragile City, 1798–1997* (New Haven, CT: Yale University Press, 2002), 128. On Lecomte's guidebook, see also Keates, *The Siege of Venice*, 45–47.

51. Jules Lecomte, *Venezia o colpo d'occhio letterario, artistico, storico, poetico e pittoresco di monumenti e curiosità di questa città* (Venice: Gio. Cecchini, 1844), 31.

52. Ibid., 468.

53. *Il gondoliere* 13, no. 34 (23 August 1845): 266.

54. Thomas G. Kaufman, *Verdi and His Major Contemporaries: A Selected Chronology of Performances with Casts* (New York: Garland, 1990), 328–33.

55. *Teatri, arti e letteratura* 45, anno 24 (30 April 1846): 72 n. 1160.

56. *Gazzetta musicale di Milano* 6, no. 3 (17 January 1847): 17–18.

57. Letter from Emanuele Muzio to Antonio Barezzi, 27 December 1847; cited in Frank Walker, *The Man Verdi* (London: J. M. Dent, 1962), 154.

NOTES TO PAGES 102–107

58. Ibid., 155.

59. Olga Jesurum, "Set Designs for Italian Operas by Romolo and Tancredi Liverani," *Music in Art* 31, nos. 1–2 (Spring–Fall 2006): 51–62, at 53.

60. *Attila* received only five performances at the Teatro La Fenice, owing to the delay caused by Verdi's illness early in the year; although it appeared in subsequent years at the Teatro Apollo and the Teatro San Benedetto, it was not staged again at La Fenice until 1972.

CHAPTER EIGHT

1. "Catahoula" is the modern spelling; the story consistently uses "Cathahoula."

2. Henry Kmen, *Music in New Orleans: The Formative Years, 1791–1841* (Baton Rouge: Louisiana State University Press, 1966).

3. Mary Grace Smith, "The Northern Tours of the Théâtre d'Orléans, 1843 and 1845," *Louisiana History: The Journal of the Louisiana Historical Association* 26, no. 2 (1985): 155–93; Sylvie Chevalley, "Le Théâtre d'Orléans en tournée dans les Villes du Nord 1827–1833," *Comptes rendues de l'Athenée Louisianais* (1955): 27–71.

4. *L'Abeille*'s list of the new troupe recorded Jobey's arrival and orchestral position on 10 November 1834. N. N. Oursel, "Jobey (Charles)," in *Nouvelle biographie normande* (Paris: A. Picard, 1886), 496, suggested that Jobey founded his own theater, but there is no evidence for this.

5. Jobey's signature appears on the letter of complaint against the theater management ("Encore quelques renseignemens utiles aux artistes qui voudraient venir à la Nouvelle-Orléans") that appeared in the Parisian *Gazette des théâtres* on 5 July 1840. His novel *L'amour d'un nègre* (Paris: Michel Lévy, 1860) also carries a damning description of the manager John Davis (37–38).

6. "Un seul d'entre nous osa d'abord mêler sa voix à cette grande symphonie de la nature." Jobey, "La Lac Cathahoula," in *L'amour d'une blanche* (Paris: E. Jung-Treuttel, 1861), 269.

7. "Et les voûtes de la forêt vierge retentirent, pour la première et sans doute pour la dernière fois, de cette belle prière de la Muette, dont le chant si simple, si large, commence par un pianissimo, ressemblant au souffle de la brise, et se terminant par un énergique point d'orgue, que l'écho nous renvoya comme un tonnerre lointain." Ibid., 272.

8. "Le fond et les détails du roman qu'on va lire sont aussi de la plus grande exactitude; les noms de beaucoup de personnages qui y figurent sont véritables," Jobey, *L'amour d'un nègre*, 3. This tactic was used regularly; for another example, see Matthew Brown, "Richard Vowell's Not-So-Imperial Eyes: Travel Writing and Adventure in Nineteenth-Century Hispanic America," *Journal of Latin American Studies* 38, no. 1 (2006): 95–122, at 100–101.

NOTES TO PAGES 107–111

9. James Duncan and Derek Gregory, "Introduction," in *Writes of Passage: Reading Travel Writing*, ed. James Duncan and Derek Gregory (London and New York: Routledge, 2002), 5; and Jennifer Yee, *Exotic Subversions in Nineteenth-Century French Fiction* (London: Legenda, 2008).

10. Brown comments similarly on "coloniser, traveller, soldier, historian or novelist" in "Richard Vowell's Not-So-Imperial Eyes," 97.

11. Rogério Budasz, "Music, Authority, and Civilisation in Rio de Janeiro, 1763–1790," in *Music and Urban Society in Colonial Latin America*, ed. Geoffrey Baker and Tess Knighton (Cambridge: Cambridge University Press, 2011), 151–70, at 162–64.

12. On musical encounters with the non-European Other, see Ruth E. Rosenberg, *Music, Travel, and Imperial Encounter in Nineteenth-Century France: Musical Apprehensions* (New York: Routledge, 2015).

13. "Nous écoutions le frémissement du feuillage . . . la note plaintive du moqueur . . . la voix adoucie du chat-tigre . . . ; les plaintes sonores des caïmans, tourmentées d'ardeurs amoureuses," Jobey, "Le Lac Cathahoula," 269.

14. Written accounts of nineteenth-century naturalists began to stress man's impact on nature, and that "the natural world itself was becoming increasingly dependent on human agency"; Paul Smethurst, *Travel Writing and the Natural World, 1768–1840* (London: Palgrave Macmillan, 2012), 2.

15. For two particularly vituperative mid-century French attacks on the United States as uncultured, see Jules Janin, "Rachel et la tragédie aux États-Unis," *Journal des débats* (15 October 1855): [1–2]; and Léon Beauvallet, *Rachel et le nouveau monde: Promenades aux États Unis et aux Antilles* (Paris: Alexandre Cadot, 1856).

16. Ban Wang, "Inscribed Wilderness in Chateaubriand's 'Atala,'" *Romance Notes* 33, no. 3 (1993): 279–87.

17. John W. Lowe, "Not-So-Still waters: Travellers to Florida and the Tropical Sublime," in *The Oxford Handbook of the Literature of the U.S. South*, ed. Fred Hobson and Barbara Ladd (Oxford and New York: Oxford University Press, 2016), 180–95.

18. "Nous voyons tous les jours des gens n'ayant jamais franchi le mur d'enceinte de Paris écrire des histoires trèsamusantes, dont les scènes se passent à quatre mille lieues de là." Jobey, *L'amour d'une blanche*, iii.

19. Jobey, *L'amour d'un nègre*, 146–61.

20. "Mademoiselle Dupuis! le rôle d'Adèle ne vous appartient pas, nous ne vous le laisserons pas jouer." Ibid., 151.

21. "A ces mots, la tempête éclate de plus belle, tout le monde est debout, à l'orchestre, au parterre, dans les loges; on s'injurie, on se menace. L'actrice, cause de tout ce bruit, prend le parti de se trouver mal sur la scène. Le désordre est à son comble, les provocations personnelles s'échangent, les bourrades, les soufflets, les coups de poings pleuvent de tous côtés. Un poignard

NOTES TO PAGES 111–115

est tiré, et, à l'instant même, vingt, cinquante, cent, deux cents poignards, couteaux, cannes à dard, pistolets, brillent dans la salle." Ibid., 151.

22. A rivalry between New Orleans's most-loved prima donna, Julia Calvé, and Mme. Bamberger resulted in scuffles in the parterre, but it did not escalate to anything like the level of Jobey's episode. See *Daily Picayune*, 30 March 1841.

23. Cormac Newark, *Opera in the Novel from Balzac to Proust* (Cambridge: Cambridge University Press, 2011), 3.

24. Newark, *Opera in the Novel*, 3.

25. Yee, *Exotic Subversions*.

26. The tradition's roots go back to Lucian's *Vera Historia* of the second century, or, more conservatively, Thomas More's *Utopia* from 1516.

27. See, for example, Auguste Robert, "Souvenirs Atlantiques: Nouvelle Orléans," *La France littéraire* (1832): 79–82; and L. Xavier Eyma, "Les femmes du Nouveau-Monde, IV," *La sylphide*, July 1848, 59.

28. On race in New Orleans's theaters, see Juliane Braun, "On the Verge of Fame: The Free People of Color and the French Theatre in Antebellum New Orleans," in *Liminale Anthropologien: Zwischenzeiten, Schwellenphänomene, Zwischenräume in Literatur und Philosophie*, ed. Jochen Achilles, Roland Borgards, and Brigitte Burrichter (Würzburg: Königshausen & Neumann, 2012), 166.

29. Jobey, *L'amour d'un nègre*, 150.

30. "Les secondes loges, réservées aux gens de couleur, n'avaient pu donner place qu'à la moitié de ceux qui s'étaient présentés au contrôle. . . . Les troisièmes, enfin, contenaient toute une population de nègres et de négresses . . . le tout tassé comme des harengs." Ibid., 149–50.

31. "Les conversations se firent à haute voix sur tous les points de la salle." Ibid., 150.

32. Ibid., 42.

33. Carl Brasseaux, *The Foreign French: Nineteenth-Century French Immigration into Louisiana*, vol. 1, *1820–1839* (Lafayette: Center for Louisiana Studies, University of Southwestern Louisiana, 1990), xi.

34. On the American Theatre, see Nelle Kroger Smither, *A History of the English Theatre in New Orleans* (New York: B. Blom, 1944).

35. Lucille Gafford, "A History of the St. Charles Theatre in New Orleans, 1835–43" (PhD diss., University of Chicago, 1930).

36. On these tensions, see Virginia R. Domínguez, *White by Definition: Social Classification in Creole Louisiana* (New Brunswick, NJ: Rutgers University Press, 1986); Juliane Braun, "Petit Paris en Amérique? French Theatrical Culture in Nineteenth-Century Louisiana" (PhD diss., Julius-Maximilians-Universität Würzburg, 2013).

37. *New Orleans and Urban Louisiana: Settlement to 1860*, ed. Samuel C. Shepherd Jr. (Lafayette: Center for Louisiana Studies, University of Louisiana at Lafayette, 2005).

NOTES TO PAGES 115–120

38. On the Palais Garnier, see Penelope Woolf, "Symbol of the Second Empire: Cultural Politics and the Paris Opera House," in *The Iconography of Landscape: Essays on the Symbolic Representation, Design and Use of Past Environments*, ed. Denis Cosgrove and Stephen Daniels (Cambridge: Cambridge University Press, 1988), 214–35.

39. Charlotte Bentley, "The Race for *Robert* and Other Rivalries: Negotiating the Local and (Inter)national in Nineteenth-Century New Orleans," *Cambridge Opera Journal* 29, no. 1 (2017): 94–112.

40. Sonia Slatin, "Opera and Revolution: *La muette de Portici* and the Belgian Revolution of 1830 Revisited," *Journal of Musicological Research* 3, nos. 1–2 (1979): 45–62.

41. Dianne Guenin-Lelle, *The Story of French New Orleans: History of a Creole City* (Jackson: University Press of Mississippi, 2016), and *The Louisiana Purchase and Its Aftermath, 1800–1830*, ed. Dolores Egger Labbé (Lafayette: Center for Louisiana Studies, University of Southwestern Louisiana, 1998).

42. See John Zilcosky, *Uncanny Encounters: Literature, Psychoanalysis, and the End of Alterity* (Evanston, IL: Northwestern University Press, 2016).

43. Yee, *Exotic Subversions*, 7.

44. Such use of the Other was common; see Yee, *Exotic Subversions*, 11.

45. Russell Goulbourne, "Satire in Seventeenth- and Eighteenth-Century France," in *A Companion to Satire: Ancient and Modern*, ed. Ruben Quintero (Malden, MA: Blackwell, 2007), 137–60.

46. Syrine Hout, "Viewing Europe from the Outside: Cultural Encounters and European Culture Critiques in the Eighteenth-Century Pseudo-Oriental Travelogue and the Nineteenth-Century 'Voyage en Orient'" (PhD diss., Columbia University, 1994).

CHAPTER NINE

1. I am grateful to all those who provided feedback on earlier drafts of this chapter; particular thanks are due to Roger Parker for turning me toward Darwin, and to Gavin Williams and Flora Willson for their feedback.

2. Entry for 1 June 1832; see R. D. Keynes, ed., *Charles Darwin's* Beagle *Diary* (Cambridge: Cambridge University Press, 1988), 69–70. The published version of the diary differs slightly, and the slight hesitancy of the original is cut: Darwin, *Journals and Remarks, 1832–1836*, vol. 3 of *Narrative of the Surveying Voyages of His Majesty's Ships "Adventure" and "Beagle" between the Years 1826 and 1836* . . . (London: Henry Colburn, 1839), 37.

3. Letter of 23 July 1834 to Charles Whitley; National Library of Australia, Canberra, MS 4260; full text available online as DCP-LETT-250, accessed 2 September 2014, www.darwinproject.ac.uk. On Europeans comparing New World voices with animals' cries, see Gary Tomlinson, *The Singing of the New World: Indigenous Voices in the Era of European Contact* (Cambridge: Cambridge University Press, 2007), and Ana Maria Ochoa Gautier, *Aurality;*

NOTES TO PAGES 120–124

Listening and Knowledge in Nineteenth-Century Colombia (Durham, NC: Duke University Press, 2014), esp. chap. 1.

4. Darwin recorded hearing the opera on 24 November 1832; Beagle *Diary*, 118. The performance of this popular work (the most successful yet in Montevideo, according to *El Universal*, 24 November, 1832) was dedicated to the Uruguayan president, Fructuoso Rivera, recently returned to the capital after some months' campaigning.

5. Felix Driver and Luciana Martins, "Views and Visions of the Tropical World," in *Tropical Visions in an Age of Empire*, ed. Driver and Martins (Chicago: University of Chicago Press, 2005), 9.

6. Emily Eden, *Letters from India*, 2 vols. (London: Richard Bentley, 1872), 1:28.

7. Ibid., 1:69.

8. Ibid., 1:91.

9. Robert Burford, *Description of a View of the City of Calcutta; Now Exhibiting at the Panorama, Leicester Square* (London: J. and C. Adlard, 1830), 3.

10. On this division, see Jeremiah P. Losty, *Calcutta: City of Palaces; A Survey of the City in the Days of the East India Company, 1690–1858* (London: British Library, 1990); Swati Chattopadhyay, "Blurring Boundaries: The limits of 'White Town' in Colonial Calcutta," *Journal of the Society of Architectural Historians* 59 (2000): 154–79; and P. J. Marshall, "The White Town of Calcutta under the Rule of the East India Company," *Modern Asian Studies* 34 (2000): 307–31.

11. *Morning Post*, 27 February 1830.

12. Burford, *Description*, 5.

13. *Examiner*, 28 February 1830. On panorama in Burford's Calcutta, see Daniel E. White, "Imperial Spectacles, Imperial Publics: Panoramas In and Of Calcutta," *Wordsworth Circle* 41 (2010): 71–81, and *From Little London to Little Bengal: Religion, Print, and Modernity in Early British India, 1793–1835* (Baltimore, MD: Johns Hopkins University Press, 2013), 35–41. On the Panorama in general, see Stephan Oettermann, *The Panorama: History of a Mass Medium* (London: Zone, 1997); Bernard Comment, *The Panorama*, trans. Anne-Marie Glasheen (London: Reaktion, 1999); Denise Blake Oleksijczuk, *The First Panoramas: Visions of British Imperialism* (Minneapolis: University of Minnesota Press, 2011); and Erkki Huhtamo, *Illusions in Motion: Media Archaeology of the Moving Panorama and Related Spectacles* (Cambridge, MA: MIT Press, 2013).

14. White, *From Little London*, 38.

15. Others have reconfigured the relationship between metropole and colony in ways useful for opera studies; see, for instance, Ana Laura Stoler and Frederick Cooper, "Between Metropole and Colony: Rethinking a Research Agenda," in *Tensions of Empire: Colonial Cultures in a Bourgeois World*, ed. Cooper and Stoler (Berkeley and Los Angeles: University of California Press, 1997); David Lambert and Alan Lester, "Introduction: Imperial Spaces,

286

NOTES TO PAGES 124–128

Imperial Subjects," in *Colonial Lives Across the British Empire: Careering in the Long Nineteenth Century*, ed. Lambert and Lester (Cambridge: Cambridge University Press, 2006), 1–31; and Miles Ogborn, *Indian Ink: Script and Print in the Making of the East India Company* (Chicago: University of Chicago Press, 2007).

16. Victor Jacquemont, letter to Adolphe de Mareste, 11 December 1828; reprinted in Jacquemont, *Letters from India: describing a journey in the British dominions of India, Tibet, Lahore and Cashmere during the years 1828, 1829, 1830, 1831, undertaken by the order of the French Government*, 2 vols. (London: Edward Churton, 1834), 1:40–41.

17. Eden, *Letters from India*, 1:101. On the Chowringhee Theatre, see Hemendra Nath Das Gupta, *The Indian Stage*, 4 vols. (Calcutta: Metropolitan, [1934–]1944), 1:246–67; Sushil Kumar Mukherjee, *The Story of the Calcutta Theatres, 1753–1980* (Calcutta, 1982); and Sudipto Chatterjee, *The Colonial Staged: Theatre in Colonial Calcutta* (London, 2007), esp. 20–24.

18. Eden, *Letters from India*, 1:107.

19. Jacquemont, *Letters from India*, 1:21–22.

20. *Bengal Hurkaru*, 20 December 1833.

21. *India Gazette*, 19 December 1833.

22. "Sketches of Indian Society: No. II; Feminine Employments, Amusements, and Domestic Economy," *Asiatic Journal and Monthly Register for British and Foreign India, China, and Australasia* 10 (January–April 1833): 115; reprinted in Emma Roberts, *Scenes and Characteristics of Hindostan, with Sketches of Anglo-Indian Society*, 3 vols. (London: W. H. Allen, 1835), 1:102.

23. See *Bengal Hurkaru*, 5 December 1833.

24. *India Gazette*, 19 December 1833.

25. *Bengal Hurkaru*, 17 January 1834. In light of other accounts of the troupe, these descriptions recall comments made by the indigo planter William Huggins about earlier Calcutta reviews: "Those who have performed well are nonpareils, and inimitable; those who have done tolerably, are admirable; and badly excellent"; Huggins, *Sketches in India* (London: John Letts, 1824), 107.

26. *Bengal Hurkaru*, 17 January 1834.

27. Ibid., 14 March 1836.

28. "The Lawfulness of Christians attending fashionable amusements," *Calcutta Christian Observer* (February 1836): 85–91.

29. *The Calcutta Christian Observer* (October 1836): 505–13, at 506. On British reaction to Durga Puja, see White, *From Little London*, 44–54.

30. "On the impropriety of Christians attending the Festivals Connected with the Durgá Pújá," *Calcutta Christian Observer* (October 1836): 505–13, at 507. Ward's original work, *A View of the History, Literature, and Mythology of the Hindoos*, published in 1811, was much reprinted.

31. Melina Esse, "Rossini's Noisy Bodies," *Cambridge Opera Journal* 21, no. 1 (2009): 27–64.

NOTES TO PAGES 128–130

32. *Bengal Hurkaru*, 14 November 1833. On Durga Puja as part of "the construction of the civic identity of Calcutta as a city," see Tithi Bhattacharya, "Tracking the Goddess: Religion, Community, and Identity in the Durga Puja Ceremonies of Nineteenth-Century Calcutta," *Journal of Asian Studies* 66 (2007): 919–62, at 948; Saugata Bhaduri, "Public Sphere and Sacred Space: Origins of Community Durga Puja in Bengal," in *Folklore, Public Sphere and Civil Society*, ed. M. D. Muthukumaraswamy and Molly Kaushal (New Delhi: Indira Gandhi Centre for the Arts, 2004), 79–91.

33. Ian Woodfield, *Music of the Raj: A Social and Economic History of Music in Late Eighteenth-Century Anglo-Indian Society* (Oxford: Oxford University Press, 2000), 5.

34. On the 1837 Calcutta census see Blair B. Kling, *Partner in Empire: Dwarkanath Tagore and the Age of Enterprise in Eastern India* (Berkeley and Los Angeles: University of California Press, 1976), 52. Out of a total of 229,705, there were just over 3,000 English, around the same number of Portuguese, 160 French, and no Italians–considerably fewer Europeans even than the one-sixteenth proportion of the city estimated by Burford (*Description*, 5).

35. *Bengal Hurkaru*, 4 December 1833. On the European construction of Italians as exotic see Robert Dainotto, *Europe (In Theory)* (Durham, NC: Duke University Press, 2007).

36. *Bengal Herald*, 21 December 1835.

37. *Calcutta Monthly Journal and General Register*, April 1836, 129.

38. See Homi Bhabha, "Of Mimicry and Man: The Ambivalence of Colonial Discourse," *October* 28 (1984): 125–33. On Shakespeare in nineteenth-century India see, for example, Jyotsna Singh, "Different Shakespeares: The Bard in Colonial/Postcolonial India," *Theatre Journal* 41 (1989): 445–58, and *Colonial Narratives/Cultural Dialogues: "Discoveries" of India in the Language of Colonialism* (London and New York: Routledge, 1996), esp. chap. 3; Sudipto Chatterjee, *The Colonial Staged*; Vikram Singh Thakur, "From 'Imitation' to 'Indigenization': A Study of Shakespeare Performances in Colonial Calcutta," *Alicante Journal of English Studies* 25 (2012): 193–208; Sarbani Chaudhury and Bhaskar Sengupta, "Macbeth in Nineteenth-Century Bengal: A Case of Conflicted Indigenization," *Multicultural Shakespeare: Translation, Appropriation and Performance* 10 (2013): 11–27; and Shormishtha Panja and Babli Moitra Sara, eds., *Performing Shakespeare in India: Exploring Indianness, Literatures and Cultures* (London: Sage, 2016).

39. "Barbarossa," "Sketches of Military Life in India," *United Services Journal and Naval and Military Magazine* (March 1837): 302–14, at 306.

40. See, for example, the "Memorandum on Native Education" reprinted in the *India Gazette* on 9 August 1833, which outlined a program of physical, moral and intellectual education: "To reading, writing, and arithmetic, should be added elementary views of natural history, of human and comparative anatomy, of the geography, chronology, and history of India and

NOTES TO PAGES 130–135

of the world, of grammar, of geometrical drawing, of geometry and survey-ing, and of practical mechanics." See Stephen Evans, "Macaulay's Minute Revisited: Colonial Language Policy in Nineteenth-Century India," *Journal of Multilingual and Multicultural Development* 23 (2002): 260–81.

41. Quoted in Zareer Masani, *Macaulay: Pioneer of India's Modernization* (London: Vintage Books, 2012), 106.

42. *Calcutta Literary Gazette*, October 1834.

43. Jacquemont, *Letters from India*, 1:189.

44. Christopher A. Bayly, "Rammohan Roy and the Advent of Constitutional Liberalism in India, 1800–1830," *Modern Intellectual History* 4 (2007): 25–41; and Partha Chatterjee, *The Black Hole of Empire: History of a Global Practice of Power* (Princeton, NJ: Princeton University Press, 2012), 155–56; see also Andrew Sartori, *Bengal in Global Concept History: Culturalism in the Age of Capital* (Chicago: University of Chicago Press, 2008), esp. chap. 3.

45. On the fund, which seems to have spurred competitive generosity between Indian and English donors, see the *Bengal Hurkaru* during May and June 1834. On Tagore's acquisition of the journals and the opera house, see Kissory Chand Mittra, *Memoir of Dwarkanath Tagore* (Calcutta: Thacker, 1870); and Kling, *Partner in Empire*.

46. *Bengal Hurkaru*, 31 January 1834.

CHAPTER TEN

1. Quoted in the *Leader*, 23 May 1914, 37.

2. *Sydney Morning Herald* (hereafter SMH), 11 October 1913, 21.

3. J. C. Williamson was the dominant theatrical agency in Australia for nearly a century, having been founded in 1879 by the American-born actor-manager James Cassius Williamson (1845–1913).

4. Jacques Malan, ed., *South African Music Encyclopaedia*, vol. 4 (Cape Town: Oxford University Press, 1996), 365.

5. Lorna Stirling, "The Development of Australian Music" [Commonwealth Literary Fund lecture given in the University of Melbourne, 21 July 1944], *Historical Studies: Australia and New Zealand* 3 (October 1944–February 1945): 58–72, at 72.

6. Katherine Brisbane, *Entertaining Australia: An Illustrated History* (Sydney: Currency Press, 1991), 12.

7. Ibid.

8. Harold Love, *The Golden Age of Australian Opera: W. S. Lyster and His Companies, 1861–1880* (Sydney: Currency Press, 1981).

9. Ibid., 203. See also Esmeralda Rocha, "Imperial Opera: The Nexus between Opera and Imperialism in Victorian Calcutta and Melbourne, 1833–1901" (PhD diss., University of Western Australia, 2012).

10. Warren Bebbington, ed., *The Oxford Companion to Australian Music* (Melbourne: Oxford University Press, 1997).

NOTES TO PAGES 135–138

11. The company later became Martin and Fanny Simonsen's Royal English and Italian Company.
12. In 1883 the Cagli Company was joined by Italians "fresh off the . . . ship." Alison Gyger, *Opera for the Antipodes: Opera in Australia, 1881–1939* (Sydney: Currency Press and Pellinor, 1990), 29.
13. Gyger, *Opera for the Antipodes*, 22. On all these companies, see Brisbane, *Entertaining Australia*.
14. Gyger, *Opera for the Antipodes*, 39.
15. Ann Blainey, *I Am Melba* (Melbourne: Black, 2008), 273.
16. See Suzanne Cole and Kerry Murphy, "Wagner in the Antipodes," *Wagner-Spectrum* 2 (2008): 237–68.
17. Katherine K. Preston, "Opera Is Elite/ Opera Is Nationalist: Cosmopolitan Views of Opera Reception in the United States, 1870–1890," in "Cosmopolitanism in the Age of Nationalism, 1848–1914," Dana Gooley, convenor, *Journal of the American Musicological Society* 66 (2013): 535–39, (at 536).
18. Michael Cannon, *Life in the Cities: Australia in the Victoria Age* (West Melbourne: Nelson, 1975), 207. See also John Rickards, "Cultural History: The High and the Popular," in *Australian Cultural History*, ed. S. L. Goldberg and F. B. Smith (Melbourne: Cambridge University Press, 1988), 178–89.
19. Love, *Golden Age*, 126, 143.
20. Ibid., 126.
21. Ibid., 140.
22. *Sydney Globe*, 26 June 1912, 9. Charles A. Hooey, "'A Voice without Equal': Caroline Hatchard," accessed 26 March 2017, http://www.musicweb -international.com/classrev/2001/june01/hatchard.htm.
23. *The Musical Standard*, 27 August 1910, 138; and John Lucas, *Thomas Beecham: An Obsession with Music* (Woodbridge: Boydell, 2011), 42–43.
24. *SMH*, 4 May 1912, 4.
25. Paul E. Bierley, *The Incredible Band of John Philip Sousa* (Urbana: University of Illinois Press, 2006).
26. Brisbane, *Entertaining Australia*, 168.
27. Information in this paragraph is drawn from Quinlan's interview with J. D. Fitzgerald, *Lone Hand*, 2 September 1912, 440. See also Gyger, *Opera for the Antipodes*; and Nicholas Tarling, *Orientalism and the Operatic World* (Maryland: Rowman & Littlefield, 2015), chap. 1.
28. *Table Talk*, 28 August 1913, 21.
29. *SMH*, 8 June 1912, 4; *SMH*, 11 October 1913, 21.
30. "Copy of Thomas Quinlan Contract," 25 April 1911, Performing Arts Collection, Southbank, Melbourne.
31. *Punch*, 29 February 1912, 41.
32. E. J. Gravestock, *Table Talk*, 25 January 1934, 26.
33. *Australian Musical News* (hereafter *AMN*), March 1913, 227.

34. *SMH*, 18 January 1924, 6. See Kay Dreyfus and Kerry Murphy, "A Passionate Paradox: Public Reception in 1920s Australia of Visiting Italian Opera Companies and Musicians," in *Italy in Australia's Musical Landscape,* ed. Linda Barwick and Marcello Sorce Keller (Melbourne: Lyrebird Press, 2012), 13–40. Australian choristers feared both loss of work and undercutting of Australian award rates, setting a deleterious precedent. See "Melba Opera Season: Imported Opera Singers," *Daily Mercury*, 8 January 1924, 4.
35. *SMH*, 12 August 1912, 5.
36. Blainey, *I Am Melba*, 275.
37. *Adelaide Advertiser*, 11 August 1913, 18.
38. *AMN*, June 1912, 330.
39. *AMN*, July 1912, 29.
40. *AMN*, February 1913, 212.
41. *SMH*, 21 May 1910, 4.
42. *Sunday Times*, 21 July 1912, 2.
43. *SMH*, 17 July 1912, 16.
44. *AMN*, July 1912, 6.
45. *Argus*, 7 October 1913, 7.
46. *AMN*, July 1912, 7.
47. *Argus*, 15 June 1912, 20.
48. *Age*, 2 November 1900, 6. See also Cole and Murphy, "Wagner in the Antipodes."
49. *SMH*, 9 August 1912, 5.
50. Ibid. The most expensive tickets for the *Ring* cycle were £1/1s., 15s., and 10s./6p., whereas the other operas were 10s./6p., 7s./6p. and 5s.
51. *SMH*, 9 August 1912, 5.
52. His ambition is mentioned in the *SMH*, 11 October 1913, 21, and his search for funding in the *New Zealand Herald*, 23 December 1913, 8.
53. See *SMH*, 4 November 1912, 3.
54. *Argus*, 24 October 1913, 9.
55. Stated in the program. See Quinlan scrapbook, Performing Arts Collection, Southbank, Melbourne.
56. Opera Scotland website, accessed 22 March 2017, http://www.operascotland.org/person/3040/Richard+Eckhold; and H. Saxe Wyndham, *Who's Who in Music* (Boston: Small, Maynard, 1913), 73.
57. *Evening News*, 18 October 1913, 10.
58. *Age*, 29 August 1913, 10.
59. *Age*, 29 August 1913, 10. Apparently, "performances suffered for nearly a week" as a result. *SMH*, 11 October 1913, 21.
60. *AMN*, June 1912, 319.
61. *Table Talk*, 16 October 1913, 36.
62. "Museums Victoria," accessed 26 March 2017, https://museumvictoria.com.au/discoverycentre/websites-mini/journeys-australia/1900s20s/.

NOTES TO PAGES 143–146

63. Gravestock, *Table Talk*, 25 January 1934, 25.
64. "A History of the Australian Railway," accessed 26 March 2017, http://www.projex.com.au/australian-railway-history/.
65. "Her Majesty's Theatre," accessed 20 March 2017, http://www.dictionaryofsydney.org/entry/her_majestys_theatre.
66. *Australasian*, 15 June 1912, 41.
67. *Table Talk*, 2 October 1913, 35.
68. Ibid.
69. "Musical Notes," *Adelaide Register*, 25 October 1913, 6.
70. *Age*, 7 October 1913, 8.
71. Ibid.
72. *Age*, 8 October 1913, 10.
73. See, for instance, comments in the *Age*, 26 August 1913, 8.
74. *Argus*, 27 September 1913, 5.
75. *Bulletin*, 28 August 1913, 8.
76. *Argus*, 26 August 1913, 8.
77. *Sun*, 14 September 1913. See Quinlan scrapbook, Performing Arts Collection, Southbank, Melbourne.
78. *Sydney Evening News*, 17 October 1913, 3.
79. Melissa Bellanta, *Larrikins: A History* (Queensland: University of Queensland Press, 2012).
80. Dirk den Hartog, "Self-Levelling Tall Poppies: The Authorial Self in (Male) Australian Literature," in *Australian Cultural History*, 226–41 at 226–27).
81. *Sydney Evening News*, 17 October 1913, 3.
82. *Sun*, 30 September 1913. See Quinlan scrapbook, Performing Arts Collection, Southbank, Melbourne.
83. Kate Darian-Smith, Patricia Grimshaw and Stuart Macintyre, "Introduction," in *Britishness Abroad: Transnational Movements and Imperial Cultures* (Carlton, Victoria: Melbourne University Press, 2007), 1–16, at 4.
84. *Sunday Times*, 14 July 1912.
85. *Referee*, 17 July 1912, 16.
86. "Mr Thomas Quinlan: The Alexander of Music," *World's News*, 18 October 1913, 5.
87. Ibid.
88. *SMH*, 12 August 1912, 5.
89. Love, *The Golden Age of Australian Opera*, 37.
90. *Adelaide Advertiser*, 12 May 1914, 12.
91. *AMN*, July 1912, 5. Similar complaints were made in the English provinces: W. J. Bowden, *Musical Standard* 36, no. 930 (28 October 1911): 280.
92. See W. J. Bowden, *Musical Standard* 36, no. 928 (14 October 1911): 248.
93. *AMN*, March 1912, 227.
94. *SMH*, 5 August 1912, 8.
95. *SMH*, 11 October 1913, 21.

NOTES TO PAGES 146–150

96. *SMH*, 13 July 1912, 16. See also conductor Tullo Voghera's comments in *SMH*, 28 November 1913, 9.
97. *Register Adelaide*, 25 October 1913, 8. *Table Talk*, 28 August 1913, 21.
98. *SMH*, 31 July 1912, 16.
99. See "Writ of Capias: Action against Opera Promoter, Melbourne," *Hobart Mercury*, 22 August 1912, 4; "Dispute over Payment: Melbourne," *SMH*, 2 October 1913, 9.
100. *Argus*, 22 June 1912, 26.
101. See note 77.

CHAPTER ELEVEN

1. Japan declared war following its attack on Pearl Harbor on 7 December 1941. The author's thanks and acknowledgments go to Suzanne Aspden, Daniel Grimley, and Takako Maeda; Hu Yianqiong of the Shanghai Symphony Archive; Xu Jinhua and Wang Feng of the Xujiahui Library (藏書樓) in Shanghai; and the Royal Musical Association for the Thurston Dart Research Grant.
2. Opera in nineteenth- and early twentieth-century Shanghai has received some critical attention, particularly with regard to touring companies and émigré musicians. See, for example, Huang Chun-zen, "Traveling Opera Troupes in Shanghai, 1842–1949" (PhD diss., Catholic University of America, 1997); and Wang Zhicheng, *Eqiao yinyuejia zai Shanghai* [Russian Musicians in Shanghai] (Shanghai: Shanghai Conservatory Press, 2007).
3. The orchestra was renamed the Shanghai Municipal Symphony Orchestra (上海市政府交響樂團) after the Second World War and then became the Shanghai Symphony Orchestra (上海交響樂團) in 1956. The Shanghai Philharmonic during the Japanese occupation is not to be confused with the present-day Shanghai Philharmonic (上海愛樂樂團), whose predecessor was the Shanghai Broadcasting Symphony Orchestra (上海廣播交響樂團).
4. Robert David Sack, *Human Territoriality: Its Theory and History* (Cambridge: Cambridge University Press, 1986), 19.
5. Yvonne Whelan, "Territory and Place," in *Key Concepts in Historical Geography*, ed. John Morrissey, David Nally, Ulf Strohmayer, and Yvonne Whelan (London: Sage, 2014), 53.
6. Fuzhou, Ningbo, Shamian Island, and Xiamen were the other ports forced to open for foreign trade under the Treaty of Nanjing.
7. The term "colonial formations" helps to convey the peculiarities of colonialism in China. Bryna Goodman and David S. G. Goodman, eds., *Twentieth-Century Colonialism and China: Localities, the Everyday, and the World* (Oxford: Routledge, 2012), 3.
8. Ibid., 1.
9. Robert Bickers and Isabella Jackson, eds., *Treaty Ports in Modern China: Law, Land, and Power* (London: Routledge, 2016).

NOTES TO PAGES 150–152

10. Tess Johnston and Deke Erh, *Frenchtown Shanghai: Western Architecture in Shanghai's Old French Concession* (Hong Kong: Old China Hand Press, 2000), 12.

11. The maps can be found in *Lao Shanghai baiye zhinan* [Directory of Businesses in Old Shanghai] (Shanghai: Shanghai Academy of Social Sciences Press, 2008), a reprint of *Shanghai shi hanghao tulu* [A Pictorial Directory of Businesses in Shanghai], 2 vols. (1937, 1947).

12. In February 1943 the British-Chinese Treaty and the US-China Treaty for Relinquishment of Extraterritorial Rights in China were signed between the Nationalist Government under Jiang Jieshi, Britain, and the United States. Although the treaties were not enforceable in Japanese-occupied Shanghai, in July of the same year the Japanese retroactively ceded the British-affiliated Shanghai Municipal Council to the Shanghai Special Municipality (which reported to Wang Jingwei's collaborationist government). In the same month, the Nazi occupation government of Vichy France handed the French Concession over to Wang Jingwei's government. The International Settlement and French Concession officially came to an end with the surrender of Japan, the conclusion of the Second World War, and the Nationalists' resumption of sovereignty.

13. Michael McClellan, "Performing Empire: Opera in Colonial Hanoi," *Journal of Musicological Research* 22 (2003): 135–66, at 135–36.

14. Ibid.

15. Evidence of French settlers' involvement in opera was not found at the time of research.

16. Laura Victoir and Victor Zatsepine, eds., *Harbin to Hanoi: The Colonial Built Environment in Asia, 1840–1940* (Hong Kong: Hong Kong University Press, 2013). Examples of touring companies performing in treaty port Shanghai include the Italian Grand Opera Company, the Stanley Opera Company, and the Russian Grand Opera Company; Huang, "Traveling Opera Troupes."

17. Three undated Shanghai Opera advertisements for productions of *Aida* (1937), *La Juive*, and *The Tales of Hoffmann* are held with the Shanghai Municipal Orchestra's (SMO) programs at the Shanghai Symphony Archive. The opera's business manager, I. I. Kounin, was also author of two books on Shanghai; one of its conductors, Alexander Sloutsky, conducted both the SMO and the Shanghai Philharmonic.

18. On the archiving in Shanghai of British- and Japanese-era materials and its implications for music research and scholarship, see Yvonne Liao, "Western Music and Municipality in 1930s and 1940s Shanghai" (PhD diss., King's College London, 2016).

19. Robert Bickers, "'The Greatest Cultural Asset East of Suez': The History and Politics of the Shanghai Municipal Orchestra and Public Band, 1881–1946," in *China and the World in the Twentieth Century: Selected Essays*, vol. 2, ed. Chang Chi-hsiung (Nankang, Taiwan: Academia Sinica, 2001), 835–75;

NOTES TO PAGES 152–155

Tang Yating, *Diguo feisan bianzouqu: Shanghai gongbuju yuedui shi* [Variations of Imperial Diasporas: A History of The Shanghai Municipal Orchestra] (Shanghai: Shanghai Conservatory Press, 2014); Wang Yanli, *Shanghai gongbuju yuedui yanjiu* [Research on the Shanghai Municipal Orchestra] (Shanghai: Shanghai Conservatory Press, 2015); and Irene Pang, "Reflecting Musically: The Shanghai Municipal Orchestra as a Semi-Colonial Construct" (PhD diss., University of Hong Kong, 2015).

20. Bickers, "Greatest Cultural Asset," 836.

21. The Shanghai Municipal Archives (hereafter SMA) holds a dossier devoted to the orchestra's services on Bastille Day. See SMA U1–1-896.

22. On the Shanghai Municipal Brass Band's sonic and aural politics, see Liao, "Western Music and Municipality."

23. Lyceum, 15 March 1936; featuring Liszt's symphonic poems *Les préludes* and *Tasso, Lamento e trionfo*, the Piano Concerto No. 1, the Mephisto Waltz, and the Hungarian Rhapsody No. 1. See SMA U1–4-928.

24. *North-China Daily News*, 2 March 1867.

25. Huang, "Traveling Opera Troupes," 12–13, from the *North-China Herald and Supreme Court and Consular Gazette,* 29 January 1874.

26. Ibid.

27. *China Press*, 19 October 1940.

28. For SMO attendance figures, see, for example, SMA U1–4-893.

29. Floria Paci Zaharoff, *The Daughter of the Maestro: Life in Surabaya, Shanghai, and Florence* (Lincoln, NE: iUniverse, 2005), 105. Strok's reputation in Asia can also be observed in his obituary of 3 July 1956 in the *New York Times*, whose headline reads "Avray Strok Dead; Impresario in East." Born in 1877 in Riga, Latvia, Strok later became an American citizen.

30. Wang Zhicheng, *Jindai Shanghai eguo qiaomin shenghuo* [The Lives of Russian Settlers in Modern Shanghai] (Shanghai: Cishu Press, 2008); and Hon-Lun Yang, Simo Mikkonen, and John Winzenburg, *Networking the Russian Diaspora: Russian Musicians and Musical Activities in Interwar Shanghai* (Honolulu: University of Hawai'i Press, forthcoming). This community included "White" Russians who fled the Bolsheviks, Ashkenazi Jews, and Soviet Jews, who arrived in northeastern China in the early twentieth century to help construct railways.

31. Shushlin remained at the conservatory after the Communist takeover and left China for the Soviet Union in 1956 (the conservatory was renamed the Shanghai Conservatory in the same year).

32. Paci stayed in Shanghai, earned a living by giving private lessons, and died in August 1946.

33. Paci was approached by Ernest F. Harris (a long-standing member and chairman of the Orchestra and Band Committee in the Shanghai Municipal Council) to provide an "autobiographical sketch" for the farewell program. Harris was the manager of Sun Life Assurance in Shanghai and president of the Shanghai Rotary Club.

NOTES TO PAGES 155–161

34. This anecdote appeared in the same "sketch."
35. Bickers, "Greatest Cultural Asset."
36. Between 1927 and 1935 there were numerous attempts at the Annual Ratepayers' Meeting to scrap the ensemble.
37. SMA, U1–4–916–2047 to U1–4–916–2052.
38. *The Land Regulations and Bye-Laws for the Foreign Settlement in Shanghai, 1845–1930*, 1. The report is held now at the Xujiahui Library.
39. SMA, U1–4–916–1978.
40. SMA, U1–4–929–1366.
41. The long-running debate around the expense of running the SMO was particularly heated in 1934 and widely reported in the English-speaking press, though less extensively in Chinese-language newspapers. See SMA, U1–4–939–0256. And on the earlier history, see Bickers, "Greatest Cultural Asset."
42. *Shanghai Times*, 9 December 1941, accessible now at the Xujiahui Library.
43. Tang, *Diguo feisan bianzouqu* [Variations of Imperial Diasporas], 197. On European Jewish musicians in Japanese-occupied Shanghai, see Yvonne Liao, " 'Die gute Unterhaltungsmusik': Landscape, Refugee Cafés, and Sounds of 'Little Vienna' in Wartime Shanghai," *Musical Quarterly* 98 (2016): 350–94. Though confined from 1943 in the Restricted Sector for Stateless Refugees, these musicians put on operetta productions such as Eysler's *Hanni geht tanzen!*, Kalman's *Die Bajadere*, Lehár's *Die lustige Witwe*, and Oscar Straus's *Ein Walzertraum*, all advertised in the *Shanghai Jewish Chronicle*, a German-language daily, copies of which can be accessed at the Xujiahui Library.
44. Advertisement for Gounod's *Faust* at the Lyceum, January 1944.
45. *Shanghai Times*, 9 December 1941.
46. John J. Stephan, *Hawai'i under the Rising Sun: Japan's Plans for Conquest after Pearl Harbor* (Honolulu: University of Hawai'i Press, 1984), 135.
47. Barak Kushner, *The Thought War: Japanese Imperial Propaganda* (Honolulu: University of Hawai'i Press, 2006), 20.
48. Held on 10 December 1944; featuring the Japanese national anthem, Yamada's sinfonia *Inno Meiji*, Beethoven's Romance in G, and Watanabe's symphonic poem *Fighting Soul*.
49. *Shanghai Times*, 9 December 1941.
50. Tang, *Diguo feisan bianzouqu* [Variations of Imperial Diasporas], 238.
51. *Shanghai Times*, 9 December 1941.
52. McClellan, "Performing Empire," 166.
53. Ibid.
54. Robert J. C. Young, *Postcolonialism: An Historical Introduction* (Oxford: Blackwell, 2001).
55. Ibid., 16–17.
56. Ania Loomba, *Colonialism/Postcolonialism* (Oxford: Routledge, 2015), 25.

NOTES TO PAGES 162–170

CHAPTER TWELVE

1. Stefan Zweig, *The World of Yesterday: An Autobiography* (London: Cassel, 1943), 23.
2. I am grateful to Gavin Williams for noting that my approach here intersects with that of Nigel Thrift on geography and capitalist cultures; see Thrift, *Non-Representational Theory: Space, Politics, Affect* (Routledge: London, 2007) chaps. 8 ("Spatialities of Feeling") and 9 ("But Malice Aforethought"), in which he considers "the affective register of cities" (171). More relevant to my preoccupations here, albeit with reference to a different location and period, would be Christopher Chowrimootoo's "Bourgeois Opera: *Death in Venice* and the Aesthetics of Sublimation," *Cambridge Opera Journal* 22, no.2 (July 2010): 175–216.
3. Theodor W. Adorno, "Bourgeois Opera," trans. David J. Levin, in *Opera through Other Eyes*, ed. Levin (Stanford, CA: Stanford University Press 1994), 25–43, opening.
4. Ibid., 28.
5. Ibid., 25.
6. Ibid., 27. Although he had supported innovative Wagner productions, like those at the Kroll Opera, in the Weimar period, Adorno here mocks the idea that the swan might be "replaced by a beam of light."
7. Ibid., 40.
8. Ibid., 29.
9. Ibid., 32, 34.
10. Ibid., 31.
11. Ibid., 30.
12. Simon Frith, "Towards an Aesthetic of Popular Music," in Richard Leppert and Susan McClary (eds.), *Music and Society: The Politics of Performance and Reception* (Cambridge: Cambridge University Press, 1987), 133–49.
13. Peter J. Martin, "Over the Rainbow? On the Quest for 'the Social' in Musical Analysis" [a review of books by Robert Adlington, Mark Anthony Neal, and Susan McClary], in *Music and the Sociological Gaze: Art Worlds and Cultural Production* (Manchester and New York: Manchester University Press, 2006), 32–55.
14. The house-as-body-or-mind metaphor emerged in K. A. Scherner, *Das Leben des Traumes* (Berlin, 1861), leading Freud to comment on it in *The Interpretation of Dreams* (*Die Traumdeutung* [Leipzig and Vienna, 1899]); see translation by James Strachey (London: Penguin [Pelican Books], 1976), 156–57.
15. Hugo von Hofmannsthal, "A Letter," trans. Joel Rotenberg, in *The Lord Chandos Letter and Other Writings* (New York: New York Review of Books, 2005), 117–28, at 121.
16. Ibid., 123–24.
17. He recalls it in his essay on Schreker in Theodor W. Adorno, *Quasi una Fantasia*, trans. Rodney Livingstone (London and New York: Verso, 1992), 131.

NOTES TO PAGES 172–177

18. Richard Sennett's work partly inspired this observation—as in *The Fall of Public Man* (1974)—but more specifically it alludes to the opening of works like Joseph Roth's novella *The Blind Mirror* (1925), describing little Fini's wanderings through Vienna, or Kafka's *The Castle* (1926), in which the castle and its attached village are, despite the rural surroundings, a city in microcosm. On de Certeau's work on walking as "the practice of making the city habitable," see Thrift, *Non-Representational Theory*, 24.

19. Relevant here is Susan Vandiver Nicassio, *Tosca's Rome: The Play and the Opera in Historical Perspective* (Chicago: University of Chicago Press, 1999).

20. Joseph Kerman, *Opera as Drama*, revised ed. (London: Faber, 1989), 205, denigrates *Tosca* as "that shabby little shocker."

21. For an absorbing, if somewhat idealistically "modernist," reading of *Tosca* as proto-fascist, see Arman Schwartz, "Rough Music: *Tosca* and *Verismo* Reconsidered," *19th Century Music* 31, no. 3 (Spring 2008): 228–44.

22. Henri Lonitz, ed., *Theodor Adorno and Alban Berg: Correspondence, 1925–1935*, trans. Wieland Hoban (Cambridge: Polity, 2005), 150. On this evaluation see Peter Franklin, *Reclaiming Late-Romantic Music: Singing Devils and Distant Sounds* (Berkeley and Los Angeles: University of California Press 2014), 144–46.

23. Bertolt Brecht, "On the Use of Music in an Epic Theatre" (1935); reprinted in *Brecht on Theatre: The Development of an Aesthetic*, ed. and trans. John Willett, 2nd ed. (London: Methuen, 1974), 84–90.

24. Thomas Mann, *Stories of Three Decades*, trans. H. T. Lowe-Porter (London: Secker and Warburg 1936), 297–319, at 309–10.

25. Max Graf, *Die Wiener Oper* (Vienna and Frankfurt am Main: Humboldt-Verlag, 1955), "Erlebnisse mit Gustav Mahler," 81–98, at 81; translations here and below are mine unless otherwise noted.

26. Ibid., 82.

27. Ibid., 81.

28. Adorno, "Bourgeois Opera," 43; the last sentence of the essay reads: "Only when the entire fullness of musical means in the face of a complaint worthy of humanity awakens something of that tension between the musical and the scenic mediums—totally lacking in the theater today—only then could opera once more match the power of the historical image."

29. For example, Mahler's repertoire at the Vienna Court Opera in 1904–5 included two new one-act-operas by Blech and D'Albert and premieres of Strauss's *Feuersnot* and Pfitzner's *Die Rose von Liebesgarten*, alongside Rossini (*Guillaume Tell*), Wagner (*Das Rheingold*) and Beethoven (*Fidelio*). See Franz Willnauer, *Gustav Mahler und die Wiener Oper* (Vienna: Löcker, 1993), 227–28.

30. This was provocatively anticipated by Ping-Hui Liao, "Opera and Postmodern Cultural Politics: *Turandot* in Beijing," in *A Night In at the Opera: Media Representations of Opera*, ed. Jeremy Tambling (London: John Libbey, 1994), 299–305. (A film of the 1998 production was issued).

NOTES TO PAGES 177–183

31. For example, Bette Davis in *Now, Voyager* (1942) leaves a concert-hall performance of Tchaikovsky's Sixth Symphony, which continues in the underscore and when she turns on her radio at home; in *Deception* (1946), Bette Davis, Paul Henreid, and Claude Raines are locked in a complex love triangle and an aural landscape setting Haydn alongside radio advertising jingles, a Salvation Army band, and music by Erich Wolfgang Korngold. See Peter Franklin, *Seeing Through Music: Gender and Modernism in Classic Hollywood Film Scores* (New York: Oxford University Press, 2011), 124–9.

CHAPTER THIRTEEN

1. In 1910 the Parisian company Le Théâtre en Plein Air produced *Le livre d'or du Théâtre en Plein Air*, detailing some twenty productions it had provided across France in its inaugural season. I am grateful to the British Academy and the *Music & Letters* Trust for funding the research underpinning this article, and to Flora Willson for astute commentary on an earlier version.
2. For a detailed case study, see Christopher Moore, "Regionalist Frictions in the Bullring: Lyric Theater in Béziers at the *fin de siècle*," *19th-Century Music* 37, no. 3 (2014): 211–41.
3. Ibid., 231–33.
4. "Le théâtre dans les ruines antiques," *Lectures pour tous: Revue universelle illustrée* [Paris] 7, no. 11 (August 1905): 981–87, at 987.
5. Anon., *La cigale* (November 1907): n.p.
6. Paul Mariéton, *Le théâtre antique d'Orange et ses chorégies, suivi d'une chronologie complète des spectacles depuis l'origine* (Paris: Éditions de la Province, 1908), 11. The text had already been published in *Le théâtre*, October 1900.
7. The Théâtre en Plein Air company offered comedies for non-classical sites only, but overall there were few such productions; see *Livre d'or du Théâtre en Plein Air*.
8. See Jean Poueigh, *Musica* 108 (September 1911): 172; and *Historique du théâtre antique de la cité de Carcassonne* (n.p., 1914), 22 (where *Samson* at Carcassonne is cited as a "fundamental error"), 26. Nevertheless, the concept of "appropriateness" was elastic: Delibes's *Lakmé*, presented in front of the reconstructed Angkor Wat at the 1922 Exposition Coloniale in Marseille, included Cambodian dancers for added "local colour."
9. Poueigh, *Musica* 108 (September 1911): 172. On other aspects of meridional authenticity in *Mireille*, see my "*Mireille*'s Homecoming? Gounod, Mistral and the Midi," *Journal of the American Musicological Society* 65, no. 2 (2012): 463–509.
10. Émile Ripert, "L'essentiel du décor est formé par toutes les maisons qui l'entourent et qui semblent attendre l'apparition de la Reine Jeanne ou du Roi René, bien plutôt que celles des Erinnyes ou de Néron." *Le petit Marseillais*, 16 August 1925, 1.

299

NOTES TO PAGES 183–189

11. See, for instance, Armand Praviel, "Le théâtre en plein air," *Le correspondant* 240 [n.s. 204], 25 July 1910, 264–87, at 276.

12. *La vie méridionale* [Montpellier] 1, no. 8, 18 August 1901, 1–2, at 1.

13. See, for instance, Marius Decavata, *Éclair de Montpellier*, 28 August 1900, 2–3, at 2.

14. Moore, "Regionalist Frictions," 221, citing Saint-Saëns's *École buissonnière* (Paris: Pierre Lafitte, 1913), 72.

15. "Elles se confondent avec celles qui forment l'horizon vrai, sans que l'œil puisse distinguer où finit la toile peinte et où commence la ligne bleue des Cévennes." Gustave Larroumet, *Le temps*, 3 September 1900, 1.

16. Denys Bourdet, *Le soleil du Midi*, 15 May 1899, 1.

17. "Le drame touche à sa fin. Prométhée est coulé sur la plus haute roche et Pandore pleure à ses pieds. Le soleil descend, et l'ombre monte lentement, comme de l'eau, dans la coupe des arènes. Le cirque et la scène y sont maintenant plongés. La roche seule émerge encore, teinte d'or et de feu par les rayons horizontaux du couchant, et le groupe du Titan et de son amante flamboie dans la clarté tragique du soleil au déclin. Cette clarté suprême s'éteint à son tour; l'ombre a tout envahi. Et le spectacle est achevé." Pierre Lalo, feuilleton, *Le temps*, 5 October 1900, 3.

18. "Ce sont là des rencontres fortuites, mais qui n'en sont pas moins prodigieuses et qui relèguent au second plan tous les effets artificiels qu'un théâtre fermé a pu jamais réaliser." *Comœdia* 16, no. 3534, 19 August 1922, 1.

19. See Sabine Teulon Lardic, "Arènes de Nîmes (12 mai 1901), lieu d'appropriation paradoxale de *Carmen* de Bizet," in *Musik—Stadt: Traditionen und Perspektiven urbaner Musikkulturen; Bericht über den XIV. Internationalen Kongress der Gesellschaft für Musikforschung, Leipzig 2008*, vol. 4, *Musik—Stadt: Freie Beiträge*, ed. Katrin Stock and Gilbert Stöck (Leipzig: Gudrun Schröder Verlag, 2012), 250–61, esp. 250–51.

20. Ibid., 251–54.

21. C. S., *La dépêche* [Toulouse], 14 August 1899, 3.

22. See Gustave Delmas, "Le théâtre en plein air," *Revue cévénole: Bulletin de la Société scientifique et littéraire d'Alais*, 2e semestre (1904): 99–119, at 116, and the *Journal du Midi* preview, as reported in Teulon-Lardic, "Arènes de Nîmes," 252.

23. C. S., *La dépêche* [Toulouse], 14 August 1899, 3.

24. *Le Monde illustré* 2948, 27 September 1913, 189.

25. "À travers cette large fresque il retrouva le modèle et, le plaçant en face de lui, il travailla d'*après nature*." *Le feu*, 1 December 1921, 290.

26. Mariéton, *Le théâtre antique*, 10. This part of the 1908 publication was originally published in 1900.

27. Letter of 16 March 1900 to the librettist Gheusi, cited in *L'art du théâtre*, 15 November 1901, 225–26.

28. Paul Dukas, *Revue hebdomadaire* (October 1900), in *Les écrits de Paul Dukas sur la musique* (Paris: Société d'éditions françaises et internationales, 1948), 505–10.

NOTES TO PAGES 189–191

29. Émile Baumann, *Camille Saint-Saëns et "Déjanire"* (Paris: A. Durand & fils, [1900]), 1. For expansion on these stylistic points see Moore, "Regionalist Frictions," 225–31.

30. *Le feu* 15, no. 14, 15 July 1921, 209–41; and Déodat de Séverac, *La musique et les lettres: Correspondance rassemblée et annotée par Pierre Guillot* (Liège: Mardaga, 2002), 341.

31. Letter of 22 April 1910 to René de Castéra. See J.-B. Cahours d'Aspry, *Déodat de Séverac (1872–1921): Musicien du soleil méditerranéen* (Anglet: Séguier, 2001), 130. *Le cœur du moulin*, which, like *La fille de la terre*, dealt with the problem of *dépaysement*, never became open-air repertoire.

32. Reviews by Raoul Davray and Marius Decavata in *L'éclair* [Montpellier], 22 August 1910, 2, contrasted *Héliogabale*'s grandeur with the "little paintings" (*tableautins*) of a "watercolorist" (*aquarelliste*) (Davray). On the "savagery" (*sauvagerie*) in *Prométhée*'s opening chorus from a "composer of delicacy" (*musicien délicat*), see Dukas, *Les écrits*, 508. See also Moore, "Regionalist Frictions," 227.

33. Shortly before writing this article, Vuillermoz had reveled in *Petrushka*'s brutal ferocity; *S.I.M.* [Société Internationale de Musique] *revue musicale mensuelle* 7, no. 7 (15 July 1911): 71–75, at 73–74.

34. *Musica* 108 (September 1911): 169.

35. *Le feu* 15, no. 14, 15 July 1921, 240.

36. "E.D." [Étienne Destranges], in *La revue musicale* 10, nos. 17–18 (1–15 September 1910): 395–99, esp. 398–99. My thanks to Steven Huebner for clarifying Destranges's authorship.

37. Review in *S.I.M. revue musicale mensuelle* 11, nos. 8–9 (August–September 1911): 65–68.

38. Ibid., 66.

39. "Où est le grand espace de la scène des arènes biterroises? Où le soleil qui lui-même semblait être de la pièce, en venant éclairer de ses rayons le bûcher d'Hercule, au moment où celui-ci demandait à Jupiter, son père, de lui envoyer le feu du ciel pour mettre fin à son horrible supplice? Où les quinze harpes à l'orchestre?—A l'Odéon nous n'en avons que deux: heureusement qu'elles sont excellentes." *Le Figaro*, 11 November 1898, front page.

40. *Le temps*, 4 September 1899, 1–2, at 1.

41. Ibid.

42. "Paris ne peut malheureusement lui donner tout ce que lui offrait Béziers. Ce n'est plus le ciel libre, la sonorité à la fois cristalline et frémissante des voix et des instruments dans l'air d'un beau jour d'été, la lumière qui joue, le vent qui passe, les vêtements qu'un souffle soulève et laisse retomber, la splendeur changeante du soleil, et le décor se confondant avec l'horizon des montagnes. C'est une salle de cirque, morne, laide et mal proportionnée; ce sont des globes électriques qui laissent tomber du plafond un rayonnement dur et froid; c'est une acoustique déplorable, qui mêle les timbres

NOTES TO PAGES 191–196

des instruments et des voix. . . . A passer ainsi d'un climat et d'un lieu dans un autre, *Prométhée* a quelque peu changé d'aspect." *Le temps*, 10 December 1907.

43. Moore, "Regionalist Frictions," 240.

44. See, for instance, the long review by Romain Rolland in *La revue de Paris*, 1 November 1901, 208–25, at 217 and 224.

45. *Mercure de France*, 1 February 1907, 449–66, at 465.

46. Ellis, "*Mireille*'s Homecoming?," 481–83.

47. On one such group pilgrimage, see Mariéton, *Revue félibréenne* (1894); repr., *Le voyage des félibres et des cigaliers (9–14 août 1894): Rhône et Vaucluse, au théâtre d'Orange, Lyon, la descente du Rhône, Tournon, Valence, Cadenet, Orange, Avignon, Vaucluse, Cavaillon; Compte rendu complet, avec tous les documents relatifs aux fêtes rhodaniennes et au théâtre d'Orange* (Paris: A. Savine, 1895).

48. "Le theâtre dans les ruines antiques," *Lectures pour tous* 7, no. 11 (August 1905): 981–87, at 987.

49. Dan Rebellato, *Theatre and Globalization* (Basingstoke: Palgrave, 2009).

50. As Teulon Lardic notes, the Toulouse and Nîmes versions of *Carmen* cried out for intercutting. Teulon Lardic, "Arènes de Nîmes," 251–52.

51. Rémy de Gourmont, *Mercure de France*, 1 September 1907, 124.

52. *Lyon mondain et sportif* 3, no. 80 (12 August 1905): 1.

CHAPTER FOURTEEN

1. Frank Howes, "Glyndebourne 1934–1952," in *Glyndebourne Festival Opera: Programme Book, 1953* (London: Glyndebourne Productions Ltd, 1953), 18.

2. Allen John Scott, "Capitalism, Cities, and the Production of Symbolic Forms," *Transactions: Institute of British Geographers*, n.s. 26 (2001): 11–23, at 12.

3. Theodor Adorno and Max Horkheimer, "The Culture Industry: Enlightenment as Mass Deception," in *Dialectic of Enlightenment*, 2nd ed. (London: Verso, 1986), 120–67. Scott, among others, has documented ways in which the culture industries have become central to the urban project over the past several decades.

4. Javier Monclús and Manuel Guàrdia, eds., *Culture, Urbanism and Planning* (Aldershot: Ashgate, 2006), xiii. See also Robert Freestone and Chris Gibson, "The Cultural Dimension of Urban Planning Strategies: An Historical Perspective," in Monclús and Guàrdia, *Culture, Urbanism and Planning*, 21–41. On opera as a marker for urban maturity and gentility, see Peter Clark, ed., *The Oxford Handbook of Cities in World History* (Oxford: Oxford University Press, 2013), 242, 300, 319, 378, 478, 485, 610, 724, 760.

5. On companies' economic problems see Michael Kennedy, *Glyndebourne: A Short History* (Oxford: Shire Books, 2010); Spike Hughes, *Glyndebourne: A History of the Festival Opera* (London: Methuen, 1965); and Jennie Jordan,

302

NOTES TO PAGES 196–198

The Buxton Festival Lifecycle: Towards an Organisational Development Model for Festivals, Discussion Papers in Arts & Festival Management 13, no.1 (Leicester: De Montfort University, 2013).

6. Adrian Tinniswood, *A History of Country House Visiting: Five Centuries of Tourism and Taste* (Oxford: Blackwell and the National Trust, 1989), 158ff. On the urban public's "thirst . . . for the natural and the historic," see Peter Borsay, "Nature, the Past, and the English Town: A Counter-Cultural History," *Urban History* 44, no. 1 (2017): 27–43, at 30; Michael Bunce, *The Countryside Ideal: Anglo-American Images of Landscape* (London and New York: Routledge, 1994), 1–4. My thanks to Peter Borsay for bringing his article to my attention.

7. For an overview of these connections see Borsay, "Nature, the Past, and the English Town," 31–36.

8. The seminal study is Raymond Williams, *The Country and the City* (Nottingham: Spokesman, 1973).

9. Bunce, *The Countryside Ideal*, 11; Borsay, "Nature, the Past, and the English Town," 35–36.

10. W. J. T. Mitchell, ed., *Landscape and Power* (Chicago: University of Chicago Press, 1994), 1.

11. Bunce, *The Countryside Ideal*, 3–4, 7–8, 34.

12. Ibid., 51.

13. Evelyn Waugh, *Brideshead Revisited* (1945; repr., London: Penguin, 1962), 216; discussed in Allan Hepburn "Good Graces: Inheritance and Social Climbing in *Brideshead Revisited*," in *Troubled Legacies: Narrative and Inheritance*, ed. Allan Hepburn (Toronto: University of Toronto Press, 2007), 239–64, at 261.

14. Williams, *The Country and the City*, 249; and Bunce, *The Countryside Ideal*, 50–51.

15. V. Sackville-West, "The Gardens," in *Glyndebourne Festival Opera: Programme Book, 1953* (London: Glyndebourne Productions, 1953), 81. Similarly, contemporary Grange Park online publicity enthuses: "You'll find the surroundings as blissfully bucolic as you could hope for . . . you just won't be able to stop yourself muttering things like *Et in Arcadia ego*"; http://www.grangeparkopera.co.uk/, accessed 10 January 2015. And the recently re-housed Garsington Opera boasts a "setting of extraordinary natural beauty. . . . which sits within the rolling landscape of the Chiltern Hills"; http://www.garsingtonopera.org/, accessed 13 November 2015.

16. "National Appeal," cited in Michael Hurd, *Rutland Boughton and the Glastonbury Festivals* (n.p.: Rutland Boughton Music Trust, 2014), 65. See also Roger Savage, *Masques, Mayings and Music-Dramas: Vaughan Williams and the Early Twentieth-Century Stage* (Woodbridge: Boydell, 2014).

17. Rolf Gardiner, *England Herself: Ventures in Rural Restoration* (London: Faber & Faber, 1943), 139–40; cited in Matthew Jefferies and Mike Tyldesley, "Introduction," in *Rolf Gardiner: Folk, Nature and Culture in Interwar Britain*, ed. Matthew Jefferies and Mike Tyldesley (Farnham: Ashgate, 2011), 5.

NOTES TO PAGES 198–200

18. Ayako Yoshino, *The Edwardian Historical Pageant: Local History and Consumerism* (Tokyo: Waseda University Press, 2010), 2; and on its commercial success, 39–66. See also Paul Readman, "The Place of the Past in English Culture c. 1890–1914," *Past and Present* 186, no. 1 (2005): 147–200.

19. The *Central Somerset Gazette*, 3 September 1920; cited in Hurd, *Rutland Boughton*, 124. The theater director Edward Gordon Craig included a brief chapter entitled "Open-Air Theatres" in *On the Art of the Theatre* (1911), ed. Franc Chamberlain (London and New York: Taylor and Francis, 2008), 142–44. His sister, Edith, transformed a seventeenth-century barn at Smallhythe Place, Kent, into a rustic theater in 1929, following the death of their mother, the actress Ellen Terry; accessed 15 February 2017, https://www.nationaltrust.org.uk/smallhythe-place/features/the-barn-theatre-at-smallhythe-place.

20. Hurd, *Rutland Boughton*, 52.

21. G. W. Foote, "Social Dreams," *Progress* 6 (May 1886): 189–94, at 190; cited in Matthew Beaumont, "To Live in the Present: *News from Nowhere* and the Present in Victorian Utopian Fiction," in *Writing on the Image: Reading William Morris*, ed. David Latham (Toronto: University of Toronto Press, 2007), 119–36, at 119. Bunce, *The Countryside Ideal*, 19–20; Martin J. Wiener, *English Culture and the Decline of the Industrial Spirit, 1850–1980* (Cambridge: Cambridge University Press, 1981).

22. William Morris, "On Some Practical Socialists" and "Architecture and History," both cited in Beaumont, "To Live in the Present," 125.

23. Grant Allen, *The British Barbarians: A Hill-Top Novel* (London: John Lane, 1895), vii. On utopian fiction see Beaumont, "To Live in the Present," 126. Peter Borsay notes the fashion for walks on "elevated sites" on the edges of towns; "Nature, the Past, and the English Town," 31.

24. Allen, *The British Barbarians*, xvii, xviii.

25. Jefferies and Tyldesley, "Introduction," *Rolf Gardiner*, 6–7.

26. Bunce, *The Countryside Ideal*, 14–17, 20–21, 33. On more general intellectual and artistic disquiet with urban life, see Morton White and Lucia White, *The Intellectual Versus the City* (Cambridge, MA: Harvard University Press and MIT Press, 1962).

27. From the "Forward" (*sic*) to *The Self-Advertisement of Rutland Boughton* (1909); cited in Hurd, *Rutland Boughton*, 34.

28. Rutland Boughton [with Reginald Buckley], *Music Drama of the Future* (1911); cited in Hurd, *Rutland Boughton*, 40.

29. Matthew Wilson Smith, *The Total Work of Art: from Bayreuth to Cyberspace* (New York: Routledge, 2007), 24–25.

30. *Daily Mirror*, 23 June 1912; cited in Hurd, *Rutland Boughton*, 52.

31. *Musical Times*, February 1915; cited in Hurd, *Rutland Boughton*, 81.

32. Charles Kennedy Scott, cited in Hurd, *Rutland Boughton*, 82.

33. Jefferies and Tyldesley, *Rolf Gardiner*.

34. Hurd, *Rutland Boughton*, 63–64, 73.

35. Ibid., 121.

NOTES TO PAGES 200–204

36. From an address to the Festival, reported in the *Central Somerset Gazette*, 8 September 1922; cited in Hurd, *Rutland Boughton*, 142.
37. Hurd, *Rutland Boughton*, 180–82.
38. Boughton's background (he was the son of a provincial grocer and a schoolmistress) was much humbler than Christie's; Hurd, *Rutland Boughton*, 3.
39. John Christie to C. L. Nelson, 30 June 1936, Glyndebourne Archive. Spike Hughes suggests that Christie's newspaper-interview statement that "English composers will be given every chance" was merely "a popular bromide of the 1930s, expressed—because demanded—with monotonous regularity. . . . a form of meaningless formality." Hughes, *Glyndebourne*, 37.
40. Cited in Hughes, *Glyndebourne*, 51.
41. The *Daily Telegraph* article for 11 November 1933 was titled: "An Opera House in Sussex. Glyndebourne: Mr. John Christie's 'Private Bayreuth.' "
42. Hurd, *Rutland Boughton*, 62–63, and 40–41 on Boughton and Buckley's *Music Drama of the Future* (1911).
43. "An Opera House in Sussex"; clipping from the Glyndebourne Archive.
44. Bunce, *The Countryside Ideal*, 53.
45. Somewhat further removed were Henry James at Lamb House in Rye (1897–1916) and Rudyard Kipling at Bateman's in Burwash (1902–36).
46. On the impact of Southern Railway's electrification on capacity and stations, see Philip Unwin, *Travelling by Train in the 'Twenties and 'Thirties* (London: Allen & Unwin, 1981), 6.
47. Ibid., 61, 8.
48. Ibid., 61.
49. Special train services were not necessarily socially exclusive; for example, there was a mid-1950s "football excursion" train; Basil Cooper, *Celebration of Steam: Kent and Sussex* (Shepperton, Surrey: Ian Allan, 1994), 13.
50. Kennedy, *Glyndebourne*, 20: "A box for nine cost 20 guineas (the equivalent of £1,080 in 2010), stalls seats ranged from £1–10s (£77 in 2010) to £2 (£103), prices unheard of in England at that time. Even Audrey [Christie's wife, a professional singer] thought they were outrageous." On "Mrs Christie's Champagne," see ibid., 29.
51. On operatic extracts at the Proms (part of the "lighter" offering) and unstaged operas at other concert venues in the late nineteenth and early twentieth centuries, see Jenny Doctor and David Wright, eds., *The Proms: A New History* (London: Thames & Hudson, 2007), 39, 97.
52. Christie added that he would offer "local performances after the 'Festspiel' " had finished, for "the poorer portion of the public." *Monthly Musical Record*, November 1933; quoted in Hughes, *Glyndebourne*, 39. Education has come to form an important part of modern Glyndebourne's offering.
53. On John Christie's legendary gregariousness, see Hughes, *Glyndebourne*, 27.

305

NOTES TO PAGES 206–212

54. Paul Kildea, "The Proms: An Industrious Revolution," in Doctor and Wright, *The Proms*, 10–31.

55. "A Sussex Opera House," *The Times*, weekly ed., 1 February 1934; clipping in Glyndebourne Archive.

56. *The Times and Tide*, 9 June 1934; clipping in Glyndebourne Archive.

57. In the 1960s George Christie called Glyndebourne's reputation for elitism "that boring old refrain"; Kennedy, *Glyndebourne*, 37.

58. Hughes, *Glyndebourne*, 28–29.

59. Ibid., 38.

60. Ibid., 41–42.

61. Marcus Binney and Rosy Runciman, *Glyndebourne: Building a Vision* (London: Thames & Hudson, 1994), 52.

62. *Glyndebourne Festival Opera: Programme Book, 1952* (London: Glyndebourne Productions, 1952), 20.

63. Peter Fowler, "Heritage: A Post-Modernist Perspective," in *Heritage Interpretation*, vol. 1, *The Natural and Built Environment*, ed. David L. Uzzell (London and New York: Belhaven Press, 1989), 57–63, at 60; and David T. Herbert, "Heritage Places, Leisure and Tourism," in *Heritage, Tourism and Society*, ed. Herbert (London: Mansell, 1995), 1–20, at 8–12.

64. Binney and Runciman, *Glyndebourne*, 45.

65. Ibid., 45.

66. Early "product placement" saw Seker's fabrics (regular advertisers) "mak[e] fabric history" by dressing singers in a new fabric for *Die Entführung aus dem Serail* in 1958, an advertisement explained. *Glyndebourne Festival Opera: Programme Book, 1958* (London: Glyndebourne Productions, 1958), 69.

67. Boughton's Glastonbury Festival fostered community by running summer schools, but even had he wanted to do so, he could not immerse audiences in the illusion of a bygone operatic era as the Christies could do.

68. Kennedy, *Glyndebourne*, 5.

69. Audrey Mildmay Christie, "Glyndebourne," in *Glyndebourne Festival Opera: Programme Book, 1957* (London: [Glyndebourne Productions], 1957), 17. Published posthumously, her words came from notes of 1949 for a lecture titled "Glyndebourne and the Idea of a Festival."

70. Siriol Hugh Jones, "And on Wednesday, We're All Going to Glyndebourne," in *Glyndebourne Festival Opera: Programme Book, 1957* (London: [Glyndebourne Productions], 1957), 64–69, at 65.

71. Kennedy, *Glyndebourne*, 24. No source is given for this quotation.

72. Ibid.

73. Incorporating dining into an entertainment began with eighteenth-century subscription concerts and was also an (informal) element of operagoing; see *The Lyric Muse Revived in Europe or a Critical Display of the Opera in all its Revolutions* (London: L. Davis & C. Reymers, 1768), 127.

74. Abercrombie and Longhurst, *Audiences*, 96, summarize the reciprocal relationship between culture and commodities in modern society: "On the one

NOTES TO PAGES 212–219

hand, all culture becomes a commodity, while, on the other, all commodities become aestheticized."

75. Michael Smith, *Glyndebourne Picnics* (Oxford: Lennard, 1989).

76. Sackville-West, "The Gardens," 81.

CHAPTER FIFTEEN

1. Theodor Adorno, "Bürgerliche Oper," *Musikalische Schriften I–III* (Frankfurt: Suhrkamp, 1978), 29.

2. Marvin Carlson, *Places of Performance: The Semiotics of Theatre Architecture* (Ithaca, NY: Cornell University Press, 1989), 105.

3. Walter Benjamin, *The Arcades Project*, trans. Howard Eiland and Kevin Mc-Laughlin (Cambridge, MA: Belknap Press, 1999). On grand opera and world's fairs, see also Anselm Gerhard, *The Urbanization of Opera: Music Theatre in the Nineteenth Century* (Chicago: University of Chicago Press, 1998), 389.

4. Walter Benjamin, "Neapel," in *Gesammelte Schriften*, ed. R. Tiedemann and H. Schweppenhäuser, vol. 4.1 (Frankfurt am Main: Suhrkamp, 1977), 309 (hereafter *GS*). Translation from *Walter Benjamin: Selected Writings*, vol. 1 (Cambridge, MA: Belknap Press, 1996), 416.

5. See also my essay "Imaginative Configurations: Performance Space in the Global City," in *Performing Architectures*, ed. Andrew Filmer and Juliet Rufford (London: Bloomsbury, 2017).

6. Alan Miller, "Opera Houses in the City: Part II; The Opéra Bastille," 9 April 2012, revised 1 August 2014, accessed 27 July 2017, http://newyorkarts .net/2012/04/opera-bastille-english/.

7. Arnold Aronson, *Looking into the Abyss: Essays on Scenography* (Ann Arbor: University of Michigan Press, 2005), 7.

8. See: Tom Wilkinson, "Typology Quarterly: Opera Houses," *Architectural Review*, 25 October 2013, accessed 27 July 2017, https://www.scribd.com /document/188740169/Opera-houses-Typology-Quarterly.

9. Benjamin, *Arcades Project*, 410 (L2a, 5).

10. Gerhard, *Urbanization*, 13.

11. Benjamin, *Arcades Project*, 844 (<Ho,3>).

12. The influence of Vicenza's horseshoe Teatro Olimpico, Sebastiano Serlio's stage plans, Michelangelo's Roman Capitol and Rome's theatrical Piazza del Popolo all attest to this. See Jonathan Barnett, *The Elusive City: Five Centuries of Design, Ambition, and Miscalculation* (New York: Harper, 1986), 10.

13. See Mary Henderson, *The City and the Theatre: New York Playhouses from Bowling Green to Times Square* (Clifton, NJ: White, 1973), 169–70.

14. Daniel Okrent, *Great Fortune: The Epic of Rockefeller Center* (New York: Penguin, 2003), 15.

15. Rem Koolhaas, *Delirious New York: A Retroactive Manifesto for Manhattan* (New York: Monacelli Press, 1994), 19–21.

NOTES TO PAGES 219–231

16. See New York's zoning laws of 1916, which regulated setback and building height.
17. Koolhaas, *Delirious New York*, 178.
18. Okrent, *Great Fortune*, 20.
19. A competing design by the Austrian architect Joseph Urban that was favored by the financier Otto Kahn was not selected.
20. Okrent, *Great Fortune*, 68.
21. Henderson, *The City and the Theatre*, 177–95.
22. Columbia University Graduate School of Architecture, *Lincoln Square: Preserving the Modern Architecture of Slum Clearance, Urban Renewal, and their Architectural Aftermath* (n.d.), accessed 27 July 2017, https://www.yumpu.com/en/document /view/11355892/lincoln-square-document-final-version-columbia-university.
23. Robert Caro, *The Power Broker: Robert Moses and the Fall of New York* (New York: Vintage Books, 1975), 1013–14.
24. Columbia University Graduate School of Architecture, *Lincoln Square*, 6–7.
25. Ibid., 10. The study notes this design's success.
26. The term was inspired by Paul Scheerbart's novel *Glasarchitektur* (Berlin: Verlag der Sturm, 1914).
27. Walter Benjamin, "Spurlos wohnen (Kurze Schatten II)," Benjamin, *GS*, 4.1: 428.
28. Annette Fierro, *The Glass State: The Technology of the Spectacle, Paris, 1981–1998* (Cambridge, MA: MIT Press, 2003), 14.
29. Ibid., 18–19.
30. Miller, "Opera Houses in the City."
31. Fierro, *Glass State*, 14.
32. Benjamin, *GS*, I.2: 693.
33. To emphasize another historical connection—to the Porte St. Antoine, a nearby entrance to the city until 1778—Ott simulated a large door in front of the opera.
34. David Dillon, *Dallas Architecture, 1936–1986* (Austin: Texas Monthly Press, 1985), 168.
35. Ibid.
36. Kevin Lynch, *The Image of the City* (Cambridge, MA: MIT Press, 1960).
37. Dillon, *Dallas Architecture*, 168.
38. Sasaki Associates, "Dallas Art District," accessed 15 July 2018, http://www .sasaki.com/project/176/dallas-arts-district/.
39. Frank Lloyd Wright, *The Disappearing City* (New York: Payson, 1932).
40. For example, the Carrée d'Art in Nîmes, France, creates continuity by juxtaposing a well-preserved Roman temple with a modern *médiathèque*; see also Deyan Sudjic, *Norman Foster: A Life in Architecture* (London: Phoenix, 2012), 148.
41. Foster + Partners, *Margot and Bill Winspear Opera House Opens in Dallas*, 15 October 2009, accessed 27 July 2017, http://www.fosterandpartners.com /projects/margot-and-bill-winspear-opera-house/.
42. Kenneth Powell, *City Transformed: Urban Architecture at the Beginning of the 21st Century* (New York: te Neues Publishing, 2000), 124–31.

308

NOTES TO PAGES 234–237

CHAPTER SIXTEEN

1. Alan Gilbert, "Latin America," in *The Oxford Handbook of Cities in World History*, ed. Peter Clark (Oxford: Oxford University Press, 2015), accessed 27 March 2017, http://www.oxfordhandbooks.com.libproxy1.usc.edu/.
2. Gilbert, "Latin America."
3. Bramante and Bernini in Rome, Palladio in Vicenza, Louis XIV's Paris and the Academy of Beaux Arts all influenced the design. Alberto Bellucci, "Arquitectura/Architecture," in *Teatro Colón: A telón abierto/In Full View*, ed. Jorge Aráoz Badí (Buenos Aires: Julio Moyano, 2000), 365–448, at 382.
4. Forty-seven works by twenty-eight Argentine composers were premiered at the Colón between 1908 and the 1970s. Malena Kuss, "Lenguajes nacionales de Argentina, Brasil y México en las óperas del siglo XX: Hacia una cronología comparativa de cambios estilísticos," *Revista Musical Chilena* 34, nos. 149–50 (1980): 61–79, at 64.
5. See Michael Scott, *Maria Meneghini Callas* (Boston: Northeastern University Press, 1991), 48, on the significance of Callas's 1949 debut at the Colón as Turandot. Leonor Plate, *Óperas, Teatro Colón: Esperando el centenario* (Buenos Aires: Editorial Lunken, 2006), 1:147–285, lists the Colón's singers and conductors from 1908 to 1999.
6. On Buenos Aires's "socially mixed" opera fans, see Claudio E. Benzecry, *The Opera Fanatic: Ethnography of an Obsession* (Chicago: University of Chicago Press, 2011).
7. For an online inventory of the Centro's productions, see the Archivos CETC at http://cetc.teatrocolon.org.ar/.
8. John King, *Sur: A Study of the Argentine Literary Journal and Its Role in the Development of a Culture, 1931–1970* (Cambridge: Cambridge University Press), 1986.
9. On Ocampo's autobiography, see Sylvia Molloy, "The Theatrics of Reading: Body and Book in Victoria Ocampo," in *At Face Value: Autobiographical Writing in Spanish America* (Cambridge: Cambridge University Press, 1991), 55–75.
10. On their friendship, see Ernest Ansermet and Victoria Ocampo, *Vies croisées de Victoria Ocampo et Ernest Ansermet: Correspondance, 1924–1969*, ed. Jean-Jacques Langendorf (Paris: Buchet/Chastel, 2005).
11. Omar Corrado, "Stravinsky y la constelación ideológica argentina en 1936," *Latin American Music Review* 26 (2005): 88–99.
12. Beatriz Sarlo, *Una modernidad periférica: Buenos Aires 1920 y 1930* (Buenos Aires: Ediciones Nueva Visión, 1988), 15.
13. Ibid.
14. Unless otherwise stated, I will use Ocampo's first name, Victoria, to refer to the character in Bauer and Sarlo's *V.O.* Like Sarlo's "modernidad periférica" is Pablo Gianera's "una modernidad inconclusa" (an unfinished modernity), applied to the musical culture of the same period. Pablo Gianera, *La música*

NOTES TO PAGES 238–247

en el grupo Sur: Una modernidad inconclusa (Buenos Aires: Eterna Cadencia Editora, 2011).

15. See Pablo Gianera, "Amor por Victoria Ocampo," review of *V.O.,* *La Nación*, 6 July 2013, accessed 20 December 2017, http://www.lanacion.com.ar/159 8568-amor-por-victoria-ocampo.

16. See Omar Corrado, "Victoria Ocampo y la música: Una experiencia social y estética de la modernidad," *Revista Musical Chilena* 61, no. 208 (2007): 37–65. Opposition to her plans led Ocampo and two other newly appointed members (Alberto Prebisch and Juan José Castro, director general) to resign from their posts after just a few months.

17. Beatriz Sarlo, *La máquina cultural: Maestras, traductores y vanguardistas* (Buenos Aires: Seix Barral, 2007).

18. I wish to express my gratitude to Beatriz Sarlo for sharing the libretto with me via email and permitting its citation.

19. Ocampo's original text reads: "Asistí, en primera fila de platea, al tumulto del *Sacre du Printemps*. Al final de la cuarta representación, creo (fui a todas), vi a Strawinsky, pálido, saludando a ese público que aplaudía *L'oiseau de feu* y silbaba despiadadamente el *Sacre*. Compré la partitura del *Sacre* y alquilé un piano para tocarla en mi salita del Meurice. No sabía bien que me atraía tanto en ese galimatías de notas y en ese ritmo brutal de cataclismo" (I attended, sitting in the front row of the orchestra, the tumult of the *Sacre du printemps*. At the end of the fourth performance, I think [I went to all], I saw a pale Stravinsky, saluting that audience, which applauded *L'oiseau de feu* and mercilessly hissed the *Sacre*. I bought the score of the *Sacre* and rented a piano to play it on in the small salon of my suite at the Meurice. I didn't really know what attracted me so much in that chaos of notes and that brutal cataclysmic rhythm). Ocampo, *La rama de Salzburgo* (Buenos Aires: Sur, 1981), 21.

20. Ocampo, *El imperio insular* (Buenos Aires: Sur, 1980), 106. Doña Sol is Victor Hugo's heroine in *Hernani*, transmuted into Elvira in Verdi's *Ernani*.

21. Sarlo, *La máquina cultural*, 99.

22. Ibid., 100.

23. In *El archipiélago*, the first volume of her autobiography, Ocampo repeatedly describes the transports of music.

24. Although Ocampo's text does not specify it, the performance must have been one of six that took place in May and June 1914; see Plate, *Óperas, Teatro Colón*, 2:212.

25. Ocampo, *La rama de Salzburgo*, 35.

26. Ibid., 39.

27. Gianera, *La música en el grupo Sur*, 23.

28. Victoria Ocampo, "Sobre un mal de esta ciudad," in *Testimonios: Segunda Serie* (Buenos Aires: Sur, 1941), 327–38, at 332–33.

29. Ibid., 337.

30. Sarlo, *La máquina cultural*, 137.

Index

acoustics, 188–90, 192
actresses, 35. *See also* women
Adam, Adolph, *Le châlet*, 179
Adam, Robert, 47, 48
Adelaide, 139
Adorno, Theodor, 163, 170, 177, 195, 233; on bourgeois opera, 163, 167, 174, 177, 213; and "massive" means of opera, 213–14, 223, 229
advertising, *205*, 206, 207, *209*, *210*
aesthetic modernity, 237, 238, 239, 246–47
Alfieri, Benedetto, 67
Algarotti, Francesco, 65; *Il saggio sopra l'opera in musica*, 65, 89
Allen, Grant, *The British Barbarians*, 199
Allen, Maud Percival, 141
Almanacchi de' teatri di Torino, 61
amateur performers, 206–7
Ambroghetti, Giuseppe, 80
Amélie-les-Bains, 187
Amsterdam, 14–15
Angelo, Henry, 46
Ansermet, Ernest, 237
Aquileia, 94–95, 98, 100, 102
architecture, 1, 7, 231; Buenos Aires, 234–35, 246–47; classical, 218; country-house, 201, 207; Dallas, 227–32; glass, use of, 214, 223, 226–27, 229–31; grid, 218–20, 222, 224, 229; historicist, 224, 226, 230, 233; identity, 229; London, 40, 45, 47–48, 53, 55; New York, 218–24; palazzo, 218, 219, 221, 223, 224; Paris, 54, 192; as performance, 229; as propaganda, 214; as scenography, 213–14, 215, 224, 227, 233; site dramaturgy, 215, 223, 224, 225, 226, 227, 232, 233; skyscraper, 219, 220, 221, 222, 229, 231; Turin, 67, 68; Venice, 89, 91. See also *Ausstellungsraum* (exhibition site)
Argentina, 234, 238, 248; Academia Argentina de Letras, 237; and Europe, 234–35
aristocracy. *See* Cavalieri, Nobile Società dei; oligarchy; social class
Arles, Théâtre Antique, 181, 183, 184, 192, 193
art, representation of, 171–76
Asahina, Takashi, 157
Auber, Daniel, *La muette de Portici*, 107, 116
audiences, 74, 88; America, 136; Australia, 133, 134, 136, 139, 140, 141, 142, 145–47; England, 27, 33, 35, 38, 43, 81, 83, 85, 123; The Hague, 19; Italy, 91, 92, 97; New Orleans, 111, 112–13, 114, 116; operatic representation of, 167, 171, 174–76; Paris, 113, 116, 124; as performers, 192–93, 208–12; Rio de Janeiro, 124; Shanghai, 153, 154; Turin, 57, 58, 60, 70, 162–63, 165, 183, 192, 194, 232, 233
Ausstellungsraum (exhibition site), 214, 215, 227, 229, 233

INDEX

Australia, 133, 137, 140, 142; critics in, 143; entertainment in, 134, 139; gold rush in, 134; Immigration Act in, 138; national opera company, 146; unionization in, 143. *See also* press

Australian character, 134, 136, 139, 140, 142, 144, 146–47

authenticity, 78, 92, 110, 182, 183, 184; of location, 198

ballet, 66, 239

ballet d'action, 54

balls, 58, 60, 62–63, 78, 79, 85, 87; at schools, 27, 28, 30, 31, 35, 36, 37, 38

Banister, Jeffrey, 30, 32

Banister, John, 37; *Musick; or a Parley of Instruments*, 37–38

Barbieri, Elisa, 124

Barnes, Edward, 229

Bartók, Béla, *Bluebeard's Castle*, 169

Bastille Opéra, 214, 215, 224–27, 231, 232, 233

Baudelaire, Charles, 245

Bauer, Martín, 238; and citation, 239; *V.O.*, 236, 237, 239, 241–42

Bayreuth, 142, 176, 201, 230

Beaujean, Philippe, 20, 24

Beauxhostes, Fernand Castelbon de, 179, 180, 190

Beecham, Thomas, 136

Beethoven, Ludwig van, 235

Bell, Vanessa, 202

Bellini, Vincenzo, 235; *Norma*, 179, 183

Benjamin, Walter, 10, 214, 215, 218, 224, 226, 228, 229, 232; *Arcades Project*, 215

Berg, Alban, 163; *Wozzeck*, 165

Berlioz, Hector, *Les Troyens à Carthage*, 183

Bernard, Oliver Percy, 137

Bertoja, Giuseppe, 97, 100, 102

Besselsleigh, 27, 29; boarding school, 27, 29, 31, 33, 35, 38; performance at, 32, 35

Bettali, Gioachino, 126

Béziers, 179, 180, 181, 183, 184, 189, 190, 191, 193–94

Bibbiena, Bernardo, *La calandria*, 90

Bizet, Georges, 235; *Carmen*, 139, 141, 157, 186, 193

Bloomsbury set, 202

Blow, John, *Venus and Adonis*, 27, 32, 33, 34

Boissy, Gabriel, 191–92

Bordeaux, Grand Théâtre de, 49

Boughton, Rutland, 197–200, 201, 202; *The Immortal Hour*, 198

Britain, 196–97, 245; country houses in, 196–97; spa towns in, 196. *See also* countryside; United Kingdom

British Broadcasting Corporation, 204

Brosses, Charles de, 67

Buck, Rudolf, 153

Buckley, Reginald, 197, 199, 200

Buenos Aires, 234–35, 236, 237, 239, 246; Ciclos de Conciertos de Música Contemporánea, 238; imported opera, 241; Teatro Politeama, 237

bullfighting, 179, 181, 186, 193

Burford, Robert: *Bay of Rio de Janeiro*, 121, *122*; *City of Calcutta*, 121–23, *122*, 125, 126

Burney, Charles, 66

Bustillo, Alejandro, 247

Cady, J. Cleaveland, 219

Caffi, Ippolito, 100, 104; *Notturno con nebbia*, 100, *101*

Calcutta, 120; Chowringhee Theatre, 129, 131

Canada, 134, 140

Canaletto, Giovanni, 100

canon, 27

capitalism, 86, 213–14, 221

Caravaglia, Margarita, 126, 128, 132

Carcassonne, 183

carnival: Italy, 57; and sociability, 57; Turin, 58, 59, 60, 61, 62, 64–66, 72–73; Venice, 89

Carr, Stephen, 228

Castiglione, Baldassare, 90

Catalini, Alfredo, *La Wally*, 158

Cavalieri, Nobile Società dei, 58, 59, 62, 63, 64, 65, 70, 71, 72, 73

censorship, 89

chamber opera, 236

Champigny, Théâtre Antique de la Nature, 181

Charles II of England, 26

Charpentier, Gustave, *Louise*, 141, 142, 174

Charpentier, Marie-Thérèse, 21–22

Charry, Joseph, 181

Chelsea, 29; boarding school, 27, 28, 30, 33, 34, 35, 38; dancing at, 33–34, 36–37; performance at, 32, 33, 36–37; staging at, 34

312

Chicago, Columbian Exposition, 218
China, 149, 154, 158, 159
chorus, 79, 83, 93–94, 95, 130, 157, 167, 192, 213; in Australia, 134, 135, 137, 138, 141, 143
Christie, John, 201, 204, 206–8
Christie (Mildmay), Audrey, 206, 208
Cibber, Colley, 41, 45
Cigna-Santi, Vittorio Amedeo, 65; *Motezuma*, 65, 66
cinema, 164, 167, 169, 170, 171, 177; as culture industry, 195; and open-air spectacle, 194; and Wurlitzer organs, 168
cinematic point of view, 171–76
citation, 238–48
citizen, 167, 168
city, representation of, 77–78, 82, 85–87, 88, 89, 90–100, 166–77, 199, 233; dystopian, 168–69, 170–71, 187, 198–99; escape from, 170–71, 198–99; as exhibition site, 218, 228, 229, 233; observer in, 171–77; and rural *dépaysement*, 187; sounds of, 169, 174, 183; urban scenography, 213–14, 215, 224, 227, 232, 233; utopian, 169. *See also* modern psyche
civic culture, 12, 58, 160, 161, 196
civilization, 90, 108–9, 120, 199; and opera, 108–9, 113–14, 115, 116, 117–18, 125–27, 129, 132, 136, 145
classicism, ancient, 180
Cocteau, Jean, *Orphée*, 167
Colbert, Jean-Baptiste, 53
colonialism, 6–7, 9–10, 108, 125, 129–30, 131, 146, 161; Argentinian, 241; British, 149–50; colonial formations, 149, 161; French, 160, 161. *See also* imperialism
commercialism, 41, 55–56, 86, 149, 164, 179
common law, 14, 22, 23. *See also* Roman law
consumerism, 7, 86, 117, 134, 136, 165, 212, 217, 232
contract law, 14, 23, 24–25
contracts, 14, 18, 22, 23–24, 25; impresarial model, 18, 23
Copenhagen, 213
cosmopolitanism, 58, 61, 86, 131, 235
Costa, Franz, 141
countryside, 196–200; and entertainments, 198; and fascism, 198, 200; and power, 196–97
Coursan, 187
court: masque, 34; Stuart, 34, 38

Craveri, Giovanni Gaspare, *Guida de' forestieri per la Real Città di Torino*, 67–68, 69, 70
Cruikshank, Robert and George, 75, 76, 76
cultural commodification, 7, 8, 26, 58, 61, 134, 166; Adorno on, 164, 195, 213
cultural geography, 2–3, 11, 74, 162, 213, 233, 237, 239; legal geography, 15; "new" cultural geography, 4; nonrepresentational geography, 4; "spatial turn," 4; urban geography, 5, 162
cultural translation, 112–16, 133, 236–37, 240–41, 247. *See also* translating experience
culture industry, 213, 229

Dallas, 213, 215, 227, 231, 233; Arts District, 215, 228, 229, 231; shops, 228
Dallas Concert Hall, 228
Dallas Museum of Art, 228, 229
Dallas Symphony Hall, 229, 232
Dallas Theater Center, 229
dancing, 30, 33–34, 37, 39, 60, 61, 63, 66, 75, 79, 83–84, 86, 124. *See also* balls
dancing masters, 29, 32
Darwin, Charles, 119–21, 123, 130, 131, 132
Debussy, Claude, 242, 247; "Hommage à S. Pickwick," 239; *Pelléas et Mélisande*, 169, 242; *The Prodigal Son*, 139–40, 141
Delhi, 130
Deseschaliers, Louis, 19–20, 22–23, 24
Dibdin, Charles, 79
Dibdin, Thomas, 81
Don Giovanni, 74, 75, 76, 77, 80–81, 85–87
Donizetti, Gaetano: *Don Pasquale*, 158; *Le convenienze ed inconvenienze teatrali*, 93; *Lucrezia Borgia*, 91–92; *Marin Faliero*, 92–93
Downton Abbey, 197
drama: biblical, 180, 183; classical, 180, 182; French, 180; mythological, 181, 197. *See also* spoken theater
Dublin, 213
Dudard, Catherine, 19–20, 21–23
Duffett, Thomas, 30; *Beauties Triumph*, 27, 30, 32, 33, 34, 35–36
Durazzo, Giacomo, 65
D'Urfey, Thomas, 30, 36; *Love for Money*, 35
Durga Puja, 127–28
Dutch Reformed Consistory, 16, 18; poor relief, 17–18, 25

313

INDEX

Ebert, Carl, 208
Eckhold, Richard, 141–42
Eden, Emily, 120–21, 123, 124, 127, 129, 130
Edinburgh, 142
Egan, Pierce, 77; *Life in London*, 74–87
Elgar, Edward, 200
English language, 237
entertainments, 27, 57, 58, 61; agricultural shows, 192; banquets, 192; gambling, 58, 63, 65; picnics, 193, 212; processions, 192; and sociability, 57; sports, 181, 186–87; street dancing, 192; timing, 58, 59, 60, 62, 63; wine festivals, 192. *See also* ballet; bullfighting
exhibition site. See *Ausstellungsraum* (exhibition site)
exoticism, 65, 89; French, 107, 110, 112, 116–17; and opera, 127–28, 140. *See also* Italian character; natural world: and exoticism

Fasciotti, Maria-Theresa, 124
fashionability, 76, 79, 84, 86, 125–26, 192
Fauré, Gabriel, 189; *Prométhée*, 181, 184, 185–86, 189, 191
félibre, 179, 183, 193
fête romaine, 179
Florence, Teatro La Pergola, 102
Foa, Arrigo, 157
Fordham University, 222
Foster, Norman, 215, 218, 228, 229, 230–32; Berlin Reichstag, 231, 232; British Museum, 231, 232
France, 20, 22, 23, 26, 52–54, 58, 91, 106, 109, 112–13, 116, 149, 187, 239, 245; *latinité*, 178; Provence, 183; republican universalism, 178; towns, 187. *See also* localism: French
Fréjus, 193
French language, 237, 240
French revolutions, 224, 225, 226

Gailhard, André, *La fille du soleil*, 181
gaming, 195
Gardiner, Rolf, 198
genre, 179, 189, 190–91
Gerhard, Anselm, 2
girls' boarding schools, 26, 27, 33; female administrators, 29–30; gender roles, 32–33; location and morality, 29, 36; performance eroticism, 34, 35; perfor-

mance morality, 27, 28, 32–34, 36, 38. *See also* schools; women
Glastonbury, 197, 198, 199–200, 201, 204, 207
Glimmerglass, 212
Glinka, Mikhail, *A Life for the Tsar*, 155
Gluck, Christoph Willibald, 183; *Armide*, 20, 23, 181, 183; *Iphigénie en Tauride*, 179, 183
Glyndebourne, 195, 200
Goodson, Richard, 31, 33; *Orpheus and Euridice*, 31, 33, 34
Gounod, Charles, 235; *Faust*, 139, 141, 157; *Mireille*, 183, 184, 192, 193
government: colonial, 150, 151; French, 179, 186; local, 14, 91, 126, 138
Graf, Max, 176
Grant, Duncan, 202
Greece, ancient, 180, 193
Greene, Graham, 237

Habsburg empire, 91, 94, 96, 104
Habsburg rulers, 93, 96
Hague, The, 12, *13*, 15, 16, 17, 20–21, 24–25
Haiphong, 151
Hamburg, Gänsemarkt Opera, 17
Hanoi, 151, 152, 160, 161
"happening," 180
Hart, Fritz Bennicke, 142
Hart, James, 30, 32
Haussmann, Georges-Eugène, 115, 218, 225
Hazon, Roberto, 135
heritage industry, 207, 212
Hofmannsthal, Hugo von, 168, 172; "Letter of Lord Chandos," 168–69
Honegger, Arthur, *Le roi David*, 237
Horkheimer, Max, 195
Howes, Frank, 195, 196
Hughes, Spike, 206–7
human geography, 3
Humperdinck, Engelbert, 165; *Hansel and Gretel*, 139, 163
hybridity, 237

identity politics, 8–9, 13, 75, 91, 156, 157; American, 222–23; Asian, 158–59; British, 196–200; civic, 213–14; French regional, 184, 187, 192–93; urban, 214, 222
immersion, 176–77, 206–12
Impekoven, Leo, 141

314

imperialism, 108, 136, 154, 160, 161, 231; American, 158–59; British, 133, 137, 145–46, 148, 152, 156, 158–59; French, 151, 160, 161; and heroism, 145; Japanese, 148, 152, 154, 158–59; and militarism, 145, 148, 159. *See also* colonialism

Indian character, 129–30, 131

indice dei teatrali spettacoli, L', 61, 63

information technology, 195

intermediality, 10–11

intermezzi, 164

Italian character, 126, 127, 128–29, 156

Jacquemont, Victor, 124, 130–31

Jambon, Marcel, 184, 190

Japan, 148, 158. *See also* imperialism: Japanese

Jobey, Charles, 106, 107–8; *L'amour d'un nègre*, 108, 110–16, 117; "Le lac Catha-houla," 105–8, 116, 117

Johnson, Ben, *To Penshurst*, 197

Jones, Siriol Hugh, 208

Juilliard School of Music, 222

jurisdiction: Dutch, 15, 16, 17, 22; French, 21; Hof van Holland, 16, 17, 20–21; local, 14–15, 16, 19, 20

Juvarra, Filippo, 67, 68

Kahn, Otto, 220

Kent, 202

Kilcullen, Bob, 229

Kit-Kat Club, 41

Komaki, Masahide, 158

Koolhaas, Rem, 215, 219, 220, 224, 229, 231–32

Korngold, Erich Wolfgang, 165; *Die tote Stadt*, 169

Kremberg, Jakob, 17

Kunc, Aymé, *Les esclaves*, 181, 190

Kusakari, Yoshito, 158

Kynaston, Francis, 37

La Lande, Joseph-Jerôme Le Français de, 66

landscape, 2–3, 5, 74, 196

Lanzhi, Gao, 158

Latin America, 236

latinité, 180

law: and culture, 12–13, 44; and gender, 22

Le Brun, Élisabeth Vigée, 67

Lecomte, Jules, *Venise ou coup-d'oeil lit-téraire . . .* , 97

legal studies, 12

Lehár, Franz, *Count of Luxembourg*, 158

Leland, John, 31

Leoncavallo, Ruggero, 164; *Pagliacci*, 141, 158

Le Texier, Anthony A., 49, 52

Lévadé, Charles, *Les hérétiques*, 181

Levant Company, 27

lieux de mémoire, 1

liminality, 107; performer, 27

Lindsey, Mary, 31

Liszt, Franz, 153

Liverpool, 140

Lloyd Webber, Andrew, *The Phantom of the Opera*, 167, 177

Lloyd Wright, Frank, 228

localism, 7, 10, 88, 92, 117, 138, 160; French, 178, 180, 182, 183, 186, 187, 192. *See also* regionalism; territoriality

locality, 75, 78, 85, 97, 124, 198, 202–4

London, 78–79, 125; Adelphi Theatre, 81; Bethnal Green, 37; City of, 28, 29; Cockney, 75, 87; court, 30; Covent Garden, 28, 37, 133, 139; dancing academy, 46; Deptford, 32; Drury Lane, 75, 76, 125; entertainments, 27, 30, 31–32, 35, 36, 39, 44, 81–82, 85; fashion, 124; fencing academy, 46; Georgian, 74–75, 86–87; Greenwich, 32; Hackney, 28, 35; hay market, 43; Holborn, 30; Islington, 29; Kensington, 35; Leicester Fields, 30; Lincoln's Inn Fields, 38; Little Theatre in the Haymarket, 43; Olympic Theatre, 75–76; operatic taste, 127; Piccadilly, 78; population of, 26; Putney, 28; Shake-spearean, 29; shops, 43–44, 46, 55, 56, 86; St. James's, 42–43; St. Martin-in-the-Fields, 42; suburbs, 28; theaters, 30, 43; Vauxhall, 78; Victorian, 86; Westminster, 28. *See also* Foster, Norman: British Museum; urban planning: London

London's Italian opera house, 39, 86, 235; depictions of, 44, 45, 46, *47*, 48, 51; as Her Majesty's Theatre, 30, 56; as King's Theatre, 41, 54, 75, 77, 78, 81; as Queen's Theatre, 45. *See also* architecture

Lucas, George, *Star Wars*, 170

Lucca, Francesco, *L'Italia musicale*, 102, *103*

Lynch, Kevin, 228

Lyster, William Saurin, 134–35, 136, 145

315

INDEX

Macao, 126
Maffei, Andrea, 96–97
Maguna, Minou, 246
Mahler, Gustav, 176
Mann, Thomas, 176
maps, 12, 42, 46, 59, 61, 150, *151*, 202, *203*; mapping, 13, 85, 162
Margot and Bill Winspear Opera House, 214, 215, 227–32, 233. *See also* Winspear, Bill
Mariéton, Paul, 183, 188, 193
Markov, Peter, 158
Marshall-Hall, G. W. L., 146–47
Mascagni, Pietro, 164; *Cavalleria rusticana*, 141, 157, 158; *Iris*, 169; *L'amico Fritz*, 155; *Le maschere*, 155
Massé, Victor, *Galathée*, 179
Massenet, Jules, *Manon Lescaut*, 141, 142
McMillan, William, 146
media archaeology, 10–11
Méhul, Étienne, *Joseph*, 179, 188
Melba, Nellie, 135, 137, 138, 139
Melbourne, 133, 134, 135, 139, 140, 141, 142, 143–44; German community in, 135; Her Majesty's Theatre, 143
melodrama, 164, 169, 179
men, 87; masculinity, 75, 77, 78–79
metropolitan culture, 166
Metropolitan Opera, 142, 214, 215, 218–24, 231, 233, 235; Metropolitan Opera board, 220, 222. *See also* New York: Lincoln Center
Meyerbeer, Giacomo, *Robert le diable*, 115
Meyerson, Morton, 229
Miami, 213
Midi, 179, 180, 181, 182, 184, 189, 191
migration, 135, 154, 157
Milan, 61, 67, 89, 94, 96, 155; La Scala, 89, 102, 154, 235; Teatro Nuovo, 89
Mitterrand, François, 215, 224, 225, 227; Grands Travaux, 215, 224
mobility, 9–10, 13, 14–15, 43, 60, 63, 74–75, 77, 78, 79, 82, 86; audience, 192–93, 232; cultural, 82, 124; and dance, 84, 86; and gender, 83, 84, 87; imaginative, 120; and music, 82–84; operatic, 76; social, 26, 75, 77, 83, 87, 131. *See also* cultural translation; translating experience
modernism, 165, 171
modern psyche, 168–69, 189, 198; dreams and, 169

Moncrieff, William Thomas, *Giovanni in London*, 75–76, 81–86
Monteverdi, Claudio, *L'incoronazione di Poppea*, 89
Montevideo, 120, 124
monuments, 55, 87, 92, 219, 220, 221
morality, 27, 127–28. *See also* schools
Moreno, Marguerite, 240
Morris, Benjamin, 221
Morris, William, 197, 198–99, 202; Arts and Crafts movement of, 199
Moses, Robert, 222
Mozart, Wolfgang, 84, 207, 208, 235; *Die Zauberflöte*, 163; *Don Giovanni*, 75, 80, 82, 84, 85, 87; *Le nozze di Figaro*, 141
Murphy, Agnes, 137
music and place, 3, 4, 83, 188–90, 193
musicians, 63; professional status, 27–28, 32, 128–29, 192
Mussorgsky, Modest, *Khovanshchina*, 155
mythology, 93, 98, 128, 181, 197–98; Arthurian, 198, 202

Naples, Teatro Nuovo, 89
Napoleon III, 215, 218
Nash, John, 49, 51, 55–56, 86
Nasher, Raymond, 229
Nasher Sculpture Garden, 229, 232
nationalism, 7, 198, 200, 201, 218
National Trust, 197
natural world, 105, 107, 108–10, 188, 201; and exoticism, 109–10, 119–23. *See also* countryside
Newman, Ernest, 141, 145
New Orleans, 105, 110, 112; American Theatre, 114, 115; St. Charles Theatre, 114; Théâtre d'Orléans, 105–6, 109, 110
New York, 106, 140, 218, 233; arcade experience, 218; Bloomingdale's, 218; Broadway, 218–19, 222; city blocks, 218, 219, 223, 224, 233; Lincoln Center, 218, 222, 223, 224, 231; Lincoln Square, 222–24; Macy's, 218; Rockefeller Center, 218, 221. *See also* architecture: New York
New York City Ballet, 222, 223
New York Philharmonic, 222, 223
New Zealand, 134, 135, 140–41
Nîmes, 181, 184, 186
Nolfi, Vincenzo, *Il Bellerofonte*, 88, 89–90, 94
nostalgia, 120, 160, 169, 172, 195, 197, 202, 212

notaries, 18–19, 21, 22, 23
Noverre, Jean-Georges, 52–53, 54, 56
Novosielski, Michael, 47–48, 51

Ocampo, Victoria, 236–38; *El imperio insular*, 240; Europhilia, 239; *Le rama de Salzburgo*, 239, 242, 244; *Le vert paradis*, 245; and Martínez, 239, 242–43; *Sur*, 236, 237, 238, 241, 247
Offenbach, Jacques, *The Tales of Hoffmann*, 139, 141, 158, 214. *See also* Paris: Bouffes Parisiens
oligarchy, 196–97, 202, 206, 208, 212, 234, 236–37, 247. *See also* Cavalieri, Nobile Società dei; social class
opera: communitarian, 199–200; and morality, 89, 127–28, 163 (*see also* girls' boarding schools); and power, 1, 4, 6–7, 116–17, 160, 166, 204 (*see also* colonialism; imperialism)
opera, country-house, 195, 200–212; and escapism, 202, 204
opera, festival, 201
opera, open-air: "slow," 188, 193; southern France, 178–94; *théâtre de la nature*, 181, 193, 198; *théâtre de verdure*, 181, 182, 187, 193; *théâtre en plein air*, 179; and weather, 184–85, 193. *See also* cinema
opera, popular, 163–66; *théâtres populaires*, 181. *See also* populism
opera, representation of, 105–6, 107, 108–12, 116–17, 119–21, 163, 173–74, 212; as art, 189; as culture industry, 195–96, 208–12, 213, 229; as democratic, 229; and Vienna, 162
opera, reputation of, 126–28, 136, 151–52, 204–6, 212; for Adorno, 163–66, 213; bourgeois, 163, 164, 165, 175, 213, 218, 224; erotic, 232, 242–43; magical, 163, 170; Romantic, 163–64, 165–66, 171, 177
opera concerts, 152, 154, 155–56, 157–58, 159, 160
opera house location, 1, 2, 55, 114–15; and civic ambition, 5–6, 15, 53, 88, 213, 215; commercialism of, 2, 6, 53, 54, 55, 56, 217; island site, 52, 54, 219, 224; monumentality of, 53, 56; symbolism of, 1, 2, 4, 115, 151, 215, 217; urban scenography of, 213–14, 215, 217, 218, 224, 232
opera studies, 4, 5, 74

operetta, 164, 169
Orange, Théâtre Antique, 178, 179, 180, 181, 182, 183, 184, 189, 192, 193; Wall at, 183, 186
oratorio, 179, 189, 191
Orchard, Arundel, 141
orchestra, 20, 45, 63, 106, 111, 124, 132, 142–43; concept of, 159; professional status of, 192; size of, 189, 213. *See also* Shanghai Municipal Orchestra; Shanghai Philharmonic Society; Shanghai Public Band
orientalism, 124. *See also* exoticism
Ott, Carlos, 224, 226, 227
Oxford, 29, 31, 32

Paci, Mario, 153, 154, 155, 156, 157
pageant, 179, 198, 199, 200
Palais Garnier, 55, 115, 181, 191, 214, 215–18, *216*, *217*, 219, 222, 224, 225, 235
panorama, 123, 214
Paris, 233, 234, 246; Arcades, 214, *216*, 217, 218; Arènes de Lutèce, 182; audiences, 112–13, 124, 174, 180; Bastille prison, 224, 225, 226, 227; Bon Marché, 218; Bouffes Parisiens, 214; Exposition Universelle, 192; fashion, 124, 192; Marais district, 225, 227; modernization, 167; Odéon, 190–91; Opéra, 53, 54–55; refinement, 189–90; and regionalism, 179, 180, 186, 189, 193; Schola Cantorum, 179; Seine, 225; shopping, 227; shops, 55; as source of performers, 192; Théâtre des Variétés, 214; Théâtre-Français du Faubourg Saint-Germain, 53; Trocadéro, Salle des Fêtes, 191–92; world fairs, 214. *See also* regionalism
patronage, 142, 179; court, 26
pedagogy, 27
Pei, I. M., 215, 229, 232
Penang, 126
performance studies, 1
performativity, 8
Petrella, Errico, 88–89
Pfitzner, Hans, 165
Phantom of the Opera, The, 167
Philippines, 153
Piano, Renzo, 215, 229, 232
Piave, Francesco Maria, 95, 96, 100, 104
Pinchoff, Thomas, 142
Pividor, Giovanni, *Souvenirs de Venise*, 98, *99*

INDEX

Pizzoni, Domenico, 126
Playford, Hannah, 29–30
Playford, John, 29
playwrights, 29
Ponchielli, Amilcare, *La Gioconda*, 155, 157
popular culture, 165
popular music, 195
population growth, 26
populism, 7, 10, 11, 181
Portzamparc, Christian de, 226
postcolonialism, 9, 131, 161
Pré Catelan, Bois de Boulogne, 181–82
press: Australia, 137, 139–40, 142, 143;
    Britain, 199, 200, 201–2, 206, 207, 208;
    Calcutta, 125, 126–30, 131–32; France,
    181, 184, 185, 186, 187–88, 189–91, 194;
    knowledgeability of, 143–44; Shanghai,
    148, 153, 154, 156, 157; Singapore, 126
Priest, Frances (Franck), 30, 32, 33
Priest, Josias, 27, 30, 32, 33, 34
Prince Regent, 40, 49, 75, 86
proms, 204
Puccini, Giacomo, 155, 164; *The Girl of the
    Golden West*, 139, 140, 141; *Il tabarro*,
    169, 170; *La bohème*, 139, 141, 155,
    174; *Madame Butterfly*, 139, 141; *Manon
    Lescaut*, 155; *Tosca*, 141, 155, 170, 173,
    177; *Turandot*, 155, 170, 177
Purcell, Henry, 27; *Dido and Aeneas*, 27,
    33–34, 36

Quesnot, Jean-Jacques, 19
Quinlan, Thomas, 133–34, 136–47; Quinlan
    International Concert Agency, 136. *See
    also* touring opera: Quinlan company

Rabaud, Henri, 194; *Le premier glaive*, 181,
    194
Racine, Jean, 240, 246; *Andromaque*, 240
realism, 92–93, 107, 140, 166, 186; verisi-
    militude, 183
regionalism, 179, 181, 182, 192, 237. *See
    also* localism
Rimsky-Korsakov, Nikolai: *The Snow Maiden*,
    155; *The Tsar's Bride*, 155
Rio de Janeiro, 108, 119, 120–21, 123–24,
    130
Risorgimento, 100
Rochelle, Pierre Drieu La, 238, 242
Rochois, Marie, 23–24
Rockefeller, John D., 220, 221, 222

Roman amphitheaters, 182, 184; ruins, 180,
    183; space, 180
Roman law: *ius commune*, 14, 22; *SC Vel-
    leianum*, 22; *societas*, 18
Rome, 25, 89, 90, 94, 100, 174, 239; Teatro
    Argentina, 96; Terme di Caracalla, 178
Rossini, Gioachino, 124, 235; *Il barbiere di
    Siviglia*, 131–32, 141; *La Cenerentola*,
    120; *La scala di seta*, 157; *L'italiana in
    Algeri*, 123, 126–27, 128–29, 130, 131;
    Moïse, 179
Ruskin, John, 199, 202
Russia, 154, 157, 158; primitivism, 189

Sackville-West, Vita, 197, 202, 212
Saigon, 151
Saintes, 193
Saint-Rémy de Provence, 193
Saint-Saëns, Camille, 188, 189, 194;
    Déjanire, 179, 180, 189, 190–91; *Les
    barbares*, 179, 183, 189, 191; *Parysatis*,
    181; *Samson et Dalila*, 141, 183
Salzburg, 201
Sarlo, Beatriz, 236, 238, 243, 245; and cita-
    tion, 238–46; *La máquina cultural*, 238,
    240; *Una modernidad periférica*, 237, 238;
    V.O., 236–48
Savoy dynasty, 58, 67
Schieroni, Teresa, 124, 126
Schillings, Max von, 165
Schoenberg, Arnold, 165
Schola Cantorum, 179
Schönbrunn, 235
schools, 5, 26; balls, 27, 30, 31, 36, 37; boys'
    boarding schools, 37; court influence,
    34; exclusivity, 28, 34; healthfulness,
    28; location of, 28; public/private na-
    ture of, 35–38; recruitment for, 36, 37,
    38; schoolmasters as performers, 32, 34,
    38. *See also* girls' boarding schools
Schott, Gerhard, 17
Schreker, Franz, 164, 174–76; *Der ferne
    Klang*, 174–76; *Die Gezeichneten*, 170–71,
    174
Sendón, Matías, 246
sensationalism, 181
Séverac, Déodat de, 190; *Héliogabale*, 181,
    184, 189, 190; *Le coeur du moulin*, 189;
    La fille de la terre, 187
Shanghai, 148–61; Bund, 152; Chinese
    Old City, 150; French Concession, 150,

318

153, 160; French Municipal Council, 151; Grand Theater, 153; International Settlement, 148, 150, 152, 156, 158, 160; Jessfield Park, 156; Koukaza Park, 157, 158; Lyceum, 151, 153, 154, 158; National Conservatory of Music, 154; Public Garden, 153; Shanghai Municipal Council, 150–51, 152, 154, 155–56, 157, 160

Shanghai Amateur Dramatic Club, 153, 154

Shanghai Ballets Russes, 158

Shanghai Municipal Orchestra, 149, 151, 152–57, 159

Shanghai Opera, 152

Shanghai Philharmonic Society, 149, 152, 157–60

Shanghai Public Band, 152, 153. *See also* Shanghai Municipal Orchestra

Shaw, George Bernard, 200

Shushlin, Vladimir, 154, 158

Sicard, Émile, *La fille de la terre*, 187

Silinskaia, Elisabetta, 155

Simonsen, Martin, 135

Singapore, 126

site-specific theater, 183, 190–92

Skelton, John, "London to Glyndebourne by road," *203*

Sloutsky, Alexander, 157, 158

Smyth, Dorothy Carleton, 137

social class, 93, 111, 124, 164–65, 177, 236, 238; America, 136; Australia, 136, 144; Britain, 28, 37, 41, 77–78, 86, 131, 136, 197, 202, 204–8; Calcutta, 131; Dallas, 231; The Hague, 16–17; New York, 222; operatic representation of, 171, 172; Shanghai, 156. *See also* mobility; schools: exclusivity

Society for the Protection of Ancient Buildings, 197

Solera, Temistocle, 95, 96, 100

South Africa, 133, 134, 138, 141

souvenirs, 193

Soviet Union, 154. *See also* Russia

Spanish language, 237

spectacle, 58, 64–72, 88, 90, 136, 170–73, 226; animals, 64, 65–66; immersive opera and, 202–12; open-air opera and, 192, 193–94; urban, 214, 215. See also *Ausstellungsraum* (exhibition site)

Spengler, Oswald, *Der Untergang des Abendlandes*, 199

spoken theater, 61, 90, 163, 179, 181, 240, 242; companies, 16, 17, 61, 222

Spontini, Gaspare, *La vestale*, 181, 183

stage lighting, 94–96, 97, 100, 102; sun, 184–85, 204

stage machinery, 16, 34, 65, 82, 88, 90, 176, 204

stage regulation, 15, 16, 58, 143

stage scenery, 19, 21, 34, 65, 66, 92–94, 97–98, 100, 102, 176, 213; in Australia, 137, 143; as metaphor, 119, 120, 121, 123; open-air, 179–80, 184

Stendhal, *De l'amour*, 242

Stratford, 197

Strauss, Richard, 165, 168; *Der Rosenkavalier*, 172

Stravinsky, Igor, 238, 247; *Perséphone*, 237; *Rite of Spring*, 239, 240

Strok, Avray, 154

Surrey, 198, 199

Sussex Downs, 201, 202

Sydney, 134, 135, 139, 140, 142–43; Her Majesty's Theatre, 143

tableaux, 193

Tagore, Dwarkanath, 131

Tagore, Rabindranath, 237

Tchaikovsky, Pyotr, *The Enchantress*, 155

Teatro Colón, 234–48; and Argentine composers, 235, 236; Centro de Experimentación (CETC), 236, 246, 247; and Europe, 235, 239; seasons, 235

territoriality, 2, 6, 17, 58–60, 90, 114–15, 121, 149–52, 161; British, 149, 150–52, 156, 159; French, 150–51; Italian, 156; Japanese, 149, 150–52, 155, 157–60; in Shanghai, 148–49, 150–52

theater, symbolism of, 91

theater finances, 17–18, 19–20, 21, 26, 43, 54, 62, 63, 181, 200, 204; touring companies, 137–38

théâtre en plein air. See opera, open-air

theatrical privileges, 15

Tom and Jerry, 74–87; and music, 78–80, 83

Torelli, Giacomo, 90

Toscanini, Arturo, 155

Toulouse, Arène des Amidonniers, 186

touring opera, 14–15, 120, 123–24, 132, 133–47, 152, 153, 234–35; Beecham company, 136; Cagli company, 135; Cagli-Pompei company, 134; Lyster

319

INDEX

touring opera (*cont.*)
    company, 134–35; Martin Simonsen
    company, 135; Melba-Williamson com-
    pany, 135, 138, 139; Montague-Turner
    company, 135; Musgrove company, 135;
    Quinlan company, 133–34, 137. *See also*
    traveling performers
tourism, 61, 67–70, 88, 92, 100, 105, 116.
    See also translating experience
translating experience, 107–8, 112–14, 116,
    237. *See also* cultural translation
transportation, 153, 235; carriage/coach,
    28, 74, 79, 83; foot, 75, 193, 218, 231,
    232; plane, 204; roads, 28, 67, 153, *203*,
    232; ship, 142–43; train, 143, 193, 202,
    204, 226
travel, 41, 43, 60, 61, 63, 74, 77–79, 88,
    239; Argentina, 235; Australia, 142–43;
    Austria, 162; France, 193; "skirting," 28
traveling performers, 5, 14–15, 19–20, 21,
    23; in Argentina, 235; in Australia,
    134–47; in Calcutta, 125, 126; Dutch,
    15; French spoken theater, 17; in New
    Orleans, 105–6; popularity of, 126; in
    Shanghai, 154, 157, 160
travel literature, 107, 109–10, 112, 116–18,
    125
treaties, 149
treaty-port, 149, 150, 151, 161
Truman, Ernest, 139
Turin, 57; as cultural center, 58, 61, 66–70;
    entertainments, 58, 59–67, 70, 72–73;
    shops, 58, 62, 65; sociability, 58, 60,
    63; Teatro Carignano, 60, 61, 62, 63,
    71; Teatro Gallo/Gallo Ughetti, 61, 62;
    Teatro Guglielmone/D'Angennes, 61,
    62; Teatro Regio, 58, 59, 60, 61, 62, 63,
    65–66, 72; Teatro Vinardi, 59, 61, 62, 72

United Kingdom, 134, 149. *See also* Britain;
    imperialism: British
United States, 134, 135, 136, 142, 145, 149.
    See also New Orleans
urban environment, 1, 2, 5, 12, 15, 74,
    166, 213, 233; Calcutta, 121; Italy, 57;
    London, 40, 43, 52, 86; New Orleans,
    106; and opera, 195–96; Paris, 54, 190,
    214, 224–25; and periphery, 28, 34, 90,
    170–71, 196, 228; pleasure gardens, 196;
    renewal, 215, 222, 224–27; and social
    spectacle, 172, 214–15; and socioeco-

nomic divisions, 28, 58, 60, 61, 62, 63,
    75, 204; Turin, 58, 59, 61, 63, 67–70;
    Venice, 88, 89, 90–91; Vienna, 162, 167,
    172. See also *Ausstellungsraum* (exhibi-
    tion site)
urbanization, 196
urban life, 219; rejection of, 198–99, 204,
    212; as transitory, 74–75, 77–78, 226
urban planning, 213; Calcutta, 121; Dallas,
    228; London, 40, 41, 49–50, 86; New
    York, 218–22; Paris, 55, 115, 214; re-
    newal, 215; Turin, 58, 59, 66, 68–70. See
    also *Ausstellungsraum* (exhibition site)
Urbino, 90
utopianism, 198–99, 200, 201–2

Valencia, 213
Valparaíso, 119, 124
Vanbrugh, John, 41, 44–45
Vela, Melchior, 153
Venice, 88; as "Adria," 93–94; Council of
    Ten, 91, 96; "dark legend," 91–92, 96;
    history of, 91, 97–98, 102–3; reputation
    of, 91–92, 96, 100, 174–75; Teatro Gallo/
    San Benedetto, 96; Teatro La Fenice, 92,
    94, 95, 100, 104; Teatro Novissimo, 89
Verande, Louis P., 137
Verdi, Giuseppe, 96, 104, 235; *Aida*, 139,
    141, 157, 234; *Attila*, 88, 94–104; *I due
    Foscari*, 96; *I Lombardi*, 89; *Il trovatore*,
    141, 163; *La Traviata*, 139, 141, 142,
    145, 158; *Rigoletto*, 139, 141, 142
verismo, 164, 169, 176
Verney, John, 28
Verona arena, 178
Versailles, 235
Vestris, Lucy, 75, 76, 79
Victor Amadeus II, 58, 67
Vienna, 162, 234; Burgtheater, 162;
    Hofoperntheater, 162, 167, 175; mod-
    ernization, 167; Staatsoper, 167, 235.
    See also urban environment: Vienna
Vinardi, Antonio, 59
Viotti, Emanuele, *Adelaide d'Aragona*, 93–94
Voight, A., "London, London town for me,"
    79–80
Voltaire, 52

Wagner, Richard, 135, 139, 140, 141–42,
    154, 199, 206–7, 208, 235, 242–43; *Die
    Walküre*, 141, 155, 176; *Götterdämmer-*

ung, 141; *Lohengrin*, 139, 163; *Meistersinger*, 141, 169, 206; and National Socialism, 170; *Parsifal*, 155, 163, 170, 242–43; post-Wagnerism, 164, 167; *Rheingold*, 141, 170; *Ring*, 140–45, 154, 176; *Siegfried*, 141; *Tannhäuser*, 139, 141, 155; *Tristan und Isolde*, 139, 140, 143, 242–43; Wagnerian heroine, 176
Wagnerians, 141, 177, 201
Ward, William, 127–28
Waugh, Evelyn, *Brideshead Revisited*, 197
Weber, Carl Maria von, *Der Freischütz*, 163
Weldon, John, 31
Wells, H. G., 199
Werner, Friedrich, *Attila, König der Hunnen*, 94

Whigs, 41
Wiedermann, Elise, 142
Williamson, J. C., 134, 135, 137–38, 145
Winspear, Bill, 229
women: dramatic representation of, 82–83, 169, 171, 174–76, 238–48; and morality, 27, 28, 29, 32–33, 136; running schools, 29–30; trouser roles, 32–33, 75, 84. *See also* mobility: and gender
Wood, Anthony, 31, 38
Woolf, Virginia and Leonard, 202
world fairs, 214, 229

Zelman, Alberto, 135, 142
Zweig, Stefan, 162, 164